CBS
3-19-97
$1 187.89

ENCYCLOPEDIA of EMULSION TECHNOLOGY

Volume 4

ENCYCLOPEDIA of EMULSION TECHNOLOGY

Volume 4

edited by
Paul Becher
Consultant
Wilmington, Delaware

Marcel Dekker, Inc. New York • Basel • Hong Kong

Library of Congress Cataloging-in-Publication Data

(Revised for volume 4)

Encyclopedia of emulsion technology.

Includes bibliographical references and indexes.
Contents: v. 1. Basic theory — v. 2. Applications —
v. 3. Basic theory, measurement, applications.
1. Emulsions. I. Becher, Paul.
TP 156.E6E62 1983 660.2'94514 82-18257
ISBN 0-8247-1876-3 (v. 1)
ISBN 0-8247-1877-1 (v. 2)
ISBN 0-8247-1878-X (v. 3)
ISBN 0-8247-9380-3 (v. 4)

The publisher offers discounts on this book when ordered in bulk quantities. For more information, write to Special Sales/Professional Marketing at the address below.

This book is printed on acid-free paper.

MARCEL DEKKER, INC.
270 Madison Avenue, New York, New York 10016

Current printing (last digit):
10 9 8 7 6 5 4 3 2 1

Printed in the United States of America

Preface

In this, the fourth volume of the *Encyclopedia of Emulsion Technology*, I have both updated earlier work, especially that originally included in Volume 1, and introduced a number of new concepts. The subject of emulsion stability has been updated by P. Walstra, and coverage of the preparation of emulsions has been updated by E. H. Lucassen-Reynders. Rajinder Pal has contributed a further discussion of emulsion rheology, with particular emphasis on the role of added polymers, both as thickeners and as stabilizers.

In the chapter combining both new and old matters, T. G. Mason et al. demonstrate the light that monodisperse emulsions can shed on, among other things, emulsion rheology and stability, as well as describing the technique for preparing such emulsions.

C. J. van Oss provides a new description of the effect of attractive forces on solubility, micelle stability, and microemulsion formation. A brief description of a new model of micro- and fine emulsions is given by Christian B. Oleksiak and Henri L. Rosano.

Finally, I have contributed the third update of the bibliography of hydrophile–lipophile balance (HLB). The fact that the literature on HLB continues to grow is illustrative of the utility of the concept.

In any collection such as this, a certain amount of duplication of discussion of basic theory is inevitable. I have striven to minimize unnecessary duplication; however, since each author may well wish to give his or her own special emphasis to the discussion, some overlapping presentations remain.

I wish to thank Sandra Beberman and Joseph Stubenrauch, of Marcel Dekker Inc., for their continued support and assistance in this work. I also express my gratitude to the authors, who have shown exemplary patience through the overlong consummation of the volume. And, of course, to Jane, my dear wife, I dedicate this latest volume.

Paul Becher

Contributors

PAUL BECHER Consultant, Wilmington, Delaware

J. BIBETTE Université de Bordeaux I/CRPP, Bordeaux, France

HU GANG Exxon Research and Engineering Company, Annandale, New Jersey

A. H. KRALL Department of Physics and Astronomy, University of Pennsylvania, Philadelphia, Pennsylvania

E. H. LUCASSEN-REYNDERS Unilever Research Laboratorium Vlaardingen, Vlaardingen, The Netherlands

T. G. MASON Université de Bordeaux I/CRPP, Bordeaux, France

CHRISTIAN B. OLEKSIAK* Department of Chemistry, The City College of the City University of New York, New York, New York

RAJINDER PAL Department of Chemical Engineering, University of Waterloo, Waterloo, Ontario, Canada

HENRI L. ROSANO Department of Chemistry, The City College of the City University of New York, New York, New York

CAREL J. VAN OSS Department of Microbiology, School of Medicine and Biomedical Sciences, State University of New York at Buffalo, Buffalo, New York

PIETER WALSTRA Department of Food Science, Wageningen Agricultural University, Wageningen, The Netherlands

DAVID A. WEITZ Department of Physics and Astronomy, University of Pennsylvania, Philadelphia, Pennsylvania

---

**Current affiliation*: Kraft Foods, Glenview, Illinois.

Contents

Contents of Previous Volumes

Contents of Volume 1: Basic Theory

Contents of Volume 2: Applications

Contents of Volume 3: Basic Theory • Measurement • Applications

1
Emulsion Stability

PIETER WALSTRA / Department of Food Science, Wageningen Agricultural University, Wageningen, The Netherlands

---

I. Introduction

The physical stability of emulsions and the underlying interfacial aspects have been extensively treated by Tadros and Vincent in the first volume of this encyclopedia [1,2]. Some additional aspects have been discussed in Volumes 1, 2, and 3 [3–5]. Since that time, new results have become available, new theories have been developed, and some new considerations may be added. This chapter is meant as an update on some salient aspects. A full treatment would not be possible; because of the vast amount of literature that has become available, it would involve an almost complete review of colloid and interface science. Use has been made of review papers, where available.

This chapter will consider only simple macroemulsions. This rules out a number of interesting new developments. But even for simple macroemulsions, numerous aspects can be discussed, not least because of the multivarious aspects of their physical stability, which often causes confusion. Another point of confusion is that between conditions governing the making of an emulsion and those governing the instability (the "breaking") of the emulsion obtained. Comparison of reviews on emulsion formation [6,7] and the present chapter will show that important variables often have very different effects on making and breaking. Taking into account that all kinds of instability are a strong function of emulsion droplet size, and that the latter is determined by conditions during formation, confusion may easily occur.

Most emulsion products need to be stable, often during prolonged storage. This is achieved mostly by making small droplets. Consequently, we will concentrate on emulsions with a droplet size of the order of a micrometer. This has important consequences. First, the Laplace pressure will generally be so high as to prevent deformation of the droplets under nearly all conditions; this is discussed further in Sec. VI. Second, the surface of the droplet will be essentially immobile in the tangential direction, since the small size will always permit development of a high interfacial tension gradient when a shearing force acts on the droplet surface, even for low surface loads; this is discussed further in Sec. II.A. Under most conditions droplets thus will behave like rigid spheres. Only for large droplets or extreme conditions (to be discussed) might the situation be different. This also implies that model studies done on large droplets are rarely relevant for real emulsions.

Figure 1 illustrates the various instabilities that may occur. *Ostwald ripening* (also called disproportionation) involves diffusion of dispersed phase material from smaller to larger droplets, due to the chemical potential of the material being higher for a smaller radius of curvature. It does not occur if the solubility of the dispersed phase material in the continuous phase is negligible. *Sedimentation*, or creaming, occurs if both phases differ in density. *Aggregation* of droplets may be said to occur when they stay very close to one another for a far longer time than if there were no attractive forces acting between them. The old distinction between flocculation and coagulation breaks down in light of the many different causative mechanisms (Sec. V), so we will use the general term aggregation (although partial coalescence may also be considered a form of aggregation). *Coalescence* occurs when the thin film of continuous phase between two closely approaching droplets breaks; the

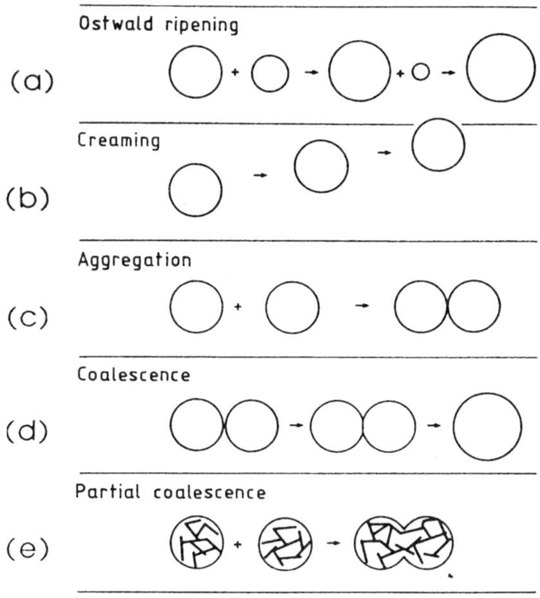

Figure 1. Types of physical instability in oil–in–water emulsions.

Laplace pressure differences then cause the droplets to rapidly fuse into one droplet. In most cases, the film between droplets is far smaller than depicted in Fig. 1. If the droplets contain solid particles, generally because part of the liquid in the droplets has become frozen, full coalescence may not occur; we may then speak of *partial coalescence*, or clumping. As discussed in Sec. VII, partial coalescence may often occur very much faster than does true coalescence. Some emulsions containing high concentrations of small-molecule surfactants may show spontaneous phase inversion [8], but this will not be discussed here.

The various changes may greatly affect each other. All changes leading to larger particles enhance the creaming rate. If the droplets have a propensity to aggregate, creaming will generally enhance the aggregation rate. Aggregation often leads to rather tenuous flocs; after flocs arrive in a cream layer or sediment, some consolidation may occur, leading to a more compact layer. In most cases, coalescence occurs only when the droplets are close to each other for a prolonged time; hence, their aggregation, and especially their presence in a cream layer or sediment, may greatly enhance coalescence. The combined effects of creaming and coalescence eventually lead to a separation into two layers: oil and aqueous phase. Most instabilities are also affected by velocity gradients in the emulsion (Secs. IV–VII).

It is often stated that emulsions are inherently unstable because of their interfacial free energy, which would be released upon breaking of the emulsion. This then suggests that the emulsion would be less stable if the surface free energy were greater, for instance, because of a higher interfacial tension or smaller droplets (i.e., larger surface area). It may be enlightening to make a few sample calculations. Assume an oil–in–water emulsion, volume fraction of oil $\varphi = 0.1$, droplet diameter $d = 10^{-6}$ m, and interfacial tension $\gamma = 0.01\,N{\cdot}m^{-1}$; an emulsifying machine would actually need about $5 \cdot 10^6\,J{\cdot}m^{-3}$ to make the emulsion. What would be the free energy gain for the various instabilities?

Aggregation would lead to an energy release of 0.5 times the coordination number (say, 8), times the droplet number ($= 6\varphi/\pi d^3$), times the free energy involved in one particle–particle contact (say, 100 kT). This about:

$0.3\,J{\cdot}m^{-3}$

Creaming over a height H (say, 0.2 m) would lead to a release of $0.5\varphi\,\Delta\rho\,gH$, where ρ is density and g is acceleration due to gravity. This is, e.g.:

$10\,J{\cdot}m^{-3}$

Coalescence would lead to release of the interfacial free energy, $6\gamma\varphi/d$, which is about

$6 \cdot 10^3\,J{\cdot}m^{-3}$

or 0.1% of the mechanical energy that was need to create the interface.

It it interesting to compare this with the energy needed to heat a square meter of the emulsion by 1 K. This turns out to be about

$4 \cdot 10^6\,J{\cdot}m^{-3}$

In other words, even full coalescence would lead to a temperature increase (in an isolated system) of only 1.5 mK. Altogether, the free energy gains are very small and there is no correlation with the stability. For instance, creaming would always occur, since there is no activation free energy preventing it. But aggregation and coalescence are generally subject to an activation free energy, and may thus be retarded for a very long time. It may be concluded that the study of emulsion stability involves primarily kinetics and, where possible, activation free energies.

Emulsion stability may also be affected by the presence of a *third phase*. Air bubbles may have significant effects: it may cause droplet disruption, or it may induce (partial) coalescence. This has been discussed elsewhere [9]. A solid phase may affect an emulsion in various ways. Large particles or equipment parts may disturb or induce (partial) coalescence, especially in a velocity gradient. Small solid particles, if adsorbing at the oil–water interface and present in sufficient concentration to fully cover the interface, may impart stability to coalescence; this Pickering stabilization has been discussed in the first volume [2]. In other cases, solid particles may induce partial coalescence (Sec. VII). The presence of a liquid crystalline phase may greatly enhance emulsion stability, either by providing the emulsion droplets with a thick "protective" layer [10] or by forming a continuous network or gel that immobilizes the droplets.

High-internal-phase emulsions, microemulsions, and multiple emulsions will not be discussed. Neither will we cover phase behavior of (mixtures of) small-molecule surfactants and its implications for emulsion stability. This is mainly because the author does not feel competent to treat these subjects. Undoubtedly, this review will be biased in other respects also.

II. Adsorption

This subject has been treated in Volume 1 [1], and here only some aspects will be discussed. We recall that a solute adsorbing onto an oil–water interface (a surfactant) lowers the interfacial tension γ with an amount Π, so that $\Pi \equiv \gamma_0 - \gamma$, where γ_0 is the interfacial tension in the absence of surfactant; Π is called the surface pressure (expressed in $N{\cdot}m^{-1}$). For one surfactant and at equilibrium conditions, Gibbs' adsorption law holds in the form

$$d\Pi = -d\gamma = \Gamma RT\, d(\ln a) \tag{1}$$

Here Γ is the surface excess (in $mol{\cdot}m^{-2}$), RT has its usual meaning, and a is the activity of the solute in the solvent, i.e., usually in the continuous phase. Thus, the activity does not equal the total concentration of surfactant; it equals the concentration in the continuous phase only if its activity coefficient is unity (which is never the case above the critical micelle concentration). If the relation between γ and ln a is completely known, so is that between Π and Γ. Nevertheless, Eq. (1) is not a surface equation of state. The simplest relation, which is valid only for a small fraction of the interface being covered by surfactant and if lateral interactions between surfactant molecules in the interface are absent, is given by

$$\Pi = \Gamma RT = \left(\frac{n}{A}\right) RT \tag{2}$$

where n is the number of moles and A is the surface area. Equation (2) is the two-dimensional analogue of the ideal gas law: $p = (n/V)RT$. See Refs [1] and [11].

Surfactants are amphipolar compounds that come in two types: small soap-like molecules and polymers (including proteins, etc.). Some workers use the word *surfactant* for the former type only. Others calls them *emulsifiers*, because of their use in making emulsions. They have, however, many functions, which is why the general term surfactant will be used here.

A. Consequences of Adsorption

The presence of a surfactant in the interface has several consequences, nearly all of which may affect the stability of emulsions.

1. The lowering of the interfacial tension *facilitates deformation* of the interface in a direction perpendicular to it. The Laplace equation states that the pressure at the concave side of a curved interface is higher than that at the convex side by an amount

$$p_L = \gamma(R_1^{-1} + R_2^{-1}) \tag{3}$$

where R_1 and R_2 are the main radii of curvature. The Laplace pressure thus opposes deformation, and for a lower pressure, i.e., a smaller γ, a smaller external stress suffices to induce a certain deformation. This has important consequences for the coalescence stability (Sec. VI).

2. If the interfacial area A is enlarged, this causes an increase in interfacial tension, as would follow qualitatively from Eq. (2). This tendency is expressed in the *surface dilational modulus*

$$E_{SD} \equiv \frac{d\gamma}{d(\ln A)} \tag{4}$$

The modulus can be determined by measuring the change in interfacial tension when changing the interfacial area, without altering the shape of the surface element (Fig. 2). See Lucassen-Reynders [12]. If the surfactant is insoluble, and its amount in the interface thus cannot change, the surface behaves elastically. The modulus then is independent of time and equal to $d\Pi/d(\ln\Gamma)$, and it can be derived from the surface equation of state. For a soluble surfactant obeying Eq. (1), exchange of surfactant with the bulk can occur and the surface behaves visco-elastically. For a plane interface in contact with a semi-infinite layer of solution the modulus is then given by

$$E_{SD} = \frac{-d\gamma/d(\ln\Gamma)}{(1+2\zeta+2\zeta^2)^{1/2}} \tag{5}$$

where

$$\zeta = \frac{dm}{d\Gamma}\left(\frac{D}{2\omega}\right)^{1/2} \tag{5a}$$

and m is the molar concentration of surfactant in the bulk. D is the diffusion coefficient of the surfactant in the solvent (say, $2 \cdot 10^{-10}\ m^2 \cdot sec^{-1}$) and $1/\omega$ is a characteristic time scale. Figure 3 gives some examples of the surface dilational modulus at the air–water interface as a function of time scale. It is seen to decrease considerably with time, owing to surfactant diffusing to the expanded interface (or away from it on contraction). The modulus is higher for a higher surfactant concentration, except at very long time scales, when it is very small anyway for soluble surfactants. (Instead of a time-dependent surface dilational modulus, one may also use a surface dilational viscosity η_{SD}.) For a sinusoidal small expansion and contraction of the interface (which means for a longitudinal surface wave), $\omega = 2\pi\nu$, where ν is the frequency. It may be mentioned that longitudinal waves cannot exist in the absence of surfactants and that such waves are usually strongly damped. For a transient expansion of the interface, we may put $\omega = d(\ln A)/dt$.

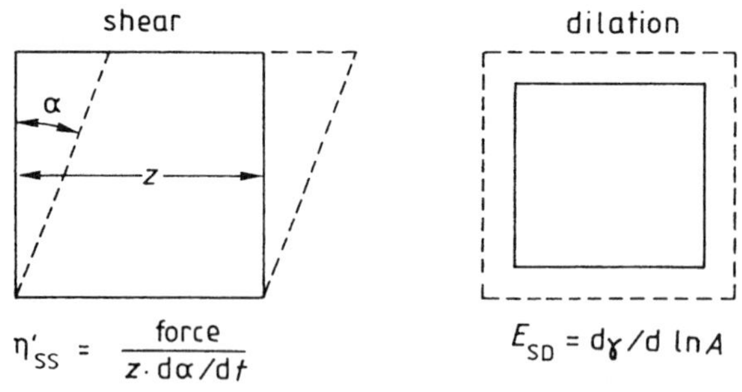

Figure 2. Illustration of the deformation of a surface element when performing interfacial rheology in pure dilation and in simple shear. (After Ref. 13.)

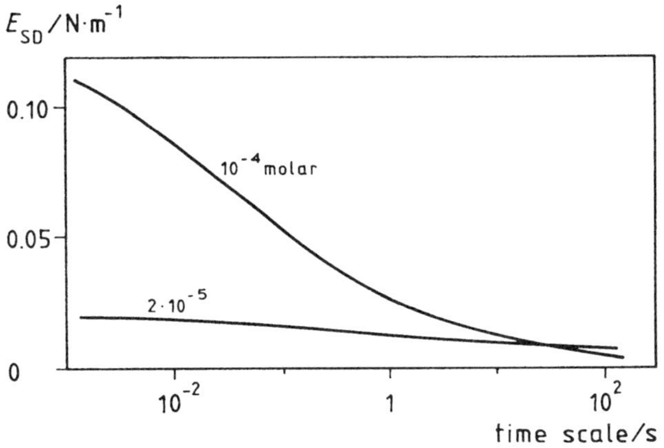

Figure 3. Surface dilational modulus E_{SD} of capric acid at the air–water interface as a function of time scale for two concentrations. (Recalculated from Ref. 12.)

These relations give, in principle, the surface dilational modulus for a variety of conditions, at least for small-molecule surfactants and for deformations that are not too large. The relations do not hold for macromolecular surfactants, because factors other than mass transport of surfactant then will play a part. Very few measurements have been made at the oil–water interface, but it can be done by generating longitudinal waves (at frequencies of 10^{-3} to $0.2\,\text{sec}^{-1}$) [14] or in an overflowing cylinder apparatus (i.e., for a fairly slow expansion of the interface) [15]. Surface dilational properties are of great importance for Ostwald ripening and for the next point.

3. An interface containing surfactant can withstand a tangential stress. This is illustrated in Fig. 4. If liquid flows along a clean interface, the liquid on the other side of the interface flows as well and the velocity is continuous across the interface (Fig. 4a). If the interface contains a surfactant, the flow tends to move the surfactant in the interface downstream, thereby creating an *interfacial tension gradient* (Fig. 4b). In the absence of other forces, the shearing stress exerted by the flowing liquid and the stress caused by the interfacial tension gradient will be equal:

$$\eta\left(\frac{dv_x}{dy}\right)_{y=0} = -\frac{d\gamma}{dx} \tag{6}$$

where η is the viscosity of the liquid, v is the linear liquid velocity, x is the coordinate in the direction of flow, and y is the coordinate perpendicular to the interface. In this way, lateral motion of the interface is stopped if a sufficiently large interfacial tension gradient can develop. Because dx is very small for an emulsion droplet, say, 1 μm, and the shearing stress exerted by the flow is rarely over 1 Pa, a $d\gamma$ of about $10^{-6}\,\text{N}\cdot\text{m}^{-1}$ wouald be sufficient to stop the interface from moving. Even the slightest amount of surfactant present would allow such a value of $d\gamma$. Of course, the gradient would depend on the value of E_{SD}, but a large shearing stress implies a large velocity gradient, hence a short

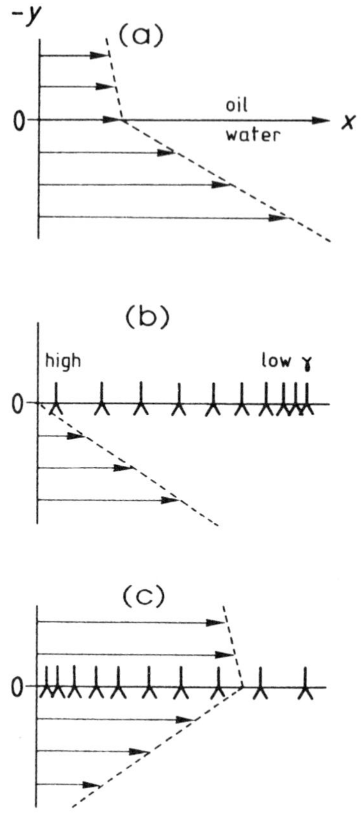

Figure 4. Interfacial tension gradients and flow. (a) No surfactant, velocity gradient continuous across interface. (b) Velocity gradient causes interfacial tension gradient. (c) Interfacial tension gradient causes flow: Marangoni effect.

time scale, hence a large modulus. Thus the surface of emulsion droplets always is immobile in the lateral direction, unless we have the combined effects of a very large droplet, a high shear stress (vigorous stirring and a high viscosity), and a very low surfactant concentration. It may be argued that Eq. (5) would predict that for a very high surfactant concentration the surface dilational modulus also becomes negligible, but this is rarely the case in practice, since commercial surfactants always contain compounds of various surface activity, and this causes the modulus to be still appreciable.

The existence of an "immovable" interface is of great importance in relation to aggregation in a shear field, sedimentation, and coalescence. Incidentally, this is all that makes existence of a foam possible, since drainage of liquid from the lamella between foam bubbles would otherwise be so fast as to cause immediate breaking of the foam [16].

The complement of the shear stress causing an interfacial tension gradient is the Marangoni effect: If for some reason (say, an uneven supply of surfactant) an interfacial tension gradient exists, it will cause a shearing flow in the adjoining liquid(s) (Fig. 4c). This is all that makes formation of an emulsion possible [7].

4. If a surface element is deformed in shear without changing its area (see Fig. 2), this consumes energy if a surfactant is adsorbed. One often measures a *surface shear viscosity* [17]; it is negligibly small for most small-molecule surfactants, but it may be appreciable for a macromolecular one. The value does not depend on exchange of surfactant with the bulk, since the area remains the same. Essentially, one measures the (bulk) viscosity of an adsorbed layer of unknown thickness; the result is expressed in $N \cdot sec \cdot m^{-1}$. This will be discussed in more detail in relation to coalescence (Sec. VI).

5. The presence of a surfactant in the interface of an emulsion droplet implies mostly that the *colloidal interaction forces* acting between two closely approaching droplets are greatly altered. If the surfactant is charged, electrostatic repulsion will occur; if flexible molecular chains protrude into the continuous phase, considerable steric repulsion may occur. This is what makes stability against aggregation—and thereby against coalescence—possible. This will be discussed in Sec. V.

6. If solid particles are present in the oil–water interface, the presence of a surfactant may alter the *contact angle* solid–oil–water (or even cause desorption of the particles), thereby affecting Pickering stabilization [2] and partial coalescence (Sec. VII).

B. Polymers as Surfactants

Over the last 15 years or so, considerable work has been done (both theoretical and experimental) on the adsorption of synthetic polymers. Here we will largely follow the recent monograph by Fleer et al. [18]. Most of the work is on adsorption at the solid–liquid interface, and the surface excess, the density profile of the polymer layer, and the colloidal interaction between two closely approaching interfaces are considered. Adsorbed polymers may, of course, provide strong steric repulsion and thereby stabilize dispersions against aggregation (Sec. V.B). For a homopolymer, important variables are the interaction energies between polymer segments, solute molecules, and adsorption sites (expressed, for instance, in Flory's χ parameters), the polymer length, and its flexibility. Many polymers have two or more different building blocks; these copolymers can be of various types, such as diblock or random copolymers. Copolymers may provide far better means of stabilization of particles, since the segments of the one type may have great affinity for the solid surface and those of the others may have affinity for the solvent, thereby protruding fairly far into the solvent.

Relatively little work has been done on polymers at the *liquid–liquid interface*, where the situation is rather different, since polymer segments can protrude for a considerable distance in both phases. Results obtained on the air–water interface may not be quite representative, since polymer segments presumably cannot protrude into the air. Figure 5 illustrates adsorption of some polymer types at the liquid–liquid interface. When using a polymer as an emulsifier, i.e., during emulsification, Bancroft's rule dictates that it must be soluble in the future continuous phase [7]. This effectively rules out the use of homo-

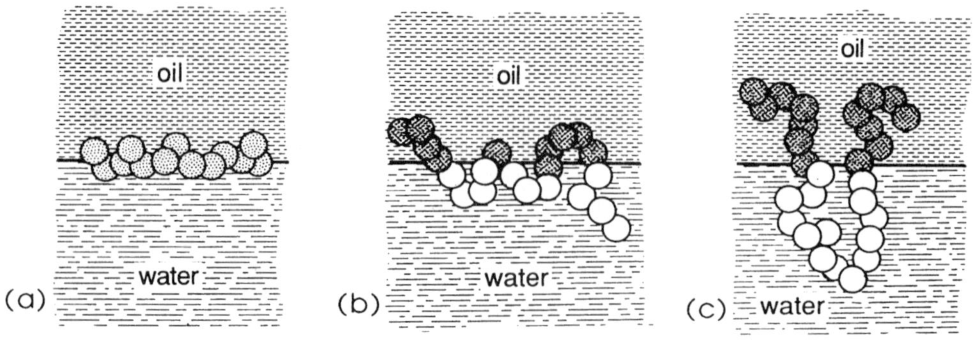

Figure 5. Polymers adsorbed onto an oil–water interface. (a) Homopolymer. (b) Random copolymer. (c) Triblock copolymer. The lighter the color of a segment, the higher its affinity for water. Highly schematic. (From Ref. 18.)

polymers, since the necessary properties of being reasonably well soluble in the continuous phase and at the same time having sufficient affinity for the disperse phase to become adsorbed are virtually incompatible. The "ideal" emulsifier may then be a diblock copolymer, with a fairly small part that has great affinity for the disperse phase (and thereby anchoring the molecule) and a longer part that has great affinity for the continuous phase (thereby ensuring sufficient solubility of the polymer to allow emulsifier action and sufficient protrusion into the continuous phase to provide steric repulsion). The density profile may then be something like that depicted in Fig. 6a, and the effective thickness of the adsorbed layer is of the order of the radius of gyration of the dissolved polymer (block). Similar profiles may occur for triblock copolymers, where "loops" as well as "tails" may protrude. Random copolymers may show a profile more

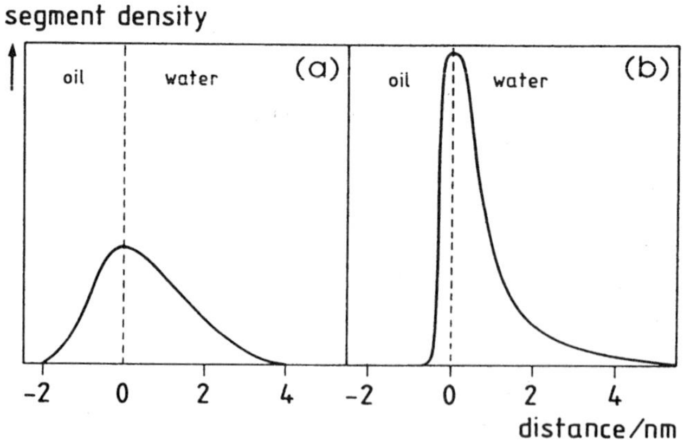

Figure 6. Density profiles of polymer molecules adsorbed at the oil–water interface. (a) Diblock copolymer with 20 hydrophobic and 40 hydrophilic segments. (b) Random copolymer of about 80 segments. Highly schematic. (Recalculated from data in Ref. 18.)

like that in Fig. 6b. The profile naturally depends on polymer length, flexibility, and solvent quality. Density profile types describes as "mushroom" (where the segment density first increases, then decreases with distance from the surface) or "brush" (where a fairly dense layer of roughly parallel polymer tails extends from the surface) are unlikely for random copolymers.

High polymers that do adsorb are always very surface active in the sense that even at very small bulk concentrations considerable adsorption occurs; see, for instance, Fig. 9, below. This may, broadly speaking, be explained by the great number of polymer segments that are simultaneously adsorbed. If the adsorption free energy would be kT per segment, a reasonable value, it would be many times kT for one molecule, which implies a very high affinity for the interface. Most polymers do not give a very low interfacial tension, presumably because tight packing of adsorbed molecules or segments at the oil–water interface cannot be attained. In accordance with this, the surface excess generally is a few milligrams per square meter, comparable to the values observed for small-molecule surfactants, despite the polymer layer being much thicker (Fig. 6).

It is generally held that Gibbs equilibrium is not reached for adsorbing polymers; i.e., Eq. (1) would not hold. This opinion may be based primarily on the observation that it is virtually impossible to desorb a polymer from emulsion droplets by "washing," i.e., by repeatedly diluting the emulsion with solvent and concentrating it, for instance, by centrifugation. It should, however, be realized that the very high surface activity (see Fig. 9, later) implies that a bulk concentration of polymer below that leading to a plateau value for adsorption, say, $10^{-5}\,\text{mol}\cdot\text{m}^{-3}$, virtually cannot be reached. Assuming that after dilution $\varphi = 0.01$, droplet diameter $= 0.6\,\mu\text{m}$, and $\Gamma_{\text{plateau}} = 10^{-7}\,\text{mol}\cdot\text{m}^{-2}$, the total concentration of polymer in the emulsion would be $10^{-2}\,\text{mol}\cdot\text{m}^{-3}$. Consequently, only 0.1% of the adsorbed polymer would have to desorb for the bulk concentration to reach $10^{-5}\,\text{mol}\cdot\text{m}^{-3}$. As a result, about 100 washings would be needed to significantly decrease Γ.

In experiments with a macroscopic interface, washing can lead to desorption, because of the much larger ratio of volume to surface. Even then, desorption may be very slow, because the very high surface activity also implies that the (absolute) concentration difference between surfactant in the bulk and adjacent to the surface is very small, which means a very small driving force for desorption. Rough calculation of diffusional transport results in desorption times of at least some days, in accordance with experimental results [108].

All this would mean that it often takes a much longer time than the duration of the experiment for equilibrium to be reached. Incidentally, the reasoning just given does not imply that any adsorbed polymer molecule will never desorb: It may, slowly and segment by segment, be displaced by another molecule, especially if the other molecule would give a lower interfacial free energy.

There is another reason why Eq. (1) would not apply to adsorption of polymers. Any synthetic polymer is polydisperse, and (other conditions being equal) the longer the molecule, the more surface active it is. The Gibbs equation in its simplest form is valid only for one surfactant, and applying it to mixtures would lead to gross errors in Γ. Random copolymers show an even greater heterogeneity, and adsorption of a polyelectrolyte, even if it is a pure protein (all molecules identical), always implies adsorption of a mixture. This is because adsorption of a polyelectrolyte is accompanied by a change in ion adsorption. Displacement of less surface-active (e.g., smaller) molecules by more surface-active ones will be a slow process, since the concentration of the latter will generally be small. It is indeed commonly observed that, upon adsorption of a polymer from a bulk phase to an oil–water interface, it takes hours or even days before the interfacial tension does not change any more, whereas the time needed for diffusion of polymer to the interface may only be a few minutes. The change in γ may also be due to slow conformational changes of molecules that remain adsorbed and possibly other aging effects. In general, any change in the system (say, altering temperature or diluting it) will cause slow changes in the interfacial layer.

Indeed, it is often observed that the interfacial tension depends on the history of the system, say, a temporary expansion of the interface, even after long waiting times. The mentioned observations would imply that emulsions may not exhibit a true adsorption isotherm, i.e., a fixed relation between surface excess and concentration of the polymer in the continuous phase. What is found is a more or less fixed relation between surface excess and c_e/A, where c_e is the concentration of polymer in the emulsion and A is the specific interfacial area created during emulsification [6]; see Fig. 7. The reader should realize that there is another reason why results for surface properties obtained after adsorption at a plane oil–water interface may differ greatly from those obtained by emulsification. In the first case, the ratio of surface area to volume may be orders of magnitude smaller than in emulsions, which implies great differences in composition of the surface layer for a heterogeneous surfactant.

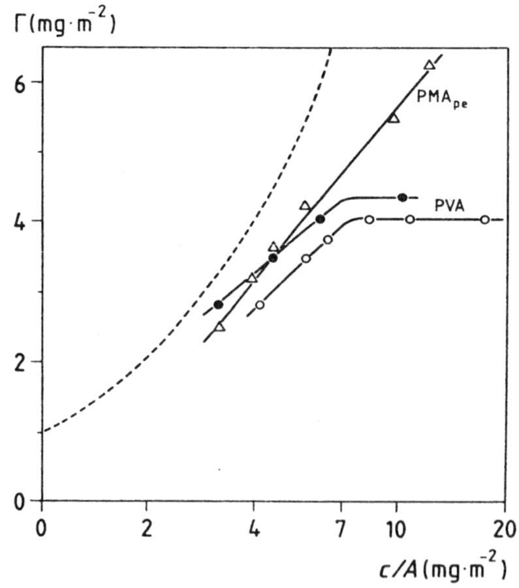

Figure 7. Surface excess Γ of polymers adsorbed onto emulsion droplets as a function of the total concentration c over the specific surface area created during emulsification A. (From data in Refs. 19 and 20.)

As an example of a suitable copolymer for making and stabilizing oil–in–water emulsions, P(VA-VAc) may be mentioned, i.e., a polyvinyl alcohol containing (mostly 2 to 12%) vinyl acetate segments, which occur in several blocks [19]. Presumably these blocks are responsible for the anchoring to the oil droplets. See also Figs. 6b and 7. It is seen that the plateau value of the surface excess is mostly 2–4.5 $mg{\cdot}m^{-2}$, depending predominantly on the molar mass. The corresponding interfacial tensions at the paraffin oil–water interface ($\gamma_0 \approx 50\, mN{\cdot}m^{-1}$) are in the range 14–25, with a higher acetate content giving a lower value.

For some polymers the surface dilational modulus was determined at the oil–water interface by application of longitudinal waves of various frequencies (see Fig. 8). It is seen that the modulus is fairly small. When expanding the interface, one would expect the surface excess to be proportionally decreased (because it would take a long time for more polymer to adsorb), which would then lead to a high modulus. However, expansion of the interface may lead to a change in conformation of the adsorbed polymer, in such a way that more segments come into contact with the interface, which would to a considerable extent compensate for the increase in interfacial tension. Thus it is not strange that the modulus would not depend on the time scale, which is observed for one of the samples within the frequency range studied. The other sample does show a frequency dependence; the explanation is unclear [19].

Some polymers are *polyelectrolytes*, e.g., polyacids. Homopolyelectrolytes [e.g., poly(metacrylic acid)] do not adsorb at all if all segments have an ionized group. If the polymer is only partly ionized (for instance, because of a lower pH), it is effectively a copolymer, and adsorption may occur, the more so for a lower degree of ionization. It may be better to use a true heteropolymer, e.g., a partially esterified PMA. Partial polyelectrolytes at the solid–water interface give a lower surface excess for a lower ionic strength; this is because it would go along with a stronger repulsion between molecules that are close to each other, as in an adsorbed layer. Results of adsorption onto emulsion droplets do show the expected dependence of the surface excess on pH (i.e., degree of ionization), but

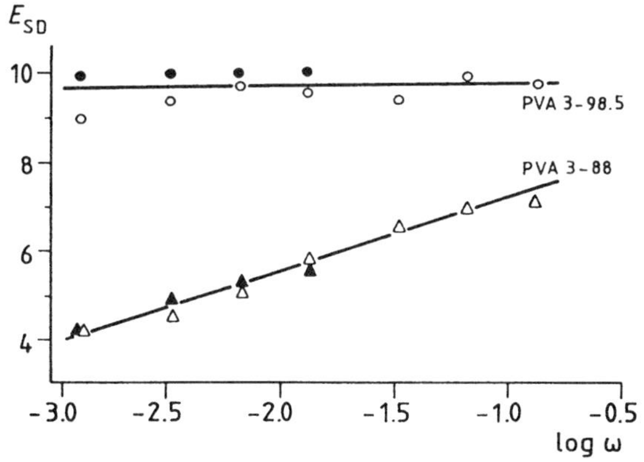

Figure 8. Surface dilational modulus E_{SD} as a function of frequency ω for 0.04% solutions of two types of PVA at the air–water interface (open symbols) and the oil–water interface (closed symbols). (After Ref. 19.)

ionic strength appears to have little effect [20]. Some results on surface excess are shown in Fig. 7.

C. Proteins as Surfactants

Proteins are often used to make and stabilize oil-in-water emulsions. Proteins have some 20 different building blocks that vary greatly in bulkiness, hydrophobicity, and charge, with positive, negative, and uncharged segments occurring in the same molecule. Moreover, most proteins have a specific, highly ordered conformation, and these so-called globular proteins do not fit into any of the molecular theories developed for polymer adsorption, although they qualitatively fit some of the observations made on synthetic polymers. A great amount of experimental work has been done on proteins at oil–water interfaces and in emulsions, too much to review here. See, for instance, two recent reviews [13,21], from which some important aspects will be discussed later.

Figure 9 illustrates the surface activity of proteins as compared to small-molecule surfactants. For reasons discussed in Sec. II.B, there appears to be no Gibbs equilibrium. The interfacial tension at the triglyceride oil–water interface varies from 8 to 13 $mN{\cdot}m^{-1}$ for most proteins; at the paraffin oil–water interface, values between 17 and 22 are observed. The surface excess follows roughly

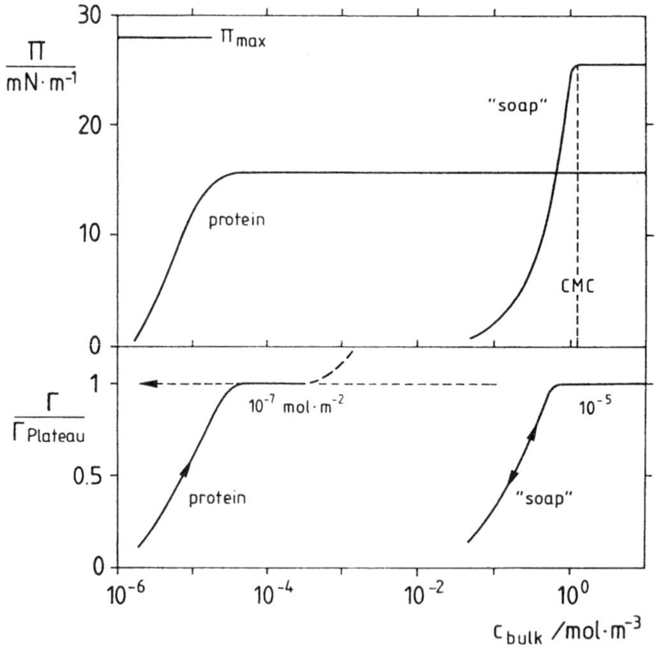

Figure 9. Surface pressure Π and surface load Γ as a function of the concentration in the continuous phase c_{bulk} of a typical protein and a typical small-molecule surfactant ("soap"). CMC = critical micelle concentration. Highly schematic examples. (From Ref. 13.)

the same trends as given in Fig. 7, and the plateau value is about $3\,mg{\cdot}m^{-2}$ for many proteins, though higher for those of a very high molar mass or for aggregated proteins. Γ tends to be higher for a pH closer to the isoelectric point. For caseinate, it has been observed that the surface excess is higher for a higher ionic strength [22], in qualitative agreement with theory.

All proteins appear to change considerably their conformation on adsorption onto hydrophobic surfaces. Globular proteins apparently exhibit a stronger conformational change at the oil–water interface than at the air–water interface. The change in conformation may be considered a denaturation of the protein (but denaturation prior to adsorption, e.g., by heat treatment, may result in still different properties of the adsorbed layer). At a surface excess of about $1\,mg{\cdot}m^{-2}$, the protein presumably is fully denatured, since this would correspond to a fully extended peptide chain. Values of Γ below 1 are usually not observed in emulsions.

Most proteins appear to be adsorbed almost irreversibly. MacRitchie [23] studied the desorption of various proteins spread at the air–water interface, by keeping the surface pressure constant. Table 1 gives some of the results in terms of a characteristic time needed for desorption, and it is seen to increase with molar mass and to decrease with increasing surface pressure. However, for protein adsorbed onto emulsion droplets, no appreciable desorption is observed, for reasons discussed in Sec. II.B; moreover, any desorption would lead to lower surface pressure and thereby to even slower desorption.

Proteins can displace one another, especially if they differ markedly in their propensity to lower the interfacial tension. For instance, casein will displace gelatin and, more specifically, β-casein and α_{s1}-casein will displace each other to form a layer of equilibrium composition. Caseins, however, are proteins with very little secondary structure, and they may form adsorption layers not unlike that depicted in Fig. 6b. Globular proteins behave differently. This may be inferred, for instance, from measurements of the surface shear viscosity at the oil–water interface. Caseins give a low value, about $2 \cdot 10^{-3}\,N{\cdot}sec{\cdot}m^{-1}$, which does not change on aging. Globular proteins give values of the order of $1\,N{\cdot}sec{\cdot}m^{-1}$, which seems to imply strong lateral interactions between adsorbed molecules. Moreover, their η_{ss} values often increase upon aging, which may

Table 1. Characteristic Times (hr) Needed for Desorption of Various Proteins (Molar Mass M) from the Air–Water Interface[a]

Protein	M (kD)	Surface pressure ($mN{\cdot}m^{-1}$) 15	20	25	30	35	40	45
Insulin	6	3	0.5					
β-Lactoglobulin	17		8	3	2			
Myoglobin	17			5	2.5	1		
γ-Globulin	160				18	8	4	
Catalase	230					5	2	1.5

[a]The proteins were spread on the surface and the surface pressure was kept constant by means of moving barriers, causing the surface area between the barriers to decrease.
Source: After results from Ref. 23.

signify that cross-linking reactions occur in the interfacial layer; especially proteins that are prone to –S–S– interchange reactions do so. This would, naturally, greatly hinder desorption.

Small-molecule surfactants may readily desorb proteins from an interface, since they tend to give significantly lower interfacial tensions (say, $3\,mN{\cdot}m^{-1}$) if present in sufficiently high concentration. Even at low concentrations they become associated with protein-covered interfaces, considerably altering surface properties (for instance, the lateral mobility of the protein). Aged layers of mixed proteins, however, may have undergone such intermolecular cross-linking that they cannot be desorbed even by large quantities of surfactant.

Few measurements of the surface dilational modulus caused by proteins have been made, and only at the air–water interface. For caseins, values of about 10–20 and for globular proteins of about $30\text{–}100\,mN{\cdot}m^{-1}$ have been observed [104,105]; there is no marked dependence on time scale. Repeated expansion and contraction of the interface may, however, greatly increase the modulus.

It may finally be noticed that all values for adsorbed proteins may depend markedly on pH, ionic strength, presence of divalent counterions, temperature, and other variables affecting solvent quality. As mentioned, values may also depend on the history of the layer.

III. Ostwald Ripening

Ostwald ripening is a well-known phenomenon in foams (where it is often called disproportionation) and solid dispersions, and it has long been realized that it may also occur in water–in–oil emulsions. More recently [24], it was shown that it may occur in emulsions of hydrocarbons in water, especially for hydrocarbons of a lower molar mass.

A. Driving Force

The driving force for Ostwald ripening is due to the increased solubility (or chemical potential) of a substance bounded by a curved interface and is generally expressed in the Kelvin equation [25], which, for a droplet of radius r, consisting of one component, takes the form:

$$s(r) = s_\infty e^{x'/r} \approx s_\infty \left(1 + \frac{x'}{r}\right) \tag{7}$$

where s is solubility of the droplet substance in the continuous phase (in $kg{\cdot}m^{-3}$) and s_∞ is the "normal" solubility, i.e., for a droplet of infinite radius. The characteristic length scale x' is given by

$$x' = \frac{2\gamma M_D}{\rho_D RT} \tag{7a}$$

where M and ρ refer to the molar mass and the mass density, respectively, of the disperse phase. It may be noted that the approximate version of Eq. (7) can also be derived from the Laplace equation (3). For many materials, x' will be 1 nm or less, implying that s(r) is higher than s_∞ by, say, 0.1% for a 1-μm droplet.

Nevertheless, Ostwald ripening can be appreciable in such a case, provided that s_∞ is finite.

Transport of material may thus occur from droplets of small radii to those of large radii. This may be called isothermal distillation. The transport takes place by diffusion, and it is generally assumed that individual molecules diffuse. In the presence of a small-molecule surfactant at a concentration above the CMC, some of the material of the disperse phase may be present in micelles in the continuous phase, and it cannot be ruled out that this would enhance Ostwald ripening; it appears, however, that molecular diffusion is the rate-limiting process, at least in the case of ionic surfactants [26]. For oil–in–water emulsions containing protein and a nonionic surfactant, transport of oil by means of association with the surfactants has been reported [27]; it is as yet uncertain whether this would induce substantial Ostwald ripening.

B. Disappearance of a Small Droplet

The classical treatment of the rate of Ostwald ripening for foam bubbles is by de Vries [28]. He envisaged a small bubble, surrounded by large bubbles (large enough to have negligible curvature) at an average distance δ. By combining the Kelvin equation with the Fickian diffusion equations, he arrived at a relation for the bubble radius r with time t, which for an emulsion droplet can be written as

$$r^2(t) = r_0^2 - \frac{2x'Ds_\infty}{\rho_D \delta} t \tag{8}$$

where r_0 is the original droplet diameter and D is the diffusion coefficient of the material of the disperse phase in the continuous phase. The equation shows that the decrease in radius of the droplet will proceed ever faster as the droplet becomes smaller (assuming that the interfacial tension remains constant). The lifetime of the droplet [i.e., t for $r(t) = 0$] is thus inversely proportional to its radius squared; some examples are given in Table 2 to illustrate the effects of interfacial tension, solubility, diffusion coefficient, etc.

Table 2. Calculated Examples of the Time t Needed for a Small Droplet to Disappear by Ostwald Ripening According to Eq. (8)

Continuous phase Disperse phase	Triglyceride Water	Water Benzene	Water Benzene	Water Decane
γ ($mN \cdot m^{-1}$)	0.01[a]	0.035	0.002[a]	0.01[a]
M_D ($kg \cdot mol^{-1}$)	0.018	0.078	0.078	0.142
ρ_D ($kg \cdot m^{-3}$)	998	880	880	730
x' (nm)	0.14	2.5	0.14	1.6
D ($m^2 \cdot sec^{-1}$)	$2 \cdot 10^{-11}$	10^{-9}	10^{-9}	$8 \cdot 10^{-10}$
s_∞ ($kg \cdot m^{-3}$)	1.4	0.7	0.7	$52 \cdot 10^{-6}$
t (sec)	$12 \cdot 10^5 \approx 14$ days	$2500 \approx 40$ min	$45,000 \approx 12$ hr	$47 \cdot 10^6 \approx 1\frac{1}{2}$ yr

[a] Surfactant present.
$R = 8.3\,J \cdot K^{-1} \cdot mol^{-1}$; $T = 293\,K$; $\delta = 10^{-5}$ m; $r_0 = 10^{-6}$ m.

C. Change in Size Distribution

In practice there is a size distribution of droplets; and small droplets will disappear, whereas large droplets will grow, as depicted in Fig. 10a. The so-called LSW treatment (after Lifshitz and Slyozov, and Wagner) is reviewed by Kabalnov and Shchukin [26], who also have done considerable experimental work on emulsions and whose review we will follow here. The theory assumes: (1) the droplets are fixed in space; (2) the distance between droplets is very much larger than their size, implying that the volume fraction of droplets $\varphi \rightarrow 0$; and (3) the concentration of the material of the disperse phase in the continuous phase is everywhere the same, except in the neighborhood of a droplet. It then turns out that Ostwald ripening eventually leads to a size frequency distribution of constant shape, given by

$$f(u) = \frac{u^2 \exp[3/(2u-3)]}{(u+3)^{7/3}(1.5-u)^{11/3}}, \qquad 0 \le u \le 1.5 \tag{9}$$

where $u = r/r_c$; for $r \gg r_c$, $f(u) = 0$. The distribution shape is given in Fig. 10b. The change in the mean droplet radius then is given by

$$\frac{d\langle r\rangle^3}{dt} = \frac{4x'Ds_\infty}{9\rho_D} \tag{10}$$

Because of the time invariance of the size distribution, the same relation holds for r_c, which at any moment is the radius of the droplets that neither grow nor shrink. An example of the change of this critical radius is in Fig. 10a. From Eq. (10) it can be derived that the time that would be needed for r_c doubling to occur is $9r^3\rho/4x'Ds_\infty$, i.e., $9r/2\delta$ times the time needed for a droplet of radius r to disappear according to the de Vries theory. This implies that even for a decane-in-water emulsion as given in Table 2, but with an average droplet radius of only 0.1 μm (about the smallest that can be achieved in a high-pressure homogenizer),

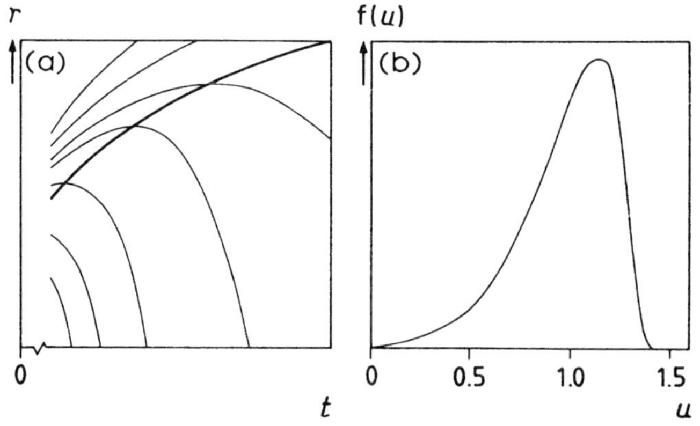

Figure 10. Ostwald ripening. Part (a) gives schematic examples of the change in radius of various droplets in an emulsion (the heavy line denotes the average radius), assuming the equilibrium size distribution f(u) depicted in part (b) exists. (Results after Ref. 26.)

the time needed to double the average radius would be only 3.4 hr. For a similar dodecane-in-water emulsion this time would be of the order of 4 wk. See, however, the following points 1 and 2.

The theory has been checked and found to be fairly correct for droplets fixed on a glass plate [26]. Nevertheless, there may be deviations for real emulsions.

1. The theory applies only for very small φ. Numerical simulations for higher-volume fractions lead to a somewhat enhanced growth rate, e.g., by a factor of 2 for $\varphi = 0.2$.
2. The droplets are not fixed in space, but show Brownian motion. In carefully controlled studies, a growth rate up to about 2.5 times that predicted by Eq. (10) was observed, presumably due to this effect.
3. Even taking these first two factors into account, it appears that Ostwald ripening in paraffin oil-in-water emulsions stabilized by nonionic surfactants may be quicker than predicted. Possibly, some transport of oil by micelles occurs.
4. It is assumed in the calculations that the size distribution is according to Eq. (9). But, of course, it takes some time before it has developed, especially if the initial size distribution is markedly different. Numerical simulations show that it may take some orders of magnitude times the critical time, implying that the average radius may already be, say, 10 times the original one before the time-invariant distribution shape is more or less attained.

Altogether, accurate prediction of the rate of Ostwald ripening will rarely be possible, but estimates may often be better than order-of-magnitude.

D. Retarding Ostwald Ripening

There are various conditions that may lead to diminishing the rate of, or even stopping, Ostwald ripening.

1. If the disperse phase consists of *two components* of different solubility in the continuous phase, this may lead to retardation. The simplest example is that of a disperse phase consisting of a main component that has some solubility, and a solute that has zero solubility in the continuous phase. Water droplets containing some salt in a triglyceride oil provide an example. The simplest way to calculate the effect is via the Laplace pressure of the droplet ($2\gamma/r$), which would increase upon shrinkage of the droplet, and the osmotic pressure, which would increase as well, thereby leading to a lower activity of the diffusible solvent (i.e., water in the present case) and thus to a driving force for back-diffusion of the solvent. In the ideal case, the osmotic pressure is given by

$$\pi_{\mathrm{osm}} = \mathrm{mRT} = \mathrm{m_0}\left(\frac{\mathrm{r_0}}{\mathrm{r}}\right)^3 \mathrm{RT} \qquad (11)$$

where m is molar concentration and the subscript 0 refers to the original droplet. No change in the size of a droplet will occur if the change in Laplace pressure with radius equals that in osmotic pressure, which implies

$$2\gamma = 3mRTr \tag{12}$$

Assuming $\gamma = 0.01\,N{\cdot}m^{-1}$ and $r = 10^{-6}$ m, an osmolarity of 2.67 mol·m^{-3} would be needed; this corresponds to less than 0.01% NaCl. Very small quantities of a suitable solute are thus sufficient to stop Ostwald ripening.

If the droplets contain two components, 1 and 2, that differ in solubility in the continuous phase, equations for the rate of Ostwald ripening can also be derived [26]. If the condition $\psi_2 < s_{2,\infty}/s_{1,\infty}$ is fulfilled, it follows that

$$\frac{d\langle r\rangle^3}{dt} \approx \frac{4x_1' D_2 s_{2,\infty}}{9\psi_2} \tag{13}$$

where ψ refers to the volume fraction of the component in the droplet. The rate is thus determined by the characteristic length scale for component 1 and the solubility and the diffusivity of component 2. Equation (13) gives the most common case. If the condition just stated is not met, the rate is determined roughly by Eq. (10), calculated for component 1.

2. The *surface dilational modulus* (see Eq. 4) of the droplet interface may play a part. When a droplet shrinks, its surface excess increases. As long as the extra surfactant does not desorb, the interfacial tension is decreased, which would lead to a lowering of the solubility of the droplet material in the continuous phase. As was already noticed by Gibbs, the condition where the increase in solubility due to a decrease in r offsets the decrease due to a decrease in γ is given [29] by

$$E_{SD} > \frac{\gamma}{2} \tag{14}$$

Since desorption would cause the surface dilational modulus to decrease rapidly with time for many surfactants (see Fig. 3), the retardation of Ostwald ripening may be negligible.

However, for macromolecular surfactants, the modulus may remain high for a long time (see, e.g., Fig. 8), since these surfactants may desorb very sluggishly from the compressed interface (see, e.g., Table 1). In such a case Ostwald ripening may be significantly retarded. Prins and coworkers. (e.g., Ref. [30]) have studied this for gas bubbles; they expressed the time-dependent surface properties by a surface dilational viscosity, defined as $-\Pi/[d(\ln A)/dt]$, and found for protein solutions a power law dependence. This led to a predictable retardation of Ostwald ripening, by about one order of magnitude for the systems studied. Although to the author's knowledge studies on emulsions have not been made, there is no reason why this approach would not be valid. It may also be mentioned that it is not uncommon for the surface dilational modulus to increase when considerable shrinkage of the interface occurs: The substances adsorbed that are most difficult to desorb will remain. An extreme example is that of adsorbed solid particles, which may not desorb at all within any reasonable time scale (cf. Sec. II.C). In such a case, complete stabilization against Ostwald ripening can be achieved.

3. Ostwald ripening can also be stopped if the continuous phase has a sufficiently high *yield stress*. (This is, in fact, similar to the situation just mentioned, where the layer of adsorbed particles provides a local yield stress.) The yield stress should be larger than the Laplace pressure of the droplet; for a

droplet of radius 0.1 μm and $\gamma = 10\,\text{mN·m}^{-1}$, this amounts to $2 \cdot 10^5$ Pa. Such high yield stresses cannot readily be attained with most gelling agents. Freezing of the emulsion, if it does not cause other damage, will, however, stop Ostwald ripening (as well as sedimentation, aggregation, and coalescence).

4. It has been assumed that adsorbed layers of some surfactants may provide a substantial *diffusional resistance* for the material in the droplets, thereby retarding Ostwald ripening in emulsions [26]. For some surfactants (phospholipids) a retardation has been observed; in the author's opnion, however, it is more likely that a fairly high surface dilational modulus was the cause of the retardation, since marked diffusional resistance caused by adsorbed monolayers has never been established in independent experiments.

IV. Sedimentation

In this section, Refs. [2,31,32] will be followed, unless indicated otherwise.

A. The Stokes Equation

According to Archimedes, the buoyancy force F_B acting on a submerged sphere is given by

$$F_B = \tfrac{1}{6}\, a\pi d^3(\rho_D - \rho_C) \tag{15}$$

where subscripts D and C refer to droplet and continuous phase, respectively. The acceleration a can be due either to gravity g ($\approx 9.81\,\text{m·sec}^{-2}$) or to centrifugation $R_e\omega^2$, where R_e is the effective radius of the centrifuge and ω is its rotation rate in rad·sec^{-1}. If $\rho_D - \rho_C = \Delta\rho \neq 0$, the droplet will move through the liquid and will feel a drag force F_S, which according to Stokes is given by

$$F_S = fv = 3\pi d\eta_C v \tag{16}$$

where f is the friction factor and v is the linear velocity of the droplet relative to the continuous phase. At equilibrium both forces will be equal and opposite, and the Stokes velocity is then given by

$$v_S = \frac{F_B}{f} = \frac{a\,\Delta\rho\, d^2}{18\eta_C} \tag{17}$$

If $\Delta\rho < 0$, as is the case in most oil–in–water emulsions, the droplet will sediment in a direction opposite to that of the acceleration; i.e., creaming occurs. For example, an oil droplet of d = 1 μm in water, $\Delta\rho = 100\,\text{kg·m}^{-3}$, $\eta_C = 1$ mPa·sec, would cream under gravity at a rate of $55\,\text{nm·sec}^{-1} \approx 5$ mm/day; an aqueous droplet of 9 μm in oil, $\eta_C = 0.08$ Pa·sec, would sediment at about the same rate.

It is often tacitly assumed that Eq. (17) can be applied to predict the sedimentation (creaming) in an emulsion in a vessel, which would imply (1) that the equation is correct, (2) that the equilibrium velocity v_S would be almost instantaneously attained, and (3) that the equation can simply be applied for the rate of arrival of droplets in a sediment layer, which then would be given by

$$q = \frac{v}{H} = \frac{a\,\Delta\rho\,d^2}{18\eta_C H} \tag{18}$$

where q is the fraction of the droplets arriving per unit time and H is the maximum sedimentation distance. Assumption (2) is always met, but the others are not always true. The various conditions for validity will now be discussed.

1. The droplets should be *perfect, homogeneous spheres*. Even a very large drop will remain spherical during sedimentation, since the shearing stress caused by the motion of the drop will be negligible compared to its Laplace pressure (Eq. 3). Very small droplets covered by macromolecules may be inhomogeneous, since the adsorbed layer may significantly affect the hydrodynamic diameter and the effective $\Delta\rho$.

2. The *interface* of the droplet should be *immobile*, since otherwise the liquid in the droplet is caused to move, thereby enhancing the sedimentation rate. As mentioned in Sec. II.A, this condition will commonly be met. By combining Eqs. (6) and (17) and taking the shear stress at the droplet surface as $3v/d$ (see below), it follows that the condition for the interfacial tension difference caused by the shearing force of the moving droplets is

$$\Delta\gamma > \tfrac{1}{6}\, a\,\Delta\rho\, d^2 \tag{19}$$

For an oil droplet of $d = 10\,\mu m$ in water, subject to a centrifugal acceleration of 10^3 times g, this means only about $\Delta\gamma > 2 \cdot 10^{-5}\,N{\cdot}m^{-1}$, a condition that will always be met.

3. The droplet *Reynolds number*, given by

$$Re \equiv \frac{d v \rho_C}{\eta_C} \tag{20}$$

should be less than 0.1, since otherwise turbulence will occur in the wake of the droplet, thereby slowing down sedimentation. Obtaining v from Eq. (17), even for a droplet of 0.1-mm diameter in water with $\Delta\rho = 100\,kg{\cdot}m^{-3}$, gives Re of only 0.06 under gravity. Consequently, the condition will nearly always be met.

4. *Brownian motion* should not disturb sedimentation. For very small droplets, Eq. (17) is invalid, and a sedimentation equilibrium is established, where the concentration c of the droplets as a function of the sedimentation distance H would be given by

$$c(H) = c_0 \exp\left(\frac{-\pi d^3\,\Delta\rho\, aH}{6kT}\right) \tag{21}$$

Assuming that the disturbance due to Brownian motion is negligible for $c(H)/c_0 < 0.02$, it follows that for the same situation as mentioned under condition 3, and a vessel height of 0.2 m, the condition comes down to $d > 50\,nm$. This will nearly always be the case.

5. There should be no disturbance due to *convection currents*, which may, for instance, be generated by even very slight temperature fluctuations. As an illustration, the free-convection flow between two vertical plates of slightly different temperature over a mutual distance b will be considered. The average velocity gradient G would then be $(\beta\rho g b\,\Delta T)/32\eta$, where β is the cubical expan-

sion coefficient [31]. For water at room temperature and $b = 0.2\,m$, G would roughly equal $10\,\Delta T$, i.e., $1\,sec^{-1}$ for a temperature difference of 0.1 K. The stress due to this gradient is ηG, and it should be much smaller than the stress due to the buoyancy, which is $4F_B/\pi d^2$. This leads to the condition

$$\frac{d}{G} \gg \frac{6\eta_C}{\pi a\,\Delta\rho} \tag{22}$$

which implies for the mentioned oil droplet in water that d in micrometers should be much larger than 2G in sec^{-1}. This means that the sedimentation of small droplets may be readily disturbed, unless the temperature is kept rigorously constant and mechanical disturbance is completely avoided. Indeed, it has been observed in creaming of oil–in–water emulsions under ambient conditions that droplets smaller than 1.5 μm hardly creamed whereas those larger than 4 μm reasonably followed Eq. (17) [32]. In a centrifuge, any effect of convection currents will be mostly negligible.

6. The continuous phase should be a *Newtonian liquid.* If this is not the case, the viscosity is a function of the shear stress or shear rate and is called an apparent viscosity η^*. Examples are given in Fig. 11. The shear stress on a sedimenting sphere is given by the buoyancy force F_B divided by the surface area of the sphere, i.e., $(1/6)ad\,\Delta\rho$, and the corresponding shear rate at the droplet surface $G_0 = 3v/d$ [33]. For all practical purposes, the stress is thus less than 10^{-3} Pa. Most polymers used as thickening agents give a type of viscosity–shear stress relation as depicted in Fig. 11, and it follows then that the relevant viscosity is the so-called zero-shear viscosity (the plateau value at low stress). If high viscosities are produced by other materials, e.g., by very small particles, a different relation is generally observed. Here, $\eta^* = kG^{n-1}$ often holds

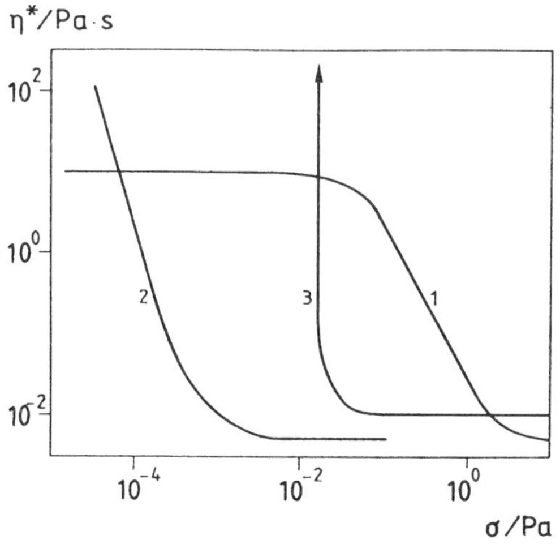

Figure 11. Schematic examples of non-Newtonian behavior: Apparent viscosity η^* (Pa·sec) as a function of shearing stress σ (Pa). Curve 1 is typical for a polymer solution, curve 2 for a slightly aggregating dispersion of very small particles. Curve 3 is for a system exhibiting a yield stress.

for a certain range in shear stress, where k is a constant and n commonly is less than 1. It has been shown that in such a case

$$v \approx \left(\frac{a \, \Delta\rho \, d^{n+1}}{18k} \right)^{1/n} \tag{23}$$

provided that the exponential relation for η^* holds for the shear stress range below G_0. There are substances (for instance, mixtures of interacting polymers) that provide the liquid with a (small) *yield stress* σ_y, which implies that it behaves as a solid (a gel) below this stress; see Fig. 11. A droplet now exerts a stress given by F_B over its cross-sectional area, i.e., $(2ad\,\Delta\rho)/3$; and if this is below the yield stress, the droplet will not sediment. In practice, a considerable margin should be realized to prevent sedimentation, since yield stresses tend to become lower for longer time scales. Moreover, very slight stirring may suffice to cause flow, and then the yield stress has to be restored, which may take some time (say, 15 min). Nevertheless, the presence of a small yield stress may effectively prevent sedimentation [33].

7. The sedimenting droplet should be surrounded by a volume of continuous phase that is very much larger than the droplet, which implies a *very low volume fraction* of the disperse phase. If φ is finite, hindered sedimentation occurs (see Sec. IV.B).

8. The droplets should be *monodisperse*. If φ is very small, it suffices for a polydisperse emulsion to replace d^2 in Eqs. (17) and (18) by the appropriate average, given by

$$d_{53}^2 \equiv \frac{\Sigma N_i d_i^5}{\Sigma N_i d_i^3} \tag{24}$$

where N_i is the number of droplets of size class i, characterized by diameter d_i. But if φ is not very small, the situation is more complicated; see Sec. IV.B.

9. There are *no colloidal interactions* between the droplets. Colloidal attraction generally leads to formation of aggregates, which tend to sediment (much) faster. This is discussed further in Sec. IV.C.

B. Hindered Sedimentation

This is a vast and complicated subject, and only some aspects important for emulsions will be discussed.

Finite volume fraction. Batchelor [35] attempted to derive a rigorous equation for the case of $\varphi \ll 0.1$. For equal-sized spheres he arrived at

$$\frac{v}{v_S} = 1 - 6.55\,\varphi \tag{25}$$

The factor −6.55 is composed of three terms: (1) an amount −5.50 for counterflow of continuous phase, due to the particles themselves displacing liquid and due to the particles dragging liquid along with them; (2) an amount −1.55 due to the overlap of the flow disturbances of two droplets sedimenting near to each other, which gives rise to a locally higher velocity gradient; and (3) an amount +0.50 due to group sedimentation (see below).

Richardson and Zaki [34] proposed the relation

$$\frac{v}{v_S} \approx (1-\varphi)^n \tag{26}$$

with $n = 4.65$. Notice that for this value of n, Eq. (26) does not approach Eq. (25) at very small φ, but that may be because Eq. (26) was derived for bubbles or droplets with a mobile surface; consequently, the exponent in Eq. (26) presumably should be 6.55.

Buscall et al. [36], by taking into account the Dougherty–Krieger relation for the viscosity of a dispersion of spheres, semi-empirically derived the equation

$$\frac{v}{v_S} \approx \left(1-\frac{\varphi}{\varphi_m}\right)^{k\varphi_m} \tag{27}$$

where φ_m is the maximum volume fraction obtainable. For $k = 6.5$ and $\varphi \to 0$, Eq. (27) indeed equals Eq. (25); for $\varphi \to \varphi_m$, $v \to 0$. Buscall et al. found good agreement between their equation and experiments with monodisperse spheres of about 1.5 μm up to $\varphi = 0.4$.

Polydispersity will, however, affect the result. Batchelor and Wen [37] considered, again for small φ, a mixture of spheres of two different sizes. As is to be expected, v/v_S increases for the smaller spheres and decreases for the larger spheres; the total sedimentation rate, however, decreases. It has also been experimentally observed in emulsions that a greater width of the size distribution for the same d_{53} resulted in somewhat slower sedimentation [32]. In the same study, with φ up to 0.1, Eq. (26) was observed to hold well, but with $n = 8.6$.

Figure 12 compares these equations, together with a semi-empirical one by Barnea and Mizrahi [38], reading

$$\frac{v}{v_S} \approx (1-\varphi)(1+\varphi^{1/3})^{-1}\exp-\left(\frac{\frac{5}{3}\varphi}{1-\varphi}\right) \tag{28}$$

which obviously is incorrect for very small φ.

There is, however, more to be discussed. Figure 13 shows results on creaming in oil–in–water emulsions (actually, mixtures of skim milk and cream) performed in such a way that virtually all the conditions stipulated in Sec. IV.A were fulfilled, except for condition 8 (monodispersity); the acceleration in the centrifuge was 200 times g, and the product of accelaration and creaming time was kept constant for all cases. It is seen that for $\varphi \to 0$, the Stokes result was obtained and that for gravity creaming, Eq. (26), with $n = 8.6$, was obeyed. The creaming in the centrifuge was, however, distinctly faster, and even surpassed the Stokes rate (by about 15%) for some φ.

A similar increase in sedimentation rate over the Stokes rate at intermediate φ has been observed by others. As an example, results by Johne [39] will be taken. For monodisperse glass spheres, $Re \approx 4 \cdot 10^{-4}$, he observed an increase over the Stokes rate by a factor of 2.3 at $\varphi \approx 0.02$. The explanation is not clear. Batchelor assumed the sedimenting particles to have a random distribution throughout the volume, but it may well be that the sedimentation itself affects the distribution, the more so for a higher φ. Johne ascribed the increase in rate to the occurrence of *group sedimentation* (this was also taken into account by Batchelor, but he

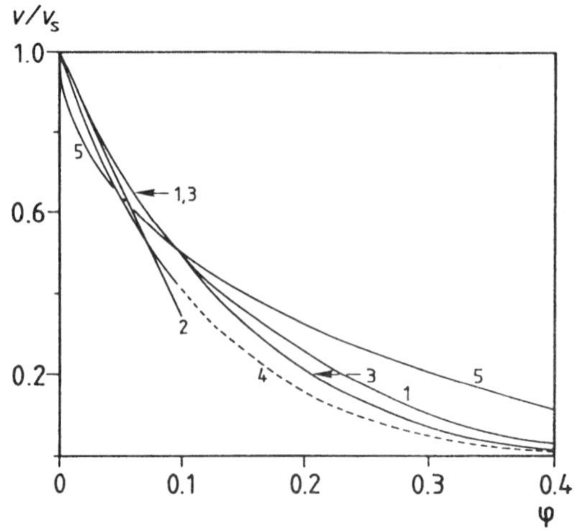

Figure 12. Calculated examples of sedimentation velocity v over the Stokes velocity v_S as a function of volume fraction φ. Curve 1: Eq. (26), n = 6.65. Curve 2: Eq. (25). Curve 3: Eq. (27), $\varphi_m = 0.6$, k = 6.5. Curve 4: Eq. (26), n = 8.6. Curve 5: Eq. (28).

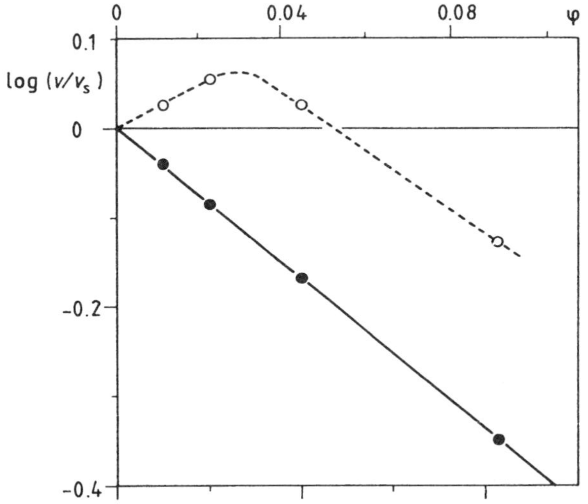

Figure 13. Creaming velocity relative to the Stokes velocity (v/v_S) of emulsions of various volume fraction φ, under gravity (●) or in a centrifuge (○); see text. (After Ref. 32.)

concluded that the effect, i.e., the term +0.50 mentioned earlier, is of limited importance for small φ). If two droplets are situated close to each other, especially one above the other, their joint resistance to sedimentation will be smaller than when they are farther apart. Diffusion will cause them to separate again, but if droplet pairs can sediment over a considerable distance during the time that the droplets are close to each other, the overall sedimentation rate may be significantly enhanced. In other words, the distribution of particles over the volume would not remain random. Here, it should be taken into account that sedimentation distance will be proportional to time, whereas the root-mean-square distance of diffusion will be proportional to the square root of time. The tendency for group sedimentation may then be expressed in a kind of Péclet number, given as the ratio of the time needed for a droplet to diffuse over a distance equal to the drop diameter over that for sedimentation (from the Einstein and Stokes equations, respectively). This is given by

$$\mathrm{Pe} = \frac{\pi \mathrm{d}^4 \mathrm{a}\, \Delta\rho}{6\mathrm{kT}} \tag{29}$$

If $\mathrm{Pe} \gg 1$, group sedimentation may be of considerable importance. In the examples of Fig. 13, Pe was about 0.5 for gravity creaming, and about 100 in the centrifuge; Johne's experimental conditions implied a very high Pe. For a very low φ, droplets would only rarely be close to each other, thereby making group sedimentation insignificant. For high φ, all droplets are always close to each other and group sedimentation would always be extensive; on the other hand, counterflow and overlap of flow disturbances would become overriding, strongly increasing with increasing φ.

It may be clear now that exact prediction of sedimentation rates will not be easy, certainly not for real (i.e., polydisperse) emulsions. The deviation from the Stokes equation will depend on volume fraction, the width of the droplet size distribution, and the Péclet number, Eq. (29). It should furthermore be realized that during sedimentation, the average droplet size d_{53} as well as φ and the distribution width outside the sedimented layer decrease; generally, the decrease in d_{53} is overriding, causing a gradual decrease in sedimentation rate.

These considerations are of especial importance for the usefulness of *centrifuge tests* to predict slow gravity creaming. Since one is interested mostly in the time needed for a little sediment (cream layer) to be formed, it follows that the centrifuge test should be done so that sedimentation here too is slight, since otherwise the mentioned change in d_{53} is too large. Another difficulty is illustrated in Fig. 13. Moreover, for non-Newtonian liquids, the effective η_C may be greatly different, mostly smaller, in a centrifuge test; see Fig. 11. Nevertheless, prediction by a centrifuge test can be fairly accurate if these pitfalls are avoided or taken care of. However, if colloidal interactions play an important part, prediction is generally impossible.

C. Further Complications

Predicting the full *sedimentation profile*, i.e., the distribution of droplets of various size over the depth of the vessel in which sedimentation occurs as a function of time, is a formidable problem. At least a suitable method is now available to determine the droplet concentration profile [40]; it is based on the effect of droplet concentration on ultrasound velocity. Several concentration profiles have now been published, e.g., Ref. [41]. A few attempts have been made at calculating concentration annex droplet-size profiles, but most of these are of a preliminary nature. In one case, creaming of pasteurized milk was modeled [42] and the full effects of droplet polydispersity, finite φ, and diffusional motion were taken into account, albeit with simplifications; the results obtained for the amount creamed reasonably agreed with Eq. (26) with $n = 8.6$.

Another complication is caused by *colloidal interaction* forces acting between the droplets. If these forces are repulsive, sedimentation will be somewhat slower, as has been studied for the case of electrostatic repulsion [103]; the main effect is a decreased importance of group sedimentation.

More interesting is the case of more or less irreversible aggregation (see Sec. V.F), which can very much enhance sedimentation rate. The aggregates will generally be of a fractal nature, and in such a case their sedimentation velocity relative to that of a single droplet will be given [43] by

$$v_{aggr} \approx 1.8 v_{drop} N_f^{(1-1/D)} \tag{30}$$

where N_f is the number of (equal-sized) drops in the aggregate and D is its fractal dimensionality. The factor 1.8, which accounts for the difference between the outer and hydrodynamic radii of the aggregate, is approximate and varies somewhat with D. For example, for $N_f = 100$ and $D = 2$ (a reasonable value), v is increased by a factor of 18. To be sure, there are other factors. If φ is not very small, the volume fraction of aggregates would soon become so large as to materially slow down sedimentation; however, the relations given in Sec. IV.B cannot be used, since the aggregates are not impermeable. On the other hand, and far more important, the difference in sedimentation rate of aggregates of different size very much enhances aggregation as big aggregates overtake small ones; this implies that sedimentation may become greatly accelerated, to such an extent that the height of the vessel has little effect on the time it takes for a sediment (cream) layer to form [9].

If φ is not small and the aggregation rate is fast, the droplets may soon form a gel [43], thereby preventing sedimentation. Such a gel may show slow *consolidation* due to the gravity forces, depending on the rigidity (or permanency) of the bonds between droplets. Also a sedimented layer of aggregates may show such consolidation with time. It may be argued that in such a case, and also for sedimentation at very high φ, a Darcy-like approach, taking an overall permeability of the disperse system into account, is the way to approach the problem, and a first attempt has been made [33].

Finally, droplets may *coalesce* after their aggregation. A full treatment of combined coalescence and creaming for polydisperse emulsions of not too high φ has been worked out by Melik and Fogler [5], for the following conditions: The occurrence of aggregation is according to the DLVO theory (i.e., only van der Waals attraction and electrostatic repulsion); the encounter rate is governed by Smoluchowski's perikinetic equations and by "overtaking" due to sedimentation; aggregation is followed immediately by coalescence. Results for a wide range of variables have been obtained and were found to reasonably agree with experimental results. The prerequisites are, however, rarely met.

V. Aggregation

Again, this is not meant as a full discussion of aggregation phenomena in emulsions, but as an update on some salient points of the earlier chapter [2]. For convenience, it is briefly recalled that the classical theory of irreversible colloidal systems (see, e.g., Ref. [44]) treats aggregation of particles as governed by their encounter frequency and the interaction forces between them. If the latter are negligible except for strong attraction at very close distance (i.e., for sticky hard spheres), the aggregation rate as caused by Brownian motion would be given by Smoluchowski's equation for (fast) perikinetic aggregation, which is, for spheres of equal diameter, d, given by,

$$-\frac{dN}{dt} = 4\pi\, dD_d N^2 = \frac{4}{3}\frac{kTN^2}{\eta} \tag{31}$$

where D_d is the diffusion coefficient of the particle and N is the number of particles per unit volume; any two particles that have aggregated are subsequently counted as one, etc. Colloidal interactions will, however, affect the rate; they are generally given as the Gibbs (free) energy V(h) needed to bring two droplets from an infinite distance to a separation h (measured between the droplet surfaces). The DLVO theory for the interaction energy takes the sum of van der Waals attraction (V_A) and electrostatic (double layer) repulsion (V_E) into account. An example is given in Fig. 14, which also gives approximate DLVO equations valid in the present case (spheres of equal size, $\kappa r \gg 1$, $h < 10\,\text{nm}$, $|\psi_0| < 40\,\text{mV}$).

In the Fuchs treatment for slow aggregation, integration over the interaction free energy with respect to h (whether given by the DLVO or by some other theory) yields a stability factor W by which the rate in Eq. (31) has to be divided to obtain the aggregation rate:

$$W = 2\int_2^\infty x^{-2} \exp\frac{V(x)}{kT}\, dx \tag{32}$$

where $x = 2(h + d)/d$.

It is now clear that this picture is greatly oversimplified, that the theory needs several new elements and refinements, and especially that the DLVO theory is valid only in some conditions. The most important new elements may be the effects of adsorbed or free polymers on the interaction free energy. Some points will be discussed shortly. Those about the interaction free energy may be relevant for coalescence in addition to aggregation.

A remark on terminology may be useful: In the classical theory, particles were said to coagulate if they would reach "the primary minimum," as near A in Fig. 14, where aggregation is presumed to be irreversible. Aggregation "in the secondary minimum," as near C in the figure, is presumed to be rather weak and reversible, and was called flocculation. As will be discussed shortly, such a simple dichotomy cannot be maintained in practice, the reason why the author prefers the general term aggregation.

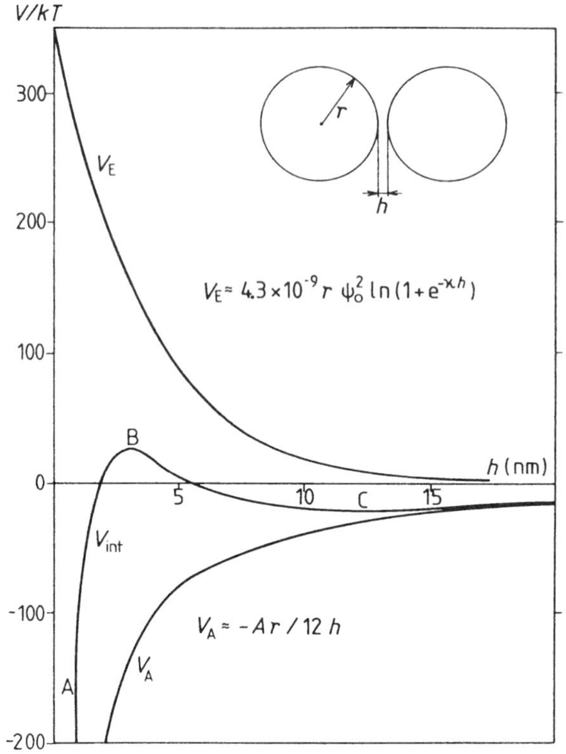

Figure 14. Interaction free energy V between two spheres of identical size at a distance h, according to the DLVO theory. Example calculated for a Hamaker constant $A = 10^{-20}$ J, droplet radius $r = 2\,\mu m$, surface potential $\psi_0 = 16\,mV$, and Debye length $1/\kappa = 3\,nm$ (ionic strength of 10 mM). $V = V_E$ (electrostatic repulsion) + V_A (van der Waals attraction). (From Ref. 102.)

A. DLVO Theory: Complications

The DLVO theory cannot be used for most water–in–oil emulsions [45]. Due to the low dielectric constant of most oils, they contain virtually no ions, causing electrostatic forces to act over distances far larger than the droplet size; in fact there is Coulombic rather than double-layer repulsion. This means that the net electrostatic force acting on a droplet is effectively zero. Consequently, other interaction forces must be present to prevent aggregation of water–in–oil emulsion droplets.

On the other hand, the theory seems to be well suited for oil–in–water emulsions stabilized by small-molecule ionic surfactants. The particles are almost perfect spheres, and important variables, such as the electrical potential and the thickness of the electrical double layer, can readily be determined and values for the Hamaker constant are tabulated: Most will be between 1 and $1.5 \times kT$. The theory certainly explains the large effects of droplet charge (e.g., via pH) and ionic strength on the stability. In some cases, good quantitative agreement is observed [5], although minor modifications may be needed [46] and in certain cases the theory by Lifshitz rather than that by Hamaker may have to be used for the van der Waals attraction. In other cases, agreement is less good. Generally, the theory is questionable at very small interparticle distances. It is assumed, for instance, that hydration forces may cause repulsion at about 1 nm [47], thereby materially diminishing the depth of the primary minimum, depicted in Fig. 14. The adsorption of fairly large molecules may also interfere with the theory. Such adsorption makes estimation of the surface potential uncertain, since the experimentally accessible electrokinetic potential may be greatly different from the potential at the droplet surface: The slipping plane, at which plane the potential is sensed, will be at some distance from the surface. The van der Waals attraction at small interparticle separations also becomes uncertain: The adsorbed layer must materially affect it, but there seems to be no theory for this effect unless the layer is compact, homogeneous, and of constant thickness.

Another point is that effects of droplet size are hardly observed in practice. Theory predicts both repulsion and attraction to be proportional to droplet diameter, which implies that for larger droplets the secondary minimum would be deeper (thus promoting "flocculation"), whereas the height of the maximum (near B in Fig. 14) would be higher (thus diminishing the chance of "coagulation"). Various modifications of the DLVO theory are now being developed, based, for instance, on incomplete desorption relaxation [48], electroviscous effects [49], or surface roughness [50]. They all predict less or no dependence on droplet size. At present, there seems to be no general agreement among colloid scientists. Any surface roughness of solid particles seems to have considerable effect on the effective colloidal interaction at distances below a few times the amplitude of the roughness [50]; whether the development of capillary waves on the surface of liquid particles may have a similar effect is yet unclear.

B. Adsorbed Polymers

If droplets are provided with a layer of protruding molecular chains, such layers will to some extent overlap on the close approach of two droplets, and this often causes strong repulsion; it is generally called *steric repulsion*. For oil–in–water emulsions, it may be caused by soaplike surfactants, which have some fairly short and flexible, hydrophilic chains, such as poly(oxyethylene) sorbitan monopalmitate. It is often cauin–sed by adsorbed polymers (see Sec. II.B), which can be very efficient stabilizers of water–in–oil or oil–in–water emulsions. A recent review is by Dickinson and Eriksson [51].

As discussed in Ref. [2], one may distinguish two main contributions to the interaction free energy $V_S(h)$. First, if $h < \delta$, where δ is the effective thickness of the layer of protruding chains, the chains are restricted in their Brownian motion, and the ensuing loss of entropy causes very strong repulsion, of the order of a few times kT per chain. However, the second contribution generally is overriding before the particles approach so close: For $\delta < h < 2\delta$, the overlap of the adsorbed layers causes a change in osmotic pressure. If the solvent quality for the protruding chains is good, the overlap would cause an increase in osmotic pressure, hence repulsion. If the solvent quality is poor, attraction may well occur. At poorer solvent quality, δ would also be thinner; hence, van der Waals attraction would be stronger at an interparticle distance of 2δ.

The theory of the polymer adsorption and the concomitant effect on $V_S(h)$ has been greatly developed at this time [18]. However, application to predict the colloidal interaction energy between emulsion droplets is fraught with uncertainties. Nevertheless, for the case of adsorbed copolymers (the only realistic situation in emulsions) it is clear that for good solvent quality of the protruding chains, repulsion is always very large, say, over a thousand times kT for distances at which van der Waals attraction still would be negligible. Figure 15 gives some rough calculations for diblock copolymers, based on results for flat surfaces [18] by applying the Deryagin approximation, and assuming "restricted equilibrium" (which means that Γ would not alter during approach); the results for $\chi = 0.5$ qualitatively agree with experimental results obtained for copolymers adsorbed at air–water interfaces (e.g., Ref. [52]). It is also seen that even a slight lowering of solvent quality may lead to very strong attraction. The latter is in agreement with observations that emulsions, whether oil-in-water or water-in-oil, stabilized with a polymer that has a theta point in the temperature range considered, show a critical aggregation temperature close to that theta temperature [53]. By and large, the use of an adsorbing copolymer that is well soluble in the continuous phase leads to very good stability against aggregation. Aggregation may occur for chains protruding over only a short distance, say, 5 nm; this may be the case for very small polymer molecules, i.e., oligomers.

If the polymer concentration is such that during emulsion formation a surface excess clearly below the plateau value is obtained, bridging between

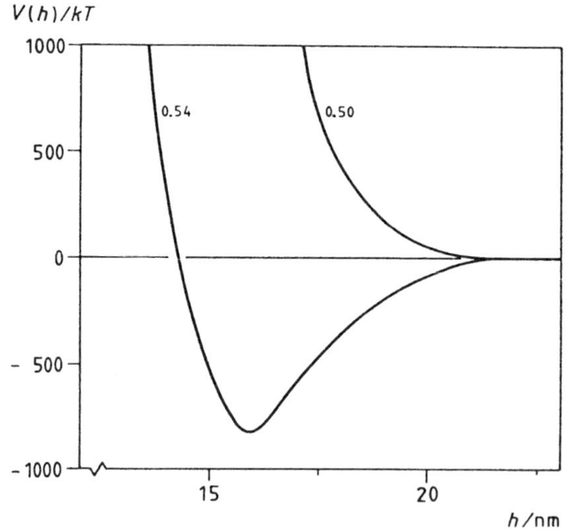

Figure 15. Steric repulsion due to polymer adsorbed onto an oil–water interface. Calculated interaction free energy V as a function of interdroplet distance h for droplets of d = 1 μm. Diblock copolymer, water-soluble block of about 50 segments, with its χ value with water indicated. Meant only to illustrate trends. (Calculated from data in Ref. 18.)

droplets may well occur and then the negative interaction energy may be high. (See Fig. 7 for information about surface excess.) Bridging flocculation may also occur after emulsification an addition of a fairly small quantity of a polymer that interacts with the surfactant (e.g., a polymer) that was used as emulsifier [54]; the interaction often is weak, implying that mild stirring would suffice to disrupt the aggregates. It may be difficult to distinguish aggregation due to this mechanism from depletion flocculation (see Sec. V.C).

Some attention should be given to the properties of charged polymers, or polyelectrolytes, often used to stabilize oil–in–water emulsions. It appears that the general theory as developed at present (e.g., Ref. [18]) is not adequately applicable to emulsions. In practice, one finds very good stability against aggregation [20]. If the average degree of dissociation of the ionogenic groups becomes small, say, less than 0.3, aggregation may occur. This is also observed for protein-stabilized emulsions near the isoelectric pH of the protein. It has been assumed (e.g., Ref. [51]) that in such cases the droplet surface may show separate patches of different charge, which would readily lead to aggregation.

Another possibility is that some kind of cross-linking is occurring between chains on different droplets; after all, steric repulsion does not prevent contact between individual chains, on the contrary. For instance, divalent counterions may provide bridges between charged groups on the adsorbed polymers. Indeed, it is observed in practice that adding Ca ions may cause aggregation at far lower concentrations than would be needed to substantially compress the electrostatic double layer. These aspects would need systematic study, however.

Incidentally, some of the mechanisms for aggregation suggested would also imply that the colloidal interaction free energy is dependent not only on h, but also on the way in which the interparticle distance is reached. More specifically, for the same h the interaction may be positive when the droplets approach each other and negative when they move away from each other. In classical colloid theory, equilibrium is mostly implied and such a hysteresis could not occur. This aspect also needs further study.

C. Depletion Interaction

Nonadsorbing polymers are depleted near an interface [18], because the center of the molecule cannot come closer to that interface than the radius of the coil. Close to the interface there is thus a layer with a polymer concentration lower than that in the bulk. This is called negative adsorption or a negative surface excess, and it goes along with an increase in interfacial tension, although that increase generally is too small to be measured. The thickness δ of the depletion layer (defined as if there were no polymer in this layer and a constant concentration beyond it) is given by the correlation length of the polymer, which for a dilute polymer solution in a good solvent is about equal to the radius of gyration R_g. Beyond the chain overlap concentration, δ will be smaller and depend on several conditions [18].

It was realized by Asakura and Oosawa [55] that depletion can lead to aggregation of particles. The depletion causes the osmotic pressure due to the dissolved polymer Π_{osm} to be higher than it would be otherwise. If two particles now come close together, parts of their depletion layers overlap (see Fig. 16),

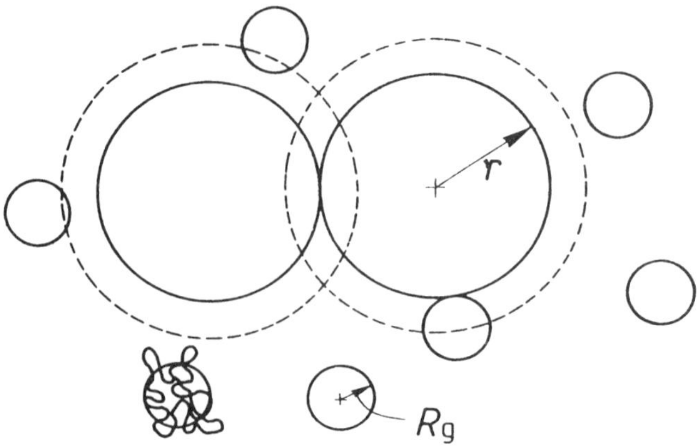

Figure 16. Schematic explanation of the depletion of macromolecules, radius of gyration R_g, from the solution by droplets, radius r, and of the resulting depletion flocculation. See text. (From Ref. 56.)

thereby decreasing both the depletion volume and the osmotic pressure. The difference may be particularly large for platelike particles, but it may also be appreciable for spheres.

For emulsion droplets with $\delta \ll r$ (the sphere radius), the interaction free energy due to the depletion effect is given approximately [18] by

$$V_D(h) \approx -2\pi r \Pi_{osm}(2\delta - h)^2 \tag{33}$$

$$\Pi_{osm} = RT\left(\frac{c}{M} + Bc^2 + \ldots\right) \tag{33a}$$

where c is polymer concentration, M is polymer molar mass, and B is the second virial coefficient. This last is higher for a better solvent quality for the polymer and may have an overriding effect on the osmotic pressure for not-too-small c. The interaction energy is proportional to the droplet diameter and increases with the size of the polymer molecules. To give an example, for $r = 1\,\mu m$, $\delta = R_g = 30\,nm$, $c = 3\,kg{\cdot}m^{-3}$ ($\approx 0.3\%$), $M = 10^3\,kg{\cdot}mol^{-1}$, and $B = 5 \times 10^{-4}\,mol{\cdot}m^3{\cdot}kg^{-2}$ (which are reasonable values for, say, a xanthan solution), it follows that $V_D \approx 25\,kT$ at $h = 0$, which means in the absence of any repulsion at short distances. This would imply strong flocculation.

Figure 17 shows some results for creaming of an emulsion with very small droplets, in the presence of xanthan at various concentrations. It is seen that even at very small concentrations, the creaming rate is greatly enhanced, undoubtedly owing to depletion flocculation. For high c, the creaming rate is slower again; this may be due partly to the increased viscosity of the continuous phase, but another factor may be a smaller δ at high concentrations, as mentioned earlier. It has indeed been observed, and also calculated from theory, that at high concentrations the depletion free energy again becomes smaller [18]. Moreover, depletion interaction may cause the droplets to form a continuous

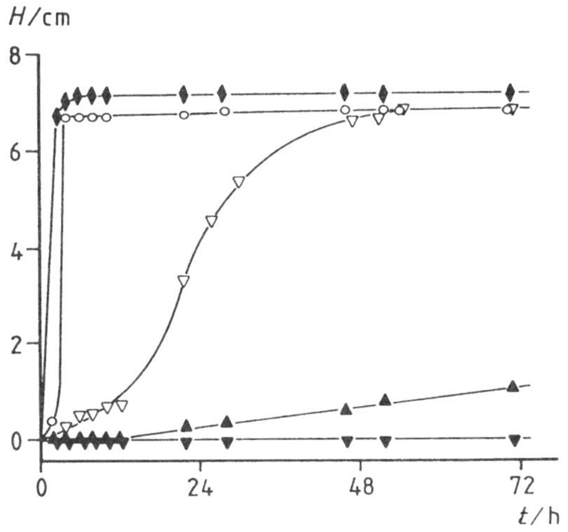

Figure 17. Creaming in emulsions containing xanthan. H is the height of the layer roughly free of droplets as a function of time t. Xanthan concentration: ▼, 0 and $\geq \frac{1}{8}\%$; ◆, $\frac{1}{40}\%$; ○, $\frac{1}{20}\%$; ▽, $\frac{1}{16}\%$; ▲, $\frac{1}{10}\%$. (From Ref. 57.)

network, leading to a very high apparent viscosity [106] or even a yield stress; see Fig. 11.

Equation (33) would apply only to undeformable ("hard") spheres. For spheres that are somewhat deformable ("soft"), the interaction energy at closest contact may be significantly higher. Deformability of the oil–water interface would generally not play a part: In the earlier example, the osmotic pressure due to the polymer would be about 20 Pa, whereas the Laplace pressure of the droplet would be of the order of 20 kPa. Only for large droplets having a very low interfacial tension may the flocculated droplets become locally flattened, thereby markedly increasing the interaction free energy. Another kind of "softness" is caused by the droplets being stabilized by a layer of adsorbed polymer. If a nonadsorbing polymer is present that induces depletion flocculation, the adsorbed layer is somewhat compressed and this causes a stronger interaction force, for instance, by a factor of 2; see Ref. [18] for a more extensive discussion. It should be mentioned, however, that specific interactions may occur between the adsorbed polymer and the nonadsorbing polymer, which may markedly affect the interaction free energy. In other words, predicting depletion flocculation is often not easy in practice.

Small particles, if present in sufficient concentration, may also cause depletion flocculation. A case in point is micelles of surfactants, as pointed out and studied by Aronson [58]. Equation (33) would apply, where M would be the molar mass of the micelle (i.e., n times the molar mass of the surfactant, where n is the aggregation number). For B the excluded volume has been taken; i.e., $B = 2/M\rho$, where ρ is the density of the micelle. It is calculated that for a concentration of surfactant in the continuous phase that is of the order of 1 percent unit higher than the critical micelle concentration, appreciable depletion flocculation may occur for many surfactants.

Bibette et al. [59] have studied depletion interaction of emulsion droplets by surfactant micelles and observed that for monodisperse droplets a kind of phase separation may occur into a "liquid" phase containing few droplets and a "solid" phase containing a close-packed array of droplets. After all, the micelles that cause the interaction are very monodisperse, meaning that δ is constant, allowing a sharp transition between a flocculated and a nonflocculated state. If the droplets are polydisperse, for a certain micelle concentration only droplets larger than a certain size will flocculate (see Eq. 33) and form a concentrated "phase"; in this way, fractionation of droplets into narrow size classes can be achieved, allowing the preparation of almost monodisperse emulsions [60]. (See also Chap. 6 of this volume.)

D. Perikinetic Capture Efficiency

In perikinetic aggregation, the particles undergo Brownian motion only. The rate of aggregation would then be given, under some conditions (e.g., spheres of equal size), by Eq. (31), divided by the stability factor W, as given in Eq. (32). The reciprocal of W is sometimes called the capture efficiency α. However, Eq. (31) should contain the translational pair diffusion coefficient D_{12}, and only at large particle separation does it equal $2D_d$, as in Eq. (31). It has been pointed out by Spielman and by Honig et al. (see [61,62] for reviews) that as the spheres come closer to each other, the diffusion coefficient becomes smaller and smaller and that Eq. (32) should be replaced by

$$W = 2\int_2^\infty \frac{D_d}{D(x)}\, x^{-2} \exp \frac{V(x)}{kT}\, dx \tag{34}$$

where for spheres of equal size [63]

$$\frac{D_d}{D(h)} \approx \frac{6u^2 + 13u + 2}{6u^2 + 4u} \tag{35}$$

where $u = 2h/d = x - 2$; equations for spheres of different sizes are also available. For $u \to \infty$, Eq. (35) $= 1$; for $u \to 0$, Eq. (35) $\to \infty$, implying that the diffusion rate becomes zero. In other words, if there were no attractive interaction between the droplets (and the theory would apply at molecular distances), aggregation would never occur. It turns out that for particles where V(h) is negative for all h, the resulting accelerating effect on aggregation and the retardation due to the additional diffusional resistance often roughly cancel each other, and Eq. (31) may thus be reasonably obeyed. In the case of a positive V(h) at small h, however, the diffusional resistance may cause considerable extra slowing down of any aggregation.

The droplet diameter d in Eq. (31) is in fact a collision radius between two droplets $r_1 + r_2$. For perfect spheres, the radius also occurs in the relation for the diffusion coefficient, and is thus cancelled (for equal-sized spheres) in the final form of the equation. However, if the particles have protruding polymer chains that cause bridging flocculation or are subject to cross-linking, the collision radius becomes larger than the diffusional (hydrodynamic) radius, and the aggregation rate may be enhanced. Moreoever, the retardation as predicted by

Eqs. (34) and (35) would not (or hardly) occur, since h would never become very small. Similar effects would occur for orthokinetic aggregation, and aggregation rates that were several times those of fast salt-induced aggregation have indeed been observed for polymer-coated particles [64].

An additional point is that the rate of bridging flocculation may be governed by the fraction of the droplet surface covered θ. In the first approximation, one would expect the rate to be slowed down by a factor of $\theta/(1-\theta)$, but it has been shown [65] that the situation is far more complicated.

E. Orthokinetic Aggregation

In a flow field, the velocity gradient G (in sec^{-1}) causes particles to encounter each other. If Brownian motion and sedimentation may be considered absent, and if the particles exhibit no interaction energy except that they stick on collision, the resulting orthokinetic aggregation rate, according to Smoluchowski (see Ref. [44]), is

$$-\frac{dN}{dt} = \frac{2}{3}\, d^3 N^2 G = \frac{4}{\pi}\, \varphi N G \tag{36}$$

where it is assumed that the flow is simple shear, implying that G equals the shear rate. Comparison with Eq. (31) shows that viscosity would have no effect (for a given shear rate) and that the effect of droplet diameter is now very large. It turns out that if Eqs. (36) and (31) were to be fully obeyed, the ratio of orthokinetic over perikinetic aggregation in water of room temperature would be approximately equal to $0.2d^3G$, where d is in micrometers. This implies that for droplets of $d = 4\,\mu m$, orthokinetic aggregation already prevails at a shear rate as low as $0.1\,sec^{-1}$; such a value would normally occur owing to slight temperature fluctuations, as discussed in Sec. IV.A, item 5.

All this is, of course, an oversimplification, since hydrodynamic and colloidal interactions come into play. This has been treated rigorously by van de Ven and Mason and is extended and reviewed by van de Ven [62]. The rate given by Eq. (36) has to be multiplied by an orthokinetic capture efficiency α_O, which may be considered the reciprocal of a stability factor W. If there is no colloidal attraction at any distance, the hydrodynamic interaction causes α_O to be zero, since the trajectories of two spheres in the shear field never lead to collision: If they come close, they merely rotate around each other before separating again. If there is attraction, some trajectories of spheres with respect to a reference sphere lead to collision, others do not. But even if particles collide, the shearing force may pull them apart again. The shearing stress is given by ηG; hence the force pulling them apart will scale as $\eta G d^2$, whereas the attractive interaction force will scale as d. This means that for a higher shear rate and larger particles, the probability of deaggregation will be larger. For van der Waals attraction only, it is found, theoretically and experimentally, that

$$\alpha_0 \approx 0.9 C_A^{0.18}, \qquad 10^{-5} < C_A < 10^{-1} \tag{37}$$

$$C_A = \frac{2A}{9\pi\eta G d^3} \tag{37a}$$

where A is the Hamaker constant. Notice that the viscosity is now a variable. For most cases, α_O is in the range 0.1 to 0.5. Notice also that the capture frequency is now proportional not to G but to $G^{0.82}$ and that for larger d the capture efficiency is smaller.

If colloidal repulsion is also taken into account, the situation is more complicated. A few examples are shown in Fig. 18, where the capture efficiency is given as a function of C_A and of the Debye length. Notice that there is a critical salt concentration for aggregation (CCC), below which no capture, hence no aggregation, would occur; for small C_A, there is a gradual transition in the capture efficiency.

Much more can be said about orthokinetic aggregation: Other kinds of colloidal interaction may occur; the droplets may differ in size and then the volume average diameter should be taken; the flow type may be different (e.g., for elongational flow at the same velocity gradient, the capture efficiency generally is greater than in shear); orthokinetic and perikinetic motion may occur simultaneously; sedimentation may occur also; etc. All these cases are discussed by van de Ven [62]. In practical situations, prediction of rates may be difficult, since flow patterns may vary with location and time, certainly where turbulence occurs. Moroever, in the author's experience with emulsions in flow fields, the presence and the nature of other surfaces, for instance, of the apparatus or of air (bubbles), may greatly affect aggregation rate.

F. Aggregation Kinetics: Further Complications

The initial aggregation rate has often been observed to be in fair agreement with Smoluchowski–Fuchs theory, albeit with (some of) the refinements mentioned earlier taken into account. In addition, the distribution of doublets, triplets, etc.,

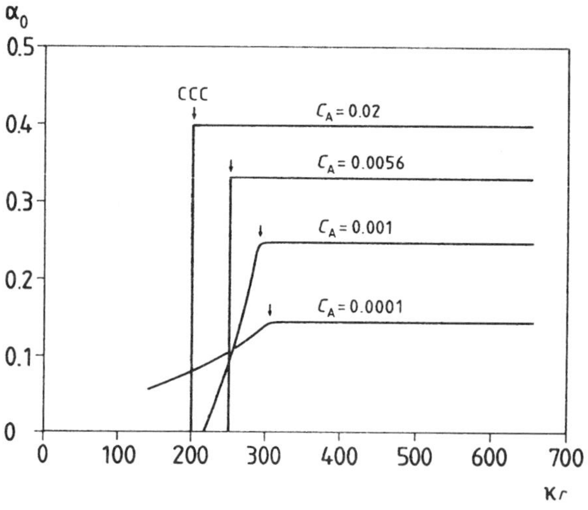

Figure 18. Orthokinetic capture efficiency α_O of spheres of radius r as a function of κr, for various values of C_A (shear rate 0.3–60 sec^{-1}). Calculated from DLVO theory, in water (ionic strength of 2–11 mM), $A \approx 1.25\,kT$, $r \approx 2\,\mu m$, $\psi_0 \approx 10\,mV$. (From Ref. 62.)

which is also given by Smoluchowski [44], may be in accordance with the theory. The only case in which agreement has been observed for further stages of aggregation is when there is *immediate coalescence* after doublet formation [5]. After all, this is the situation that Smoluchowski apparently had in mind, since his theory assumes all particles to be perfect spheres. Immediate coalescence may occur if emulsion droplets aggregate in the primary minimum, i.e., if only van der Waals attraction acts, or if electrostatic repulsion is very slight owing to a high ionic strength. (Incidentally, most studies on aggregation rates have been done with electrostatically stabilized dispersions to which a high concentration of salt was then added.) But in all other cases, aggregation of emulsion droplets will lead to aggregates in which the original droplets are still present. To be sure, slow coalescence may take place in the aggregates, and a kinetic theory has been worked out for combined aggregation and coalescence by van den Tempel [2,66], but it will not be discussed here, partly because the complications to be mentioned below seriously interfere with it; nevertheless, it may be useful in the early stages of aggregation if the stability factor W is fairly high and coalescence occurs fairly fast.

If particles can aggregate unhindered and the mutual aggregation of aggregates is also taken into account ("cluster–cluster aggregation"), *fractal aggregates* are obtained; see e.g., Refs. [43,67,68]. These are ramified and self-similar structures; i.e., at scales larger than that of the primary particle, they exhibit a similar structure at any magnification. In other words, they are stochastic fractals in three dimensions. The number of particles in such an aggregate is given by

$$N_f \approx \left(\frac{R}{r}\right)^D \tag{38}$$

where R is the radius of the sphere that can just contain the aggregate and D is the fractal dimensionality, which is less than 3. The latter implies, that the aggregate becomes ever more tenuous as it becomes larger. Equation (38) has been observed to hold very well over a wide range of sizes, both from simulations and experimentally. For spheres in Brownian motion that immediately stick to one another after touching, without allowing rearrangement afterwards ("diffusion-limited aggregation"), it turns out that $D \approx 1.8$. In many situations there will be a significant stability factor for aggregation ("chemical-limited aggregation"), and some rearrangement may occur in the aggregate as well. These factors lead to higher D values; in many situations in practice, D may not be far from 2.1. The theory needs several refinements, such as averaging over sizes if they show scatter, but these will not be considered since they do not materially affect the conclusions.

Keeping in mind the preceding considerations as well as some others, we ask: What might affect the *aggregation rate*, other than the variables involved in Eqs. (31) through (36)? According to Refs. [43,56,69], several complications could arise.

1. In Smoluchowski's equations, the volume fraction of particles is ignored. To rigorously take it into account, intricate numerical simulations would be needed. As an approximation, the diffusional distance z

of the droplets may be considered. For $\varphi \to 0$, it would equal $N^{-1/3}$. For finite φ, z is smaller, and the square of $N^{-1/3}/z$ would in first approximation yield the factor by which the perikinetic aggregation rate would be increased; this factor then is about $(1-\varphi^{1/3})^{-2}$. For $\varphi = 0.025$, this equals 2; for $\varphi = 0.32$, a factor 10 is obtained.

2. According to Eq. (38), φ would increase during aggregation, thereby further increasing the aggregation rate. In Fig. 19 a calculated example for perikinetic aggregation is given, and it is seen that the effect may be large. (In the figure, the reciprocal of the growth rate is plotted, and the area under the curve would be proportional to the aggregation time.)
3. Polydispersity may increase in the course of aggregation. Although according to Eq. (31), the average particle size does not affect the rate, the size ratio of the colliding particles does. The aggregation is faster by a factor $(r_1 + r_2)(1/r_1 + 1/r_2)/4$ [44]; this would then lead to an increasing average size ratio during the process. In perikinetic fractal aggregation, the effect appears to be small or even absent, presumably because of points 4 and 5, next.
4. For any aggregate, the collision radius tends to be larger than the hydrodynamic radius, causing an increased aggregation rate. For fractal aggregates, the ratio between both radii, which is the factor by which the rate is increased, is about $1.5(D+2)/D$, i.e., for $D = 2$ about a factor of 3. If rotational diffusion is taken into account, the difference may well be larger.
5. The stability factor W may alter as aggregation proceeds. Simulation of chemically controlled cluster–cluster aggregation generally shows that it tends to change into diffusion-limited aggregation after the

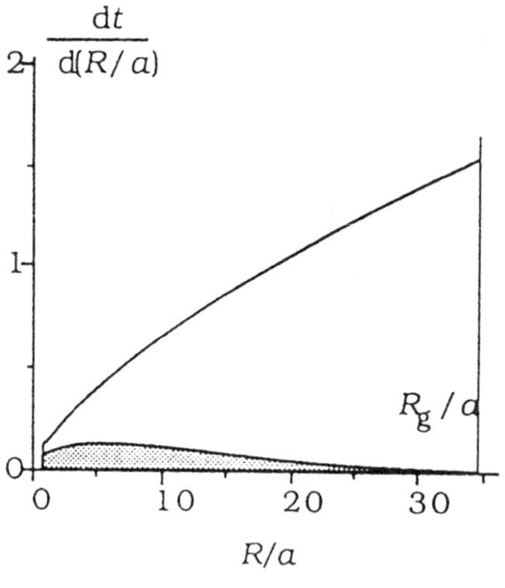

Figure 19. The reciprocal change in aggregate radius R as a function of R. The upper curve is calculated without correction for the increase in volume fraction with increasing R; the lower curve is calculated with the correction. Original $\varphi = 0.004$, fractal dimensionality is 1.7, particle radius a = 100 nm, water is at room temperature. (From Ref. 43.)

aggregates have grown fairly large. In terms of Smoluchowski, the aggregation changes from "slow" to "fast." Point 4 may play a part, according to the simulation method. Another point is that the extra diffusional resistance as given by Eq. (35) would become relatively unimportant as the aggregates grow large; after all, the aggregates are permeable for the solvent. Most important may be that the Fuchs treatment breaks down. Two aggregates can to some extent interpenetrate, because of their tenuous structure, to a greater extent when they are bigger. Because of rotational and sideways lateral diffusion, individual strands of the aggregates may now frequently touch before the aggregates have diffused away from each other. If they touch a number of times greater than W, they will nearly always form a bond upon approach, which implies fast aggregation. See, however, the next point.

6. If the aggregates become large, even a slight velocity gradient would cause orthokinetic aggregation to override perikinetic aggregation; see Sec. V.E. Also, this factor would cause aggregation to be accelerated. Moroever, it would cause the range of aggregate sizes to increase, because larger aggregates grow faster than smaller ones.
7. If the aggregates grow large and the density difference between the two phases is not very small, sedimentation of aggregates would occur, and this also may greatly enhance aggregate rate; see Sec. IV.C.

Points 1 to 5 would also apply to orthokinetic aggregation, though the governing equations would be different. Most of the points would not apply in the situation of immediate coalescence, but points 1, 6, and 7 do.

It seems clear that prediction of aggregation rate may be far from easy. Nevertheless, some approximate expressions for *aggregation times* are given next, for perikinetic aggregation leading to fractal aggregates, largely following the approach by Bremer and co-workers [43,56,69]. By integration of Eq. (31) until the time that the number of particles N is halved, the halving time is given by

$$t_{0.5} = \frac{\pi \eta d^3 W}{8kT\phi} \tag{39}$$

This often is called the coagulation time, but that is clearly a misnomer in many situations. One generally is interested in the time elapsed until some desired or undesired change in the emulsion as a whole has occurred, which implies the time after which some critical aggregate radius R_c is reached. This may be the radius of a visible aggregate or the average radius at which the system forms a gel or at which it shows separation into layers. By rewriting Eq. (31) as dR/dt and by integrating it from d/2 to R_c, one obtains for the critical time

$$t_c = t_{0.5}\left(\frac{2R_c}{d} - 1\right)^D, \qquad 2R_c > d \tag{40}$$

It then turns out that for different-type R_c (but the same $t_{0.5}$), the critical time may differ greatly, by up to several orders of magnitude [43,56].

Following Ref. [69], the critical time may be calculated for various changes, whereby the shortest of these has to be taken. Of course, these are approximations. If the relevant $R_c \geq d/2$, it may be assumed that the aggregation time is in any case very short. The following situations may be relevant for emulsions.

1. The fractal aggregates grow until the volume they occupy fills the whole system; at that moment a *gel* is formed. From Eq. (38), the critical radius is $\frac{1}{2}d\varphi^{1/(D-3)}$. Moreover, the increase in aggregation rate caused by the rapidly decreasing diffusion distance has to be introduced into Eq. (40). This leads [43] to

$$t_c = t_{gel} \approx \left(1 - \frac{6D}{2D+3} + \frac{3D}{D+6}\right) t_{0.5}\varphi^{D/(D-3)} \tag{41}$$

Here, the term in Eq. (40) is negligible.

2. After the aggregates have attained a certain size, *sedimentation* becomes the rate-determining step; and, moreover, the rate then is so accelerated that the additional time needed for a gel to form or for a separation into layers to occur generally would be negligible. The resulting expression,

$$R_c = R_{sed} \approx \left(\frac{kT}{d^{(3-D)} g\, \Delta\rho}\right)^{1/(D+1)} \tag{42}$$

can be inserted into Eq. (40).

3. *Orthokinetic* aggregation may also become the rate-determining mechanism, again leading to very fast aggregation afterwards. In that case, the expression

$$R_c = R_{ortho} \approx \left(\frac{kT}{4G\eta}\right)^{1/3} \tag{43}$$

should be inserted in Eq. (40). G is the velocity gradient, and it may be remembered that even very slight temperature fluctuations or other disturbances can lead to $G \approx 0.1\,\text{sec}^{-1}$.

Some sample calculations are illustrated in Fig. 20. Of course, there are several uncertainties. W may not remain constant and would probably increase during aggregation. The decrease in the diffusion distance also is not taken into account, except in Eq. (41), where it would be most pronounced. In Eq. (43), $G = 0.1\,\text{sec}^{-1}$ has been used, as a kind of practical minimum; a larger G would lead to a shorter critical time. If the viscosity of the continuous phase were higher, the same density differences would cause a lower G but about the same ηG. This means that the resulting R_{ortho} would still be applicable, although t_c would be higher in proportion to η. The same applies to the other critical times. It is seen that different critical times should be used, primarily according to the values of d, φ, and $\Delta\rho$. Also, the effective fractal dimensionality influences the result. If aggregation is immediately followed by coalescence, the same equations can be used with $D = 3$; then in Eq. (41), the first factor of the equation should be omitted.

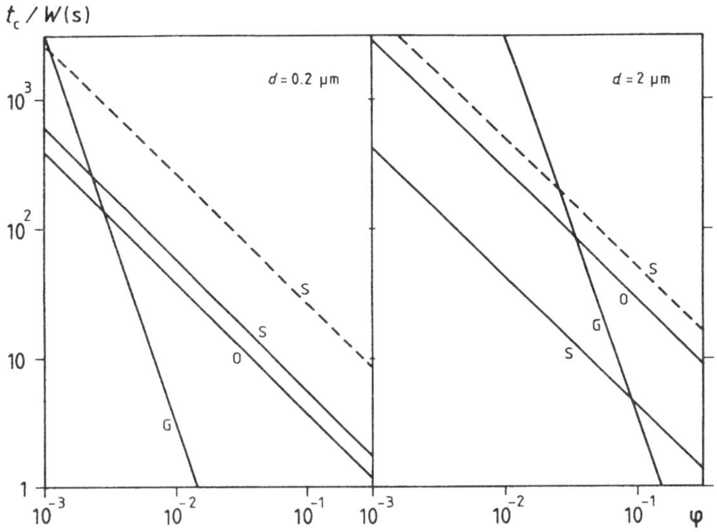

Figure 20. Critical time for aggregation t_c over the stability ratio W as a function of volume fraction φ for two droplet diameters d. G, gel formation; O, onset of orthokinetic aggregation ($G = 0.1\,sec^{-1}$); S, onset of sedimentation ($\Delta\rho = 100\,kg{\cdot}m^{-3}$). Calculated for water at room temperature, fractal dimensionality = 2. See text.

VI. Coalescence

Coalescence is induced by rupture of the film between two droplets that are very close to each other. For this to occur, the droplets must be either aggregated or packed in a sediment (cream) layer. Coalescence generally is a first-order rate process, and generally is fairly or even extremely slow. This implies that the rupture of a film between droplets is a chance event. The theory of coalescence is in a state of confusion. Consequently, it makes little sense to extensively review the literature on factors affecting it. Instead, some considerations will be given that, hopefully, will stimulate further systematic research.

There seems to be a number of reasons for the confusion. First, the coalescence rate often is correlated with other changes, such as the aggregation and sedimentation mentioned previously, and these changes may depend on variables (such as volume fraction, droplet size, type of surfactant) in a different way than does coalescence itself. Second, making and breaking of the emulsion tend to become mixed up. For instance, a surfactant giving a lower interfacial tension generally yields smaller droplets on emulsification [7], and this may in itself cause slower coalescence. Moreover, such tests are sometimes done under conditions where the surfactant concentration is the limiting factor for the droplet size obtained, implying that the surface excess may be lower than its plateau value. If one wants to study the effect of the surface layer of the droplets on coalescence, one should do it at equal size distribution, and this may be very hard to achieve.

The most important cause of confusion may be that two situations have to be distinguished, that is, whether or not two droplets form a flat film between them; see Fig. 21. This will be governed by the balance between any external stress acting on the droplet pair and the internal Laplace pressure (Eq. 3) [70]. We may define a Weber number for the ratio between these:

$$\mathrm{We} \equiv \frac{\Lambda\sigma_{\mathrm{ext}}}{\mathrm{p_L}} \approx \frac{(\mathrm{r/h})\sigma_{\mathrm{ext}}}{\mathrm{p_L}} = \frac{\mathrm{r}^2\sigma_{\mathrm{ext}}}{2\gamma\mathrm{h}} \tag{44}$$

where Λ is a stress concentration factor that would apply only as long as the droplets are spherical and that is of the order of droplet radius (r) over interdroplet distance (h). The importance of this will be illustrated with a few examples. For a small droplet (say, $\mathrm{r} = 0.2\,\mu\mathrm{m}$) provided with a polymeric surface layer ($\gamma = 10\,\mathrm{mN{\cdot}m^{-1}}$), we may have $\mathrm{p_L} = 10^5$ Pa; for a droplet 10 times larger and with a small-molecule surfactant ($\gamma = 1\,\mathrm{mN{\cdot}m^{-1}}$) this would be 10^3 Pa. The following external stresses may act.

1. Van der Waals attraction, where it is envisaged that at a distance h a hard-wall repulsion acts. We then have

$$\Lambda\sigma_{\mathrm{A}} \approx \frac{(\mathrm{r/h})(\mathrm{dV_A/dh})}{\pi\mathrm{r}^2} \approx \frac{\mathrm{A}}{12\pi\mathrm{h}^3} \tag{45}$$

Assuming the Hamaker constant A to be $6 \cdot 10^{-21}$ J, the external pressure would be 160 Pa for h = 10 nm and 6 kPa for h = 3 nm.

2. With a shearing stress due to a velocity gradient G, the pressure, in first approximation, would be

$$\Lambda\sigma_{\mathrm{S}} = \frac{\eta\mathrm{Gr}}{\mathrm{h}} \tag{46}$$

For $\eta\mathrm{G} = 100$ Pa, which is high, and $\mathrm{r/h} = 100$, the pressure would be 10 kPa.

3. In a cream layer (or in a sediment), the droplets in the top (or bottom) layer feel a pressure due to buoyancy

$$\Lambda\sigma_{\mathrm{C}} = \left(\frac{\mathrm{r}}{\mathrm{h}}\right)\varphi_{\mathrm{C}}\,\Delta\rho\,\mathrm{aH} \tag{47}$$

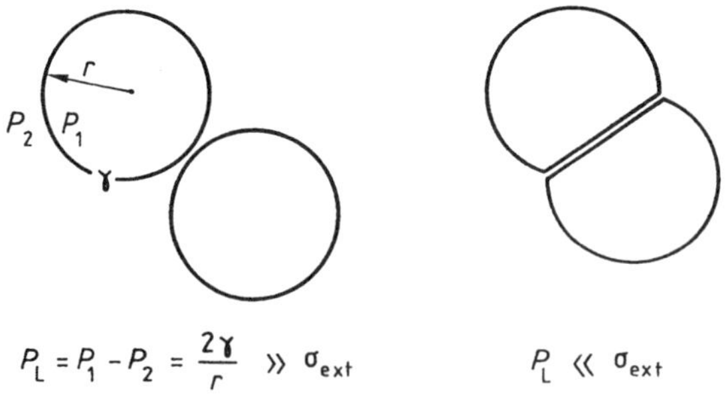

Figure 21. Closely approaching droplets for the cases that the Laplace pressure $\mathrm{p_L}$ is much higher or much lower than the external stress σ_{ext} acting on the droplet pair.

Assuming r/h = 100, the volume fraction of droplets in the cream layer $\varphi_C = 0.5$, the density difference $\Delta\rho = 100\,\text{kg·m}^{-3}$, a = g, and the height of the cream layer H = 2 cm, the pressure would be 1 kPa. In a centrifuge, a may be very much higher, e.g., 10^4 g, which would readily yield We $\gg$ 1.

It may be concluded that in most situations in practice, We is less than 1 and often even much less than 1, implying that the drops remain spherical. This would lead to fairly rapid approach of two droplets (e.g., of the order of a millisecond in Brownian motion) and in a very small "film" between the droplets, with a radius of the order of $\sqrt{rh}$; for r = 1 μm and h = 5 nm, this would imply only 70 nm. If We $\gg$ 1, a relatively large flat film develops; see, e.g., Refs. [71,72]. The drainage of a large film is fairly slow, and it may even involve dimple formation [2].

Film rupture would always occur if the film is very thin. The simplest theory considers the free energy needed to create a small hole in the film (see Fig. 22a), which would be of the order of γh^2. It is envisaged that the Brownian motion of the molecules in and near a small part of the film may provide the kinetic energy (i.e., an activation energy) to create a hole. For $\gamma = 2\,\text{mN·m}^{-1}$ and h = 3 nm, this energy would be 4.5 times kT, implying that rupture would readily happen. More elaborate theory considers the development of varicose capillary waves in the film, which may provide local thin (i.e., unstable) spots; see Fig. 22b. Capillary waves would always develop due either to Brownian motion or to vibration, but they will be damped. In a theory by Vrij [72], the effect of the interaction free energy in the film is taken into account and a criterion for stability was derived, i.e.,

$$\frac{d^2V(h)}{dh^2} > -\frac{2\pi^2\gamma}{a^2} \tag{48}$$

where the interaction potential V(h) is in J·m^{-2} and a is the radius of the film. A qualitative explanation for the effects of γ and film radius is that at a lower

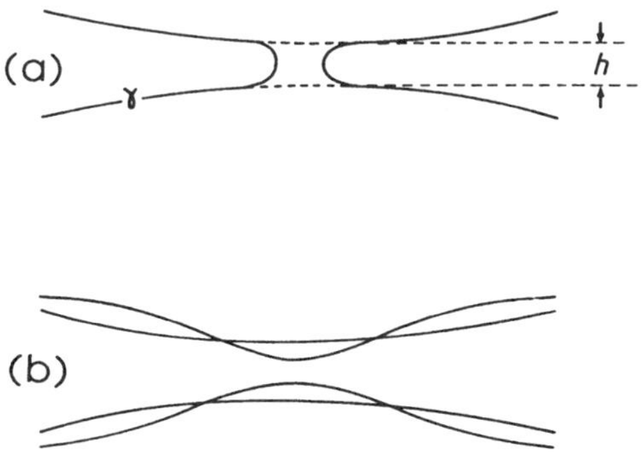

Figure 22. Rupture of a film. (a) Formation of a hole by chance. (b) Development of varicose waves in the film. γ = interfacial tension, h = film thickness.

interfacial tension and for a larger film, rupture would occur more easily, because then the same wave amplitude goes along with a lower local Laplace pressure and the latter, naturally, has a strong damping effect. In an extension of the theory [73], the role of the surface dilational modulus has been considered, since in the absence of a modulus considerable lateral motion of the interface can occur, thereby greatly enhancing film instability. It was concluded, however, that even a very small modulus, say, $10^{-5}\,\mathrm{N{\cdot}m^{-1}}$, would prevent such lateral motion, and a very low surfactant concentration would thus be sufficient.

A somewhat different view is held by Scheludko and co-workers (see, e.g., Ref. [74]). They suggest that the process of attaining a state of instability rather than the develoment of the instability itself determines the rate of film rupture. They give a theory for the rate of development of nonhomogeneities in the film that appears to fit experimental results on rather large flat films fairly well; the diameter of the nonhomogeneities is, however, at least 50 μm.

Considering first the case of We $\ll 1$, it has been remarked, e.g., in Ref. [74], that none of the theories would apply to small films bounded by strongly curved interfaces. In the author's view, Vrij's theory may possibly be modified for this case. The theory involves the first derivative of the stress in the film; for spheres, a first approximation may be the introduction of the stress concentration factor Λ (Eq. 44). By further assuming the film to extend to a thickness of less than 2h, we obtain for the size of the film $a^2 = rh$, and the stability criterion would become

$$\frac{d^2V(h)}{dh^2} - \frac{dV(h)}{h\,dh} > -C\,\frac{\gamma}{r^2} \tag{49}$$

where C is an unknown constant (about 10?); γ/r^2 would be, for instance, $10^9\,\mathrm{N{\cdot}m^{-3}}$. It would follow that a low γ, a large r, and also a weak repulsion between the droplets would favor coalescence. Applying Eq. (49) to DLVO-type interaction profiles reasonable for emulsion droplets, it follows that instability would always and only occur if the droplet pair is "in the primary minimum." Coalescence rate would then be governed by the probability of overcoming the maximum in the interaction free energy (see Fig. 14). It may be noticed that even for droplets that would never aggregate when diffusing freely, the buoyancy forces in a cream layer may be sufficient to overcome the repulsion; for instance, for a cream layer thickness of 2 cm and droplets of 5 μm, a force of almost 10^{-10} N would act between two droplets, comparable with the force calculated (by DLVO theory) from the maximum (negative) slope of V versus h curves. For steric interaction caused by adsorbed copolymers of not very small molar mass, coalescence would never occur according to Eq. (49). However, if the droplets were close to each other for a long time, as in a cream layer or when aggregated, rearrangements of the polymer layers might occur; whether this would eventually lead to coalescence is uncertain. Altogether, further study of these aspects would be very useful.

For We $\gg 1$, it is qualitatively expected that coalescence would (far) more readily occur, since the film between droplets is much larger; moreover, the droplets may stay close together for a far longer time. A similar dependence of coalescence on interfacial tension and interaction energy would be expected as for We < 1. However, the time needed for drainage of the film may now be rate-determining [2], and larger droplets may give larger films; the same may hold for a lower interfacial tension. Moreover, polymeric surfactants may impede drainage. Variation in surface dilational modulus may also play a part, since a far larger modulus would now be needed to arrest lateral motion of the interfaces, again because of the size of the film; see Sec. II.A, point 3. It may be mentioned that the factors determining the stability of macroscopic films are far from clear; see, e.g., Refs. [75,76] for protein-stabilized films. The pressure in a film as a function of thickness is in good agreement with theory, but prediction of rupture is often impossible. There is always considerable scatter in rupture time, and it increases on average with increasing film thickness, although variation among surfactants often cannot be explained.

Several studies have been done on coalescence in emulsions being subject to high-speed centrifugation or even ultracentrifugation, where large thin films are formed between droplets [2]. Generally, the correlation with coalescence in the same emulsions at rest is very poor. This may be no surprise, since it is to be expected that for $\mathrm{We} \gg 1$ (centrifuge) and $\mathrm{We} \ll 1$ (at rest), the stability would be very different. The same holds true for studies by Bibette et al. on osmotic deswelling of emulsions [77], where the droplets are also greatly deformed against each other. These studies are highly interesting for the stability of thin films, but they do not yield information on the stability of emulsions under practical conditions. Incidentally, although there is great practical need for fast tests predicting coalescence stability, these are not available for most systems. The best solution seems to be a very sensitive test for changes in average droplet size, e.g., from turbidity, so that a coalescence rate can be determined within, say, a day and used for prediction. Even then, caution may be needed, since coalescence rate is not necessarily constant with time.

Results on coalescence stability are confusing. By and large, emulsions with small droplets that neither aggregate nor cream (the latter because of a very small density difference or slow rotation of the vessel containing the emulsion) are very stable to coalescence. The idea mentioned earlier that for a DLVO-type interaction energy, coalescence would occur (and then fast) only if the droplets were to reach the primary minimum seems to agree with many observations. Most emulsions made with well soluble adsorbing copolymers are stable for several years; this may not be surprising, knowing that γ is fairly large and steric repulsion acts over a relatively large distance (Sec. V.B). Most protein-stabilized emulsions are also very stable, provided the droplets are small and the surface load has reached the plateau value; if the droplets are aggregated, very slow coalescence may occur. If a small-molecule surfactant is added in sufficiently high concentration, it will desorb the protein from the droplets (because it gives a lower γ; cf. Fig. 8), and this may be very detrimental to coalescence stability.

As expected, a very low γ goes along with rapid coalescence, as is clear, for instance, from the studies by Shinoda and co-workers on the effects of temperature on "phase inversion" and γ for some nonionic surfactants (e.g., Ref. [78]). Others have found a similar effect of γ in other situations [79,80]. The latter is a kinetic study, where a greatly enhanced rate was observed for very low γ, which was related to the formation of flat films between droplets; in other words, a transition from $\mathrm{We} < 1$ to $\mathrm{We} \gg 1$ went along with much faster coalescence. It may be mentioned that some nonionic surfactants may give a very low γ, down to 10^{-5} N·m^{-1}, at high salt concentration (e.g., Ref. [80]).

The coalescence stability in shear field may depend in a different way on the variables mentioned [81,82]. Here, a lower γ, a higher shear rate, and sometimes also a larger droplet size may lead to greater stability, because the time during which the droplets are close together may be too short for film rupture to occur. In these cases, We > 1. Several variables play a part, and they may affect each other in different ways; this would need a separate review.

Results on the effects of droplet size seem to be conflicting. Often, emulsions with small droplets are observed to be more stable, but this may be due to (1) more coalescence events being needed for the result to be macroscopically observable; (2) less creaming (sedimentation) occurring, where creaming may cause coalescence in the cream layer; and (3) We being <1 rather than >1. The only systematic study into coalescence of single droplets for We $\ll$ 1 of which the author is aware is by Dickinson et al. [83]. Here, small droplets in a dilute protein solution were allowed to cream to a planar oil–water interface of varying age, and the time needed for coalescence was observed. There was considerable scatter (relative standard deviation was about 25%), but the trend of greater stability for a smaller droplet was very clear; see Fig. 23. (The authors ascribe the effect of droplet size to the difference in buoyancy, but this appears unlikely: For a droplet diameter of 4 µm, the gravitational energy for movement over 1 nm would be about 0.01 kT; even for very weak repulsion, the change in colloidal interaction energy over 1 nm would be several times kT at its maximum.) The results of Fig. 23 were obtained under conditions (protein concentration and adsorption time) where the surface excess of protein would have been less than 0.5 mg·m^{-2}, whereas even a fully unfolded peptide chain would give about 1 mg·m^{-2}; in other words, repulsion must have been relatively weak. In

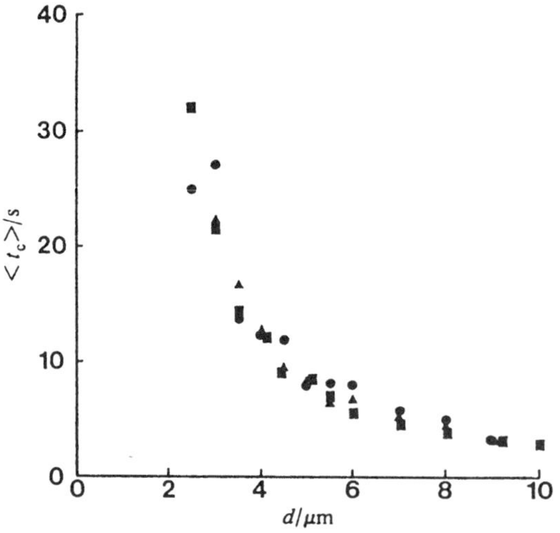

Figure 23. Coalescence time t_c for coalescence with a planar oil–water interface of droplets of various diameters d in 20-min-old 1-ppm protein solutions. ▲, β-casein; ●, κ-casein; ■ lysozyme. (From Ref. 83.)

cases where a thicker adsorbed layer was allowed to form at the planar interface (with the droplets still having a very low surface excess), Dickinson et al. found very little coalescence for small droplets.

Besides the variables γ, d, and V(h), there seem to be other factors determining coalescence, even for We $\ll$ 1. It has been mentioned by several authors that a thick adsorbed layer, especially of a polymeric substance, may provide a "skin" that would prevent coalescence. Possibly, the presence of a multilayer of small-molecule surfactant in the form of a liquid crystalline phase around a droplet, which appears to greatly enhance coalescence stability [10,84], acts in a similar way. In the author's view, the stability may arise because a thick layer keeps the oil–water interfaces farther apart (and then it may possibly fit into the Vrij theory), or because it hinders curvature of, thus wave formation on, the oil–water interface. In the latter case, this should show up in a bending modulus of the oil–water interface, but such measurements have not been made, to the author's knowledge. Instead, several workers have studied the surface shear modulus (see Sec. II.A, point 4), of, for instance, protein layers [83] or of high-molar-mass aromatic components in crude oils [85] and have observed correlations with coalescence stability. Unfortunately, the surface shear viscosity has from its inception [86] been called film strength, or even mechanical film strength [75], while it has clearly nothing to do with a mechanical strength. The correlations with coalescence stability often are poor and sometimes are absent [87,13]. For instance, the protein casein generally gives a low viscosity (say, $2\,\mathrm{mN{\cdot}sec{\cdot}m^{-1}}$) but very good coalescence stability, whereas for several globular proteins a viscosity of about $1\,\mathrm{N{\cdot}sec{\cdot}m^{-1}}$ is observed, although the coalescence stability certainly is not better. Small-molecule surfactants give negligible surface shear viscosity. It may well be that in the cases where a correlation with coalescence stability is observed, this is via a correlation with the effective thickness of the adsorbed layer.

Some other mechanisms of coalescence have been suggested. Hazlett and Schechter [88] proposed that in emulsions with very high surfactant concentration, microemulsion droplets may be present as well as "normal" droplets, and that the former may form a network between the latter; such a network would then induce coalescence. It should, however, be mentioned that the emulsions studied had a very high φ and that γ was very low. Davies et al. [89] studied what they called heterocoalescence. Small-droplet emulsions of water in cyclohexane and brine in cyclohexane were each observed to be very stable, but mixtures of the two showed marked coalescence. They concluded that film rupture probably was closely associated with transport of water, and possibly hydrated ions, across the oil film between two different droplets.

Before ending this section, it may be useful to comment on what may be called the interfacial-tension paradox. A surfactant is needed to make an emulsion and it lowers interfacial tension, thereby lowering the interfacial free energy of the system. Consequently, one might expect the emulsion to be more stable for a lower γ (other parameters being constant). However, the opposite is generally observed. As mentioned in Sec. II.A, the surfactant plays several roles. It acts during making of the emulsion by inducing the Gibbs–Marangoni effect [7]. It is generally needed to provide repulsion between droplets, and thereby imparts stability. It adsorbs only because it lowers the interfacial tension, but to obtain good coalescence stability, the lowering should not be so large as to produce a very low γ.

Summarizing, four cases of coalescence stability may be distinguished.

1. During emulsification. Here, no equilibrium surface layer exists, the colloidal repulsion provided by adsorbed surfactant is generally too weak to prevent coalescence (since the droplets may "collide" with very high kinetic energy), and the Gibbs–Marangoni effect is essential. This last plays no part in a "finished" emulsion.
2. We $\ll$ 1. Here, a high γ, a small d, and strong repulsion go along with stability, though other factors may also play a part. Emulsions can be very stable.
3. We $\gg$ 1. Here, the emulsion often is fairly unstable, but the situation is very complicated and several conditions determine how the mentioned variables affect stability.
4. Solid particles are present in or between the emulsion droplets and they may trigger coalescence. In the following section partial coalescence is discussed, but it may be envisaged that the presence of a few suitable particles may induce true coalescence in certain cases. This has been studied in foams [90], but not in emulsions.

VII. Partial Coalescence

If the emulsion droplets contain a network of solid particles, rupture of the film between closely approaching droplets cannot lead to full coalescence. Instead, an anisometric clump is formed, in which the original droplets (if still discernable) are held together by a "neck" of oil and possibly crystals; see Fig. 1e. The system thus has three phases and is, strictly speaking, not an emulsion. The clumping process has been called partial coalescence. In practice, it may occur if the dispersed phase becomes partially crystallized after making the emulsion. Such a formation of crystals in the droplets may happen on cooling of oil–in–water emulsions if the oil is a triglyceride mixture or a paraffin mixture. This case has been studied in some detail, especially in the author's laboratory [9,91–98] and by Darling [99], and this section will be based on those studies. Presumably, partial coalescence may also occur in water–in–oil emulsions, if the aqueous phase has a high concentration of solute, so partial freezing of a droplet may occur.

Partial coalescence is like aggregation in the sense that ragged aggregates are formed. The viscosity of the emulsion thus increases, which it would not do if true coalescence occurred. The aggregates may even grow into a solid network throughout the system under some conditions (high φ, fairly low shear). Important aspects are that partial coalescence may proceed very much faster, for instance, by six orders of magnitude, than coalescence would in the same emulsion without crystals. In many food emulsions, partial coalescence is by far the most important physical instability. It is all but certain that the rupture of the film between two droplets is caused by a crystal protruding from the droplet surface and puncturing the film, thereby making contact with the oil in the

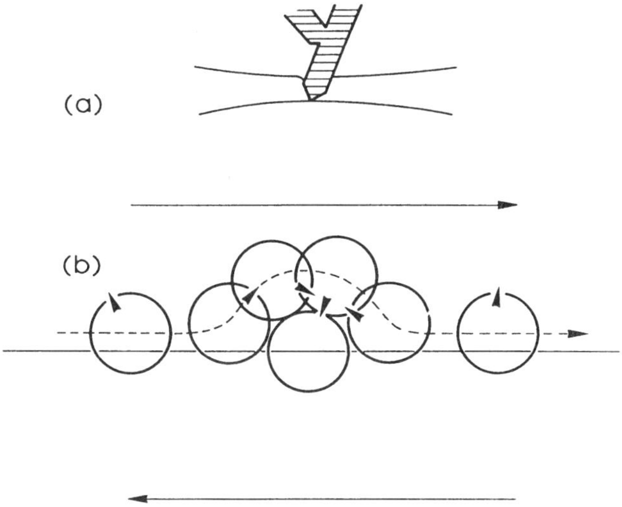

Figure 24. Illustration of the onset of partial coalescence. (a) Protruding crystal puncturing the film between two droplets. (b) Enhanced possibility for puncture of the film if droplets roll around each other in a shear field.

other droplet (see Fig. 24a); this then leads to oil–oil contact due to wetting phenomena, and thus to partial coalescence.

Typically, partial coalescence occurs in, or is at least very much enhanced by, a *velocity gradient* (G), and this is primarily the situation that will be discussed shortly. The flow may be of any type, but Couette flow, i.e., almost pure simple shear, has been studied most. The reasons for the overriding effect of shear may be the following:

The encounter rate of droplets will be about proportional to shear rate; see Sec. V.E.

The shearing stress presses the droplets closer together, depending on the colloidal interaction between them, thereby enhancing the probability that a protruding crystal may reach the other droplet. It should be realized that the size of the protruding part of the crystal may be very small, of the order of some nanometers. This also implies that the stress concentration factor Λ may be very large in this case, say, 10^5, which would greatly facilitate puncturing a fairly thick adsorbed layer.

Presumable, the chance that a crystal protrudes far enough from the surface is small, implying that the encounter of droplets in Brownian motion would rarely be such that a suitable protrusion is present at the region of closest approach. In a shear field, encountering droplets roll around each other, greatly enhancing the chance for a protruding crystal to be in the film. This is illustrated in Fig. 24b.

That it is sufficient for one droplet in the encounter to have protruding crystals is supported by the results in Fig. 25, where mixtures of emulsions with and without crystals were studied. If the fraction of emulsion with crystals

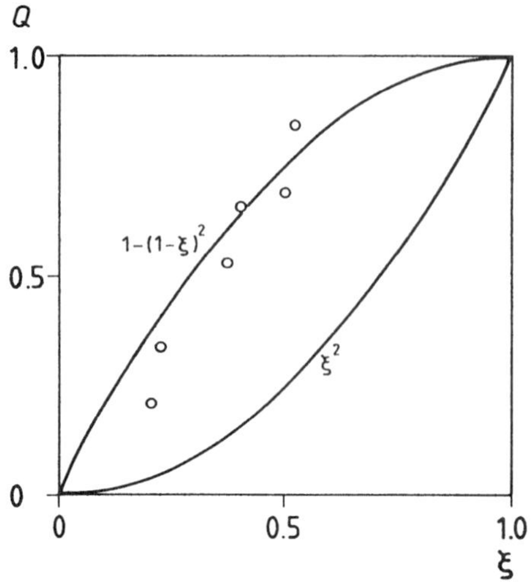

Figure 25. Partial coalescence rate relative to the rate at $\xi = 1$, Q, for mixtures of emulsions with and without crystals in the droplets, where ξ is the proportion with crystals. Various emulsions, d = 2.7–3.9 μm, $\varphi - 0.2$, G = 300 sec^{-1}. (From results in Ref. 92.)

is ξ, the partial coalescence rate would be proportional to ξ^2 if both droplets in an encounter need to have crystals, and proportional to $\{1 - (1 - \xi)^2\}$ if it is only one of the droplets; clearly, the latter model fits the results far better.

Differences in *fat crystallization* have a large effect on partial coalescence. Fat crystallization in itself is an intricate subject, and crystallization in droplets adds several complications; thus the subject is far too wide to discuss here. Only a few salient points will be recalled [100]:

Natural fats are mixtures of many components and show a melting range, which may span several times 10 K. In many fats, temperatures as low as about −40°C are needed to ensure complete crystallization. Most fats have final melting points between 0 and 40°C.

In emulsion droplets, considerable supercooling can occur, between a few and about 30 K below the final melting point. As soon as a droplet contains at least one crystal, other triglycerides (or paraffins) crystallize, with little supercooling. Supercooling is generally less in larger droplets and it may decrease with time. Thus, in an emulsion, some droplets may be partially crystalline and others may be fully liquid.

Crystal size may vary greatly, depending on several, partly unknown, factors, but cooling rate often has a fairly small effect. Triglyceride crystals tend to be small (< 1 μm), and even small droplets may contain numerous crystals. Paraffin crystals tend to be larger, and droplets may contain only one or two crystals. Most crystal shapes are fairly long platelets.

Crystals always aggregate, generally forming a continuous network. The strength of the network depends on crystal size and on the presence of some surfactants, notably monoglycerides.

Fat crystallization is slow and involves several unstable states (monotropic polymorphs, compound crystals), and this is the main cause of sintering between aggregated crystals. This implies that the crystal network may greatly increase in strength on aging.

In sum then, the proportion of fat crystallized, the size of the crystals, and their arrangement in the droplets depend on fat composition, the presence of other substances, droplet size, temperature, and the temperature history of the emulsion.

The *arrangement of the crystals* in the droplets may vary widely. In relation to partial coalescence, three cases can be distinguished; these are illustrated in Fig. 26.

(a) Some crystals are oriented at the oil–water interface in a parallel orientation. For most emulsifiers used, the contact angle θ, as measured in the aqueous phase, is between 120° and 160°, implying that the crystal is oriented preferentially in the interface, sticking out further into the oil than into the water. However, crystals may not reach the interface because they are held in the crystal network. Even if there are crystals in the interface, there may be only a few and, moreover, they may not stick out very far. Low orthokinetic capture frequencies α_O are observed, say, 10^{-6}. This implies slow though measurable partial coalescence in a shear field.

(b) From a network of crystals that extends throughout the droplet, some stick out partly in a roughly perpendicular orientation. The protrusion distance presumably depends on (1) θ at the moment the crystal becomes oriented in the interface; (2) the size and shape of the crystal; and (3) the extent of shrinking of the perimeter of the droplet due to ongoing crystallization of its contents. This is a fairly common case for several triglyceride mixtures. α_O values of, say, 10^{-3} are observed, implying a halving time of about 10 sec for $\varphi = 0.2$ and $G = 300\ \text{sec}^{-1}$.

(c) Crystals stick out far due to "growing out" of the droplet. Whether this happens depends on the combination of fat composition and the type of emulsifier; it has never been observed with polymeric emulsifiers. Two subtypes may be distinguished. In the first, only a single crystal sticks out very far, and in this case $\alpha_O \approx 1$; pouring the emulsion once suffices

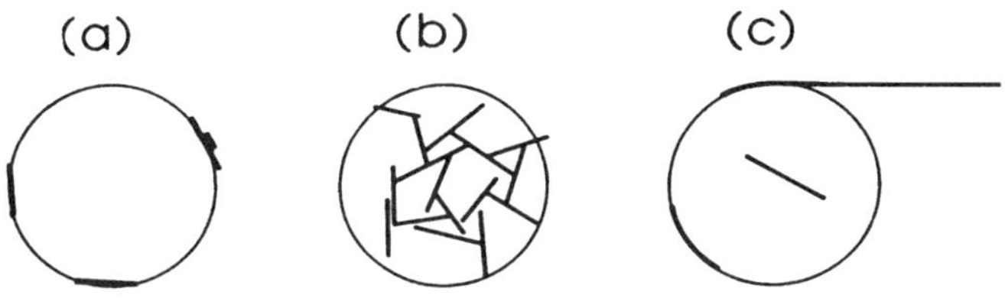

Figure 26. Different arrangements of crystals in oil droplets in relation to partial coalescence. See text.

to produce a gelled mass. In the other subtype, several crystals stick out but not very far; also, these emulsions are extremely unstable, and strong clumping may occur during quiescent cooling. Such partial coalescence without substantial shear presumably occurs when some of the droplets already contain (protruding) crystals while others are still liquid.

Unfortunately, the conditions determining whether and to what extent protruding crystals occur are not well understood. Moreover, other arrangements can occur, of which the most obvious may be the absence of protruding crystals.

Besides the mentioned size and arrangement of the crystals in the droplet, the following *factors affecting the rate of partial coalescence* have been identified. The effects of some of them are depicted in Fig. 27.

Flow conditions. In simple shear, the rate often is approximately proportional to shear rate (Fig. 27a), although the minimum G observed for partial coalescence to occur varied from a few to over 100 sec^{-1}. (In some emulsions, even shear rates of 1000 sec^{-1} or more may have no effect.) In the case of other flow types, such as Taylor vortices and turbulence, the rate may be much higher.

Volume fraction of droplets (φ). Generally the rate is approximately proportional to φ^2, as is to be expected (Fig. 27b).

Proportion of fat solid (ψ). The rate generally increases with ψ, presumably because more crystals are protruding from the droplet. For some types of fat, the partial coalescence rate decreases again for very high ψ, presumably because the network of crystals tenaciously holds the oil left, thereby leaving insufficient oil

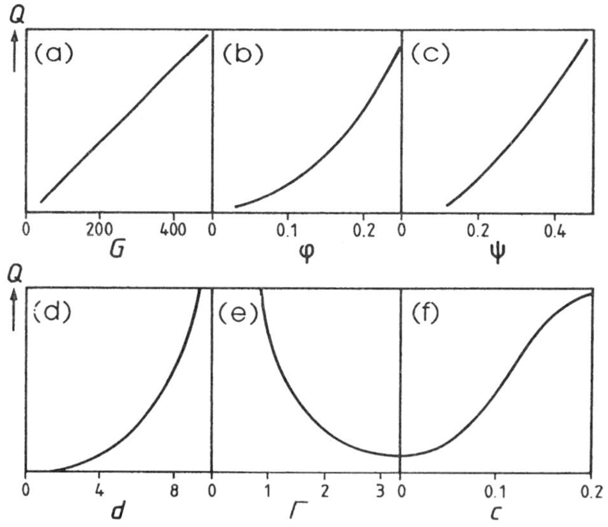

Figure 27. Effects on partial coalescence rate Q of (a) shear rate G, (b) volume fraction φ, (c) proportion of fat solid ψ, (d) droplet diameter d, (e) surface excess of emulsifier Γ, and (f) concentration of added surfactant c. In (e) and (f) the emulsifier was a protein. High schematic; meant only to illustrate trends. (After various sources.)

to produce "necks" between droplets. In most cases, a minimum ψ is needed to allow partial coalescence (Fig. 27c). This must be because a certain amount of crystalline material is needed to produce a crystal network extending throughout the whole droplet; if the network were smaller than the droplet, crystals would not stick out from the surface. The theory of fractal flocculation may be helpful. Equation (38) gives the size of a fractal aggregate. In the derivation of Eq. (41), the critical diameter of an aggregate for a continuous network to form was given as $d\varphi^{1/(D-3)}$, where φ is the volume fraction and d is the diameter of the aggregating particles. Applying the same relations to fat crystals of size y, the critical aggregate size for network formation can be found as a function of ψ. However, this critical size should be smaller than the droplet diameter (d), for a continuous network to form. This leads to the condition

$$\frac{d}{y} > \psi^{1/(D-3)} \tag{50}$$

Assuming $d = 10y$ and a fractal dimensionality of $D = 2$, this would yield a minimum ψ of about 0.1. This calculation is rather approximate, but serves to illustrate a trend. (For gelation to occur in a macroscopic vessel, condition (50) is virtually always fulfilled.)

Droplet size. Figure 27d gives an example, and such a strong dependence on droplet size is always found, albeit the curve may be shifted to smaller or larger diameters. The effect may be because condition (50) is more readily fulfilled for larger droplets or, alternatively, because larger droplets have larger crystals, which may thus protrude farther, other conditions being equal. The main cause may be that the force pressing the droplets together is larger for larger droplets (in a given flow field). Notice that the strong dependence of the partial coalescence rate on particle size would cause the rate to increase during the process; on the other hand, the shearing forces may be high enough to disrupt the larger aggregates.

Surface layer. The emulsifier used may affect the contact angle and (if it is also soluble in the dispersed phase) the crystal network. A smaller θ (as measured in the aqueous phase) generally goes along with faster partial coalescence. The emulsifier will also affect colloidal repulsion and thus the distance of closest approach between droplets (in a given flow field) and thereby the probability that a protruding crystal can puncture the film. Proteins generally give good stability against partial coalescence, unless the surface load Γ is low (Fig. 27e). If a small-molecule surfactant is added afterwards, stability is much diminished (Fig. 27f), presumably because the protein is displaced from the oil–water interface.

Composition of the continuous phase. The concentration of emulsifier, if it is a small-molecule surfactant, may affect θ. Other conditions, like ionic strength, pH, and solvent quality, may greatly affect the repulsion distance between droplets. High quantities of salt may also affect θ.

Since it is often very difficult to vary just one parameter, and since the effects of the variables may be interdependent, the results just given may be subject to exceptions.

Another complication is the intricacy of the *kinetics of partial coalescence*. Although some kinetic models have been developed [93,94,98], here only some trends will be discussed. The kinetics vary widely in type and appear to be related to the cases as depicted in Fig. 26. If the emulsion droplets are as in case (a), a regular increase of average particle size with time is observed, where the process is first order with respect to time; after long times, the process may be accelerated. For case (b), mostly a rapid formation of very big clumps is observed, so big that they rapidly cream out of the emulsion. This points to a strong increase of reactivity as bigger aggregates are formed. The average size of the particles remaining in the lower layer then often decreases with time. Intermediate cases have also been found, and a first attempt has been made to relate the differences in kinetics to some clearly defined parameters [94]. A full explanation would, however, need far more study.

Finally, it may be mentioned that emulsion droplets often readily "coalesce," i.e., make contact, with a solid surface. This may be similar to the induction of partial coalescence, since surface roughness may provide protrusions that can puncture the film between the droplet and the solid surface. Also, the presence of fat crystals in oil droplets may greatly enhance the probability that the droplets make contact with air bubbles [90].

VIII. Epilogue

It may be clear from the discussion in this chapter that even for a relatively simple colloid like a (macroscopic) emulsion, much about it stability is insufficiently known. There are uncertainties about the colloidal interaction energy and its effect on aggregation rate. Coalescence especially needs further research, where the different variables should be studied separately. Also, partial coalescence would need more study, although the main variables have been identified and their importance has been established. Prediction of combined instabilities, which will often be encountered in practice, is far more difficult, although it has been tried in some cases [5,101].

Nevertheless, much is known in a qualitative sense. Moreover, for most situations at least order-of-magnitude calculations can fruitfully be made about the changes to be expected in a certain time. However, it often is essential (1) to determine the droplet size distribution of the emulsion, and (2) to establish experimentally to what kind of instability the emulsion is subject. Often, simple light microscopy may suffice for the latter. Other tests can be derived from the nature of the variables affecting the various instabilities. Aggregated droplets can often be dispersed again, for instance, by altering pH or by decreasing salt content (DLVO-type interactions); by adding a suitable small-molecule surfactant (bridging); or by diluting the emulsion with solvent (depletion flocculation). Ostwald ripening or coalescence cannot be undone. Partially coalesced "droplets" fully coalesce when the crystals present are melted by increasing the temperature.

It may also be useful to consider that aggregation is mostly a second-order rate process, and often increases in rate with time; coalescence is first order; and Ostwald ripening goes slower and slower. However, especially in the case of aggregation, the rate may be determined by some slow change in solvent conditions, say, a decrease of pH. To distinguish between Ostwald ripening and coalescence, observation of the mode of change of the size distribution may be helpful. Ostwald ripening leads to the loss of the smaller droplets: The size frequency distribution tends to become (or remain) unimodal and negatively skewed. Coalescence often affects the largest droplets the most: The size distribution broadens, becoming positively skewed or even bimodal. Interesting examples are given by Dickinson et al. [107].

Acknowledgments

I am greatly indebted to Dr. E. Dickinson for discussions on several of the points raised in this chapter. Also, Drs. B. Bergenståhl, M. Cohen Stuart, D. G. Dalgleish, and T. van Vliet kindly scrutinized various parts of the manuscript. I am also indebted to Ms. I. Smulders and Mr. L. Taisne for help in finding references, correcting errors, and the like.

Symbols

a	activity (mole fraction)
	acceleration $(m \cdot sec^{-2})$
	radius of film (m)
A	surface area (m^2)
	specific surface area (m^{-1})
	Hamaker constant (J)
B	2nd virial coefficient $(mol \cdot m^3 \cdot kg^{-2})$
C	a constant
c	concentration $(kg \cdot m^{-3})$
D	diffusion coefficient $(m^2 \cdot sec^{-1})$
	fractal dimensionality (–)
d	droplet diameter (m)
E_{SD}	surface dilational modulus $(N \cdot m^{-1})$
F	force (N)
f	friction factor $(N \cdot sec \cdot m^{-1})$
G	velocity gradient (sec^{-1})
g	acceleration due to gravity $(m \cdot sec^{-2})$
H	vertical distance (m)
	sedimentation distance (m)
h	interparticle distance (m)
k	a constant
	Boltzmann constant $(1.38 \times 10^{-23}\, J \cdot K^{-1})$
M	molar mass $(kg \cdot mol^{-1})$
m	molar concentration $(mol \cdot m^{-3})$
N	number concentration (m^{-3})
N_f	number of particles in aggregate (–)
p	pressure (Pa)
p_L	Laplace pressure (Pa)
Q	(partial) coalescence rate

R	gas constant ($8.314\,J{\cdot}K^{-1}{\cdot}mol^{-1}$)
	radius of aggregate (m)
R_g	radius of gyration (m)
r	droplet radius (m)
s	(relative) solubility
T	absolute temperature (K)
t	time (sec)
$t_{0.5}$	halving time (sec)
v	velocity ($m{\cdot}sec^{-1}$)
W	stability ratio (–)
x,y,z	distances (m)
x'	characteristic length (m)

Dimensionless Groups

Pe	Péclet number
Re	Reynolds number
We	Weber number

Greek

α	capture rate (= 1/W) (–)
α_O	perikinetic capture efficiency (–)
Γ	surface excess (surface load) ($mol{\cdot}m^{-2}$ or $kg{\cdot}m^{-2}$)
γ	interfacial tension ($N{\cdot}m^{-1}$)
γ_0	γ in absence of surfactant ($N{\cdot}m^{-1}$)
δ	layer thickness (m)
η	viscosity ($Pa{\cdot}sec$)
η^*	apparent viscosity ($Pa{\cdot}sec$)
η_{SS}	surface shear viscosity ($N{\cdot}sec{\cdot}m^{-1}$)
θ	surface fraction (–)
η^*	contact angle (degrees)
κ	reciprocal Debye length (m^{-1})
Λ	stress concentration factor (–)
π	surface pressure ($N{\cdot}m^{-1}$)
κ_{osm}	osmotic pressure (Pa)
ρ	mass density ($kg{\cdot}m^{-3}$)
σ	stress (Pa)
φ	volume fraction of droplets (–)
φ_m	φ at close packing (–)
χ	Flory's interaction parameter (–)
ψ	electric potential (V)
	volume fraction of component (–)
	mass fraction solid (–)
ω	frequency ($rad{\cdot}sec^{-1}$)
	reciprocal time scale (sec^{-1})

Subscripts

A	attractive
B	buoyancy
C	continuous phase

c	critical
D	dispersed phase
	depletion
d	droplet
E	emulsion
i	of size class i
S	according to Stokes
	steric

References

1. T. F. Tadros and B. Vincent, in *Encylopedia of Emulsion Technology*, Vol. 1 (P. Becher, ed.), Dekker, New York, 1983, pp. 1–56.
2. T. F. Tadros and B. Vincent, in *Encyclopedia of Emulsion Technology*, Vol. 1 (P. Becher, ed.), Dekker, New York, 1983, pp. 129–286.
3. K. Shinoda and H. Kunieda, in *Encyclopedia of Emulsion Technology*, Vol. 1 (P. Becher, ed.), Dekker, New York, 1983, pp. 337–368.
4. V. B. Menon and D. T. Wasan, in *Encyclopedia of Emulsion Technology*, Vol. 2 (P. Becher, ed.), Dekker, New York, 1985, pp. 1–75.
5. D. H. Melik and H. S. Fogler, in *Encyclopedia of Emulsion Technology*, Vol. 3 (P. Becher, ed.), Dekker, New York, 1988, pp. 3–79.
6. P. Walstra, in *Encyclopedia of Emulsion Technology*, Vol. 1 (P. Becher, ed.), Dekker, New York, 1983, pp. 57–128.
7. P. Walstra, *Chem. Eng. Sci. 48*:333 (1993).
8. J. L. Salager, in *Encyclopedia of Emulsion Technology*, Vol. 3 (P. Becher, ed.), Dekker, New York, 1988, pp. 79–134.
9. H. Mulder and P. Walstra, *The Milk Fat Globule: Emulsion Science as Applied to Milk Products and Comparable Foods*, PUDOC, Wageningen, the Netherlands, 1974.
10. S. Friberg, L. Mandell, and M. Larsson, *J. Colloid Interf. Sci. 29*:155 (1969).
11. E. H. Lucassen-Reynders, in *Anionic Surfactants: Physical Chemistry of Surfactant Action* (E. H. Lucassen-Reynders, ed.), Dekker, New York, 1981, pp. 1–54.
12. E. H. Lucassen-Reynders, in *Anionic Surfactants: Physical Chemistry of Surfactant Action* (E. H. Lucassen-Reynders, ed.), Dekker, New York, 1981, pp. 173–216.
13. P. Walstra and A. L. De Roos, *Food Rev. Intern. 9*:503 (1993).
14. J. Lucassen, *Trans. Faraday Soc. 64*:2221 (1968).
15. D. J. M. Bergink-Martens, H. J. Bos, H. K. A. I. van Kalsbeek, and A. Prins, in *Food Colloids and Polymers: Stability and Mechanical Properties* (E Dickinson and P. Walstra, eds.), Royal Society of Chemistry, Cambridge, UK, 1993, pp. 376–386.
16. P. Walstra, in *Principles of Foam Formation and Stability* (A. J. Wilson, ed.), Springer, London, 1989, pp. 1–15.
17. E. Dickinson, B. S. Murray, and G. Stainsby, *J. Colloid Interf. Sci. 106*:259 (1985).
18. G. J. Fleer, M. A. Cohen Stuart, J. M. H. M. Scheutjens, T. Cosgrove, and B. Vincent, *Polymers at Interfaces*, Chapman & Hall, London, 1993.
19. J. M. G. Lankveld and J. Lyklema, *J. Colloid Interf. Sci. 41*:475 (1972).
20. J. T. C. Böhm, *Meded. Landbouwhogeschool Wageningen 74–5* (1974).
21. E. Dickinson, *J. Chem. Soc. Faraday Trans. 20*:2973 (1992).
22. E. Tornberg, *J. Sci. Food Agr. 29*:867 (1978).
23. F. MacRitchie, *J. Colloid Interf. Sci. 105*:119 (1985).
24. S. S. Davis and A. Smith, in *Theory and Practice of Emulsion Technology* (A. L. Smith, ed.) Academic Press, London, 1976, pp. 285–306.
25. W. Thomson, *Proc. Roy. Soc. Edinburgh 7*:63 (1871).

26. A. S. Kabalnov and E. D. Shchukin, *Adv. Colloid Interf. Sci. 38*:69 (1992).
27. D. J. McClements, S. R. Dungan, J. B. German, and J. E. Kinsella, *J. Colloid Interf. Sci. 156*:425 (1993).
28. A. J. de Vries, *Rec. Trav. Chim. Pays-Bas 77*:209, 283 (1958).
29. J. Lucassen, in *Anionic Surfactants: Physical Chemistry of Surfactant Action* (E. H. Lucassen-Reynders, ed.), Dekker, New York, 1981, pp. 217–265.
30. A. D. Ronteltap, B. R. Damsté, M. de Gee, and A. Prins, *Coll. Surf. 47*:269 (1990).
31. R. B. Bird, W. E. Stewart, and E. W. Lightfoot, *Transport Phenomena*, Wiley, New York, 1960.
32. P. Walstra and H. Oortwijn, *Neth. Milk Dairy J. 29*:263 (1975).
33. T. van Vliet and P. Walstra, in *Food Colloids* (R. D. Bee, P. Richmond, and J. Mingins, eds.), Royal Society of Chemistry, Cambridge, UK, 1989, pp. 206–217.
34. J. F. Richardson and W. N. Zaki, *Trans. Inst. Chem. Eng. 32*:35 (1954).
35. G. K. Batchelor, *J. Fluid Mech. 52*:245 (1972).
36. R. Buscall, J. W. Goodwin, R. H. Ottewill, and T. F. Tadros, *J. Colloid Interf. Sci. 85*:78 (1982).
37. G. K. Batchelor and C. S. Wen, *J. Fluid Mech. 124*:495 (1982).
38. E. Barnea and J. Mizrahi, *Chem. Eng. J. 5*:171 (1973).
39. R. Johne, *Chem. Ing. Techn. 38*:428 (1966).
40. A. M. Howe, A. R. Mackie, and M. M. Robins, *J. Dispersion Sci. Technol. 7*:213 (1986).
41. P. A. Gunning, D. J. Hibberd, A. M. Howe, and M. M. Robins, *Food Hydrocolloids 2*:119 (1988).
42. X. D. Chen and G. C. Wake, *Food Hydrocolloids 7*:123 (1993).
43. L. G. B. Bremer, "Fractal Aggregation in Relation to Formation and Properties of Particle Gels," Ph.D. thesis, Wageningen Agricultural University, Wageningen, the Netherlands, 1992.
44. H. R. Kruyt (ed.), *Colloid Science*, Vol. I, Elsevier, Amsterdam, 1952.
45. W. Albers and J. T. G. Overbeek, *J. Colloid Sci. 14*:501, 510 (1959), *15*:489 (1960).
46. E. Hantz, A. Cao, P. Depraetere, and A. Taillandier, *J. Phys. Chem. 89*:5832 (1985).
47. J. N. Israelachvili, *Intermolecular and Surface Forces*, 2nd ed., Academic Press, London, 1992.
48. S. S. Dukhin and J. Lyklema, *Faraday Disc. Chem. Soc. 90*:261 (1990).
49. P. Warzyński and T. G. M. van de Ven, *Faraday Disc. Chem. Soc. 90*:313 (1990).
50. S. Yu. Shulepov and G. Frens, *J. Colloid Interf. Sci., 170*:44 (1995).
51. E. Dickinson and L. Eriksson, *Adv. Colloid Interf. Sci. 34*:1 (1991).
52. J. Lyklema and T. van Vliet, *Faraday Trans. Chem. Soc. 65*:25 (1978).
53. G. C. March and D. H. Napper, *J. Colloid Interf. Sci. 61*:383 (1977).
54. B. Bergenståhl, in *Gums and Stabilizers for the Food Industry*, Vol. 4 (G. O. Phillips, P. A. Williams, and D. J. Wedlock, eds.), IRL Press, Oxford, 1988, pp. 363–372.
55. S. Asakura and F. Oosawa, *J. Polymer Sci. 33*:183 (1958).
56. P. Walstra, in *Food Colloids and Polymers: Stability and Mechanical Properties* (E. Dickinson and P. Walstra, eds.), Royal Society of Chemistry, Cambridge, UK, 1993, pp. 1–15.
57. Y. Cao, E. Dickinson, and D. J. Wedlock, *Food Hydrocolloids 4*:185 (1990).
58. M. Aronson, *Langmuir 5*:494 (1989).
59. J. Bibette, D. Roux, and F. Nallet, *Phys. Rev. Letters 65*:2470 (1990).
60. J. Bibette, *J. Colloid Interf. Sci. 147*:474 (1991).
61. L. A. Spielman, in *The Scientific Basis of Flocculation* (K. J. Ives, ed.), Sijthoff & Noordhoff, Alphen aan den Rijn, the Netherlands, 1978, pp. 63–88.
62. T. G. M. van de Ven, *Colloidal Hydrodynamics*, Academic Press, London, 1988.

63. E. P. Honig, G. J. Roebersen, and P. H. Wiersema, *J. Colloid Interf. Sci. 33*:562 (1971).
64. Y. Adachi, M. A. Cohen Stuart, and R. Fokkink, *J. Colloid Interf. Sci. 165*:310 (1994).
65. G. J. Fleer and J. Lyklema, *J. Colloid Interf. Sci. 52*:228 (1974).
66. M. van den Tempel, *Rec. Trav. Pays-Bas 72*:433, 442 (1953).
67. F. Family and D. P. Landau (eds.), *Kinetics of Aggregation and Gelation*, North Holland, Amsterdam, 1984.
68. P. Meakin, *Adv. Colloid Interf. Sci. 28*:249 (1988).
69. L. G. B. Bremer, P. Walstra, and T. van Vliet, *Colloids Surfaces 99*:121 (1995).
70. J. A. de Feijter, J. B. Rijnbout, and A. Vrij, *J. Colloid Interf. Sci. 64*:258 (1978).
71. M. P. Aronson and H. M. Princen, *Colloids Surfaces 4*:173 (1982).
72. A. Vrij, *Disc. Faraday Soc. 42*:23 (1966).
73. A. Vrij, F. T. Hesselink, J. Lucassen, and M. van den Tempel, *Proc. Kon. Akad. Wetensch. B 73*:124 (1970).
74. B. P. Radoev, A. D. Scheludko, and E. D. Manev, *J. Colloid Interf. Sci. 95*:254 (1983).
75. L. R. Fisher and E. E. Mitchell, in *Food Colloids* (R. D. Bee, P. Richmond, and J. Mingins, eds.), Royal Society of Chemistry, Cambridge, UK, 1989, pp. 123–137.
76. V. G. Babak, *Food Hydrocolloids 6*:45 (1992).
77. J. Bibette, D. C. Morse, T. A. Witten, and D. A. Weitz, *Phys. Rev. Letters 69*:2439 (1992).
78. H. Saito and K. Shinoda, *J. Colloid Interf. Sci. 32*:647 (1970).
79. R. D. Hazlett and R. S. Schechter, *Colloids Surfaces 29*:71 (1988).
80. J. A. M. H. Hofman and H. N. Stein, *J. Colloid Interf. Sci. 147*:508 (1991).
81. H. Mousa and T. G. M. van de Ven, *Colloids Surfaces 60*:19, 39 (1991).
82. F. Groeneweg, F. van Voorst Vader, and W. G. M. Agterof, *Chem. Eng. Sci. 48*:229 (1993).
83. E. Dickinson, B. S. Murray, and G. Stainsby, *J. Chem. Soc. Faraday Trans 1., 84*:871 (1988).
84. S. Friberg, *J. Colloid Interf. Sci. 37*:291 (1971).
85. P. Becher, in *Interfacial Phenomena in Apolar Media* (H. F. Eicke and G. D. Parfitt, eds.), Dekker, New York, 1987, pp. 257–280.
86. J. A. Serralach, G. J. Jones, and R. J. Owen, *Ind. Eng. Chem. 25*:816 (1993).
87. L. M. Baldauf, R. S. Schechter, W. H. Wade, and A. Graciaa, *J. Colloid Interf. Sci. 85*:187 (1982).
88. R. D. Hazlett and R. S. Schechter, *Colloids Surfaces 29*:53 (1988).
89. R. Davies, D. E. Graham, and B. Vincent, *J. Colloid Interf. Sci. 126*:616 (1988).
90. P. R. Garret, *Chem. Eng. Sci. 48*:367 (1993).
91. M. A. J. S. van Boekel, *Agr. Res. Reports Wageningen 901*:1 (1980).
92. M. A. J. S. van Boekel and P. Walstra, *Colloids Surfaces 3*:109 (1981).
93. J. P. Melsen, "The Stability of Recombined Milk Fat Globules," Ph.D. thesis, Wageningen Agricultural University, Wageningen, the Netherlands, 1987.
94. K. Boode, "Partial Coalescence in Oil-in-Water Emulsions," Ph.D. thesis, Wageningen Agricultural University, Wageningen, the Netherlands, 1992.
95. K. Boode, C. Bisperink, and P. Walstra, *Colloids Surfaces 61*:55 (1991).
96. K. Boode and P. Walstra, *Colloids Surfaces 81*:121 (1993).
97. K. Boode, P. Walstra, and A. E. A. de Groot-Mostert, *Colloids Surfaces 81*:139 (1993).
98. K. Boode and P. Walstra, in *Food Colloids and Polymers* (E. Dickinson and P. Walstra, eds.), Royal Society of Chemistry, Cambridge, UK, 1993, pp. 23–30.
99. D. F. Darling, *J. Dairy Res. 49*:695 (1982).

100. P. Walstra, in *Food Structure and Behaviour: Stability and Mechanical Properties* (J. M. V. Blanshard and P. J. Lillford, eds.), Academic Press, London, 1987, pp. 67–85.
101. B. Bergenståhl, "Topics in Food Emulsions," Ph.D. thesis, University of Lund, Sweden, 1994.
102. P. Walstra, in *Food Structure and Behaviour* (J. M. V. Blanshard and P. J. Lillford, eds.), Academic Press, London, 1987, pp. 87–106.
103. E. Dickinson, *J. Colloid Interf. Sci. 73*:578 (1980).
104. D. E. Graham and M. C. Phillips, *J. Colloid Interf. Sci. 76*:227 (1980).
105. G. A. van Aken and M. T. E. Merks, *Voedingsmiddelentechnologie 27(9)*:11 (1993).
106. E. Dickinson, M. I. Goller, and D. J. Wedlock, *Colloids Surfaces 75*:195 (1993).
107. E. Dickinson, V. B. Galazka, and D. W. Anderson, *Carbohydrate Polymers 14*:373 (1991).
108. J. C. Dijt, M. A. Cohen Stuart, and G. J. Fleer, *Macromolecules 25*:5416 (1992).

2
Dynamic Interfacial Properties in Emulsification

E. H. LUCASSEN-REYNDERS / Unilever Research Laboratorium Vlaardingen, Vlaardingen, The Netherlands

I. Introduction

Emulsification can be defined loosely as the breakup of large drops of one liquid in another, immiscible, liquid, leading to an average drop size in the 0.2–50-μm range. The interfacial area created by breakup into, say, 6-μm drops is as large as 10^6 m^2 per cubic meter of dispersed liquid. The excess energy associated with this large interfacial area renders the emulsion thermodynamically unstable against phase separation; yet dynamic stability over quite long time spans can be achieved by the addition of stabilizing surface-active materials as emulsifiers. Much practical use of emulsions depends on the possibility of controlling and manipulating their drop size and stability by the judicious addition of emulsifiers. Thus, the basic functions of emulsifiers can be described simply as, first, to help make drops small and, second, to keep them small by retarding coalescence.

This chapter will focus on the mechanisms by which emulsifiers can fulfill these functions under the dynamic conditions of emulsification. Since emulsifiers can function only if adsorbed at the interface, the main emphasis will be on their capacity (1) to reduce the interfacial tension and (2) to impart viscoelasticity to the interface. The essential mechanism by which the surface activity of emulsifiers aids emulsification, as first clearly expounded by van den Tempel in 1960 [1], is not that emulsifiers produce a low interfacial tension, but that their presence can lead to gradients in interfacial tension that enable the interface to resist tangential stresses; it is this resistance that can be expressed in interfacial elasticity and viscosity.

II. Interplay of Hydrodynamics and Interfacial Dynamics

A. General Description

The mechanisms by which various types of liquid agitation can provide the energy needed for breakup of the disperse phase have been reviewed in an earlier volume in this series [2]. Three stages can be distinguished in any emulsification process:

1. Extension of initially spherical drops of the disperse phase in the applied flow field. In simple shear, drop extension tends to be moderate because of the rotational component of the flow, especially when the disperse phase is highly viscous. In elongational flow, drops are more easily extended into very long cylindrical threads.
2. Breakup of the extended drops, into either two relatively large drops or a multitude of tiny droplets.
3. Recoalescence of drops, which will occur especially in emulsions with low concentrations of poor emulsifiers.

Only the first two stages have been studied extensively with single drops, mainly in emulsifier-free systems [3–7]. In the first stage, extension of drops by an imposed flow regime is opposed by the capillary (Laplace) pressure, ΔP, which aims to preserve the spherical shape:

$$\Delta P = \frac{2\sigma}{R} \tag{1}$$

where σ is the interfacial tension and R is the radius of the undeformed drop. With a few notable exceptions, current hydrodynamic treatments of such drop breakup ignore the effects of emulsifiers other than their direct effect on the Laplace pressure. Thus, these treatments assume (1) that the interface transmits tangential stresses undiminished (i.e., that interfacial elasticity/viscosity is negligible) and (2) that the interfacial tension is constant and uniform.

If reduction of capillary pressure were the only function of emulsifiers, one would expect the size of emulsion drops to vary linearly with the interfacial tension, σ, during emulsification in shearing flow, at given conditions of rate and bulk viscosities. At least two factors arising from the dynamic character of any emulsification process can disturb this simple picture:

1. During breakup, the interfacial tension of the drops is neither constant nor uniform. The average value of σ will be higher than at equilibrium because of the necessary overall extension of the interface, but some parts of the drop may have much lower values than other parts, since the deformation is not homogeneous.
2. In many emulsifying machines, conditions of flow are such that zones of low shear exist side by side with highly sheared regions. After breakup in a high-shear region, drops almost always spend some time in zones where they can recoalesce if their dynamic stability is low. This leads to a larger average drop size, particularly in a relatively inefficient emulsifying apparatus.

As a result, a simple proportionality of average drop size with equilibrium interfacial tension is unlikely even in the simplified case of laminar flow (see Sec. IV).

It is clear that dynamic effects will be negligible only if (1) transport of emulsifier between bulk solution and interface by convective diffusion is slower than both the surface extension of the drops and their breakup and (2) interfacial rheological properties (dilational elasticity/viscosity) do not interfere with the breakup process.

As to the third stage, no general theory is available for recoalescence of drops. The role of emulsifiers in this stage can be expected to be crucial, because the rheological properties that emulsifiers impart to an interface play a key role in the stabilization of the films separating emulsion drops [1,2,8], as foaming agents do in the films separating air bubbles [9,10].

B. Hydrodynamics of Drop Breakup in Emulsifier-Free Systems

Deformation and breakup of a drop occur when the stress exerted on the drop by the flow is high enough to overcome the Laplace pressure given by Eq. (1). The ratio of these opposing stresses is known as the capillary number (Ca), defined by Eq. (2) if the applied flow is laminar [4,6,7]:

$$\mathrm{Ca} = \frac{\eta_c \dot{\gamma} R}{\sigma} \tag{2}$$

(See the symbols list at the end of this chapter for the meaning of symbols.) Drops eventually break up into two or more main fragments, accompanied by much smaller satellites, if the capillary number exceeds a critical value, $\mathrm{Ca_{cr}}$. Numerical values of $\mathrm{Ca_{cr}}$ are not predicted by simple hydrodynamic theory; experimentally, $\mathrm{Ca_{cr}}$ is found to depend on the type of flow and on the viscosity ratio of the two liquids. The classical work by Grace reproduced in Fig. 1 shows that the breakup in simple shear is easiest if the drop phase is somewhat less viscous than the continuous phase [4,6]. Drops are generally found to break up into two large fragments (binary breakup) if the applied flow is only slightly stronger than that corresponding to $\mathrm{Ca_{cr}}$; if much stronger, the flow draws the drops out into long cylindrical threads before they burst [5]. Figure 2 gives a schematic illustration of the two events as described by Rumscheidt and Mason [5].

A different picture is observed if the applied flow is elongational or hyperbolic [3,5,6]. The critical capillary number in this flow is generally lower (see Fig.

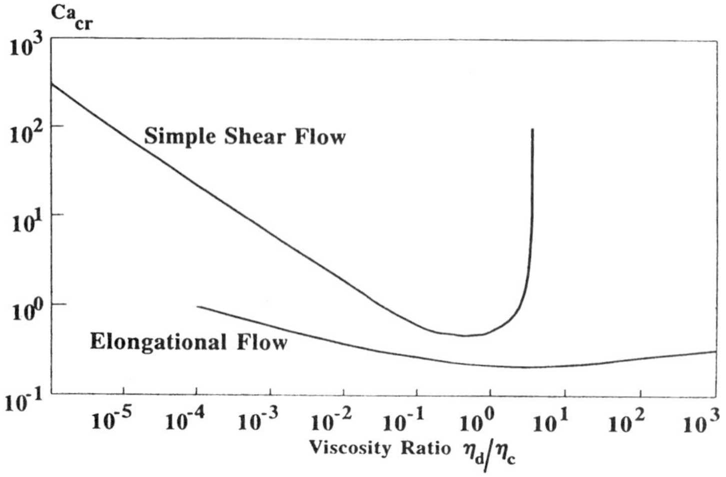

Figure 1. Critical capillary number for drop breakup as a function of the viscosity ratio, in simple shear flow and plane elongational flow for surfactant-free systems. (After Ref. 6.)

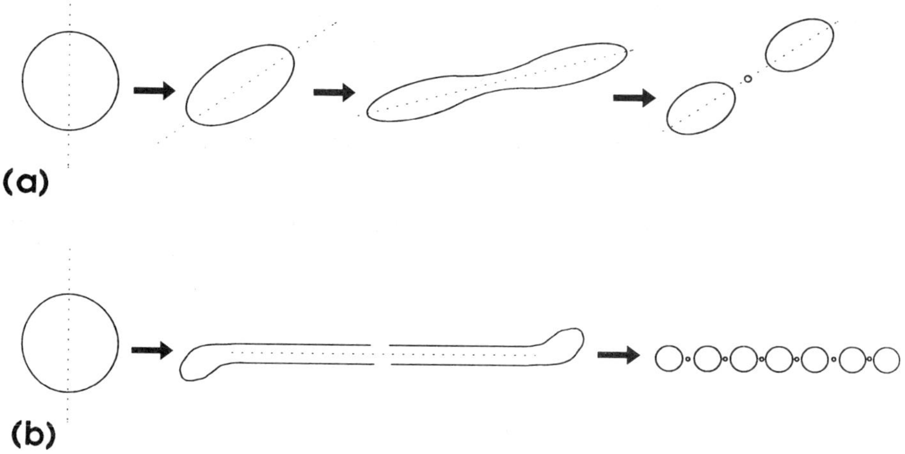

Figure 2. Deformation and breakup of drops by simple shear. Viscosity ratio ≈ 1. (a) Capillary number just above Ca_{cr}; (b) $Ca \gg Ca_{cr}$. (Adapted from Ref. 4.)

1), and long cylindrical threads are more easily formed than in simple shear. These differences mean that elongational flow is the more efficient type of flow for emulsification purposes [11,12].

Final drop size after the breakup of cylindrical threads depends on the ratio L/r of cylinder length L to radius r, and the question arises as to which factors determine the value of this ratio. The qualitative answer is that the extension of a viscous cylinder in a less viscous medium will go on as long as the flow is accelerated [3]. When the flow is stopped, the cylinder will break up spontaneously into N droplets, because this reduces its interfacial area (Rayleigh instability):

$$N = \frac{L}{2\pi r} f_1 \tag{3}$$

with the numerical factor f_1 dependent on the viscosity ratio η_d/η_c. Breakup in terms of the critical capillary number occurs under quasistatic conditions; i.e., the flow conditions must be kept constant over a time span longer than the time scale of the breakup process. This time scale, τ_b, can be approximated by

$$\tau_b = \frac{\eta_d r}{\sigma} f_2 \tag{4}$$

where the factor f_2, like f_1 in Eq. (3), is a function of the viscosity ratio. Available evidence suggests an inverse-square-root dependence on η_d/η_c for f_2 [6].

Equation (4) is valid for both the breakup of an elongated cylinder into many fragments and binary breakup. The essential difference between the two situations is that binary breakup occurs only if the flow is maintained for a time τ_b, whereas the breakup of long cylinders takes place spontaneously after the flow is stopped.

Summarizing, simple hydrodynamic theory for laminar flow describes drop breakup under quasistatic conditions in terms of shear rate ($\dot{\gamma}$), viscosities (η_c and η_d), a constant uniform interfacial tension (σ), and a number of empirical coefficients (Ca_{cr}, f_1, f_2). The presence of emulsifiers obviously requires accounting for the effects of the interfacial tension being neither constant nor uniform under the dynamic conditions of emulsification.

C. Interfacial Dynamics of Emulsifiers in Drop Breakup

The preceding treatment for pure liquids [3–7] ignores the effects of emulsifiers, where it assumes the following:

1. The interfacial tension is constant and uniform.
2. The interface transmits tangential stresses undiminished; i.e., interfacial elasticity/viscosity is neglected.

Emulsifiers can have two main effects on the deformation and breakup of drops:

1. Changes in interfacial tension, e.g., by convective diffusion of emulsifier from either liquid phase to the expanding interface
2. Viscoelastic behavior of the interface

The effects of interfacial viscoelastic behavior on the deformation and orientation of drops in simple shear flow and extensional flow were analyzed by Flumerfelt [13]. His hydrodynamic theory accounts for the effect of interfacial tension gradients on the tangential stress boundary condition at the interface in terms of interfacial elasticities and interfacial viscosities, for the case of negligible solution/interface interchange of surfactant, i.e., for insoluble surfactants. An important result of his theory is that nonzero interfacial rheological parameters are described as effectively increasing the viscosity of the dispersed phase. Thus, interfacial rheology hinders the breakup of drops more viscous than the continuous phase in simple shear, with reference to Fig. 1. It also suppresses internal circulation, even in low-viscosity drops.

Milliken et al. [14] also considered the effects of insoluble surfactant on drop deformation and breakup in extensional flow for a range of drop viscosities, using an explicit surface equation of state (Frumkin). Their analysis in terms of interfacial tension gradients (which is equivalent to one in terms of dilational elasticity) demonstrates that during drop stretching, surfactant is convected to the ends of the drop. This reduces the local interfacial tension and so facilitates the formation of pointed ends, leading to tip streaming, illustrated in Fig. 3. The effect of surfactant is found to be particularly pronounced at low drop viscosity, and it decreases with increasing viscosity ratio η_d/η_c. This analysis, too, shows that gradients in interfacial tension retard the interfacial motion and cause the drop to appear as though it were more viscous. Moreover, the authors study the interfacial-tension-driven motion of an elongated drop suspended in an otherwise-quiescent fluid. Whereas previously stretched drops without surfactant tend to fragment by "end pinching," the same drop with surfactant does not fragment: Larger extensions of the drop can be achieved without producing end pinching. Surfactant is also shown to affect the critical ratio L/r (see Eq. 3) at which long cylinders break up, and generally to slow down the process of retraction, on account of the lower interfacial tension. The reduced interfacial tension should also slow down the growth of the capillary instabilities leading to the breakup of a cylinder. In this way, the presence of surfactant results in fewer drops. These findings on stretched drops after flow is stopped have obvious practical relevance for emulsification conditions where zones of intense flow coexist with fairly quiet regions.

D. Stabilization of Liquid Films Between Drops by Emulsifiers

The essential stabilizing function of surface-active molecules during emulsification is that their presence can lead to *gradients in interfacial tension*, which enable the interface to resist tangential stresses from the adjoining flowing liquids [1]. Thus, an emulsifier-covered interface can behave as a two-dimensional body with its own two-dimensional rheological properties. These provide the liquid layers separating emulsion drops with a mechanism for

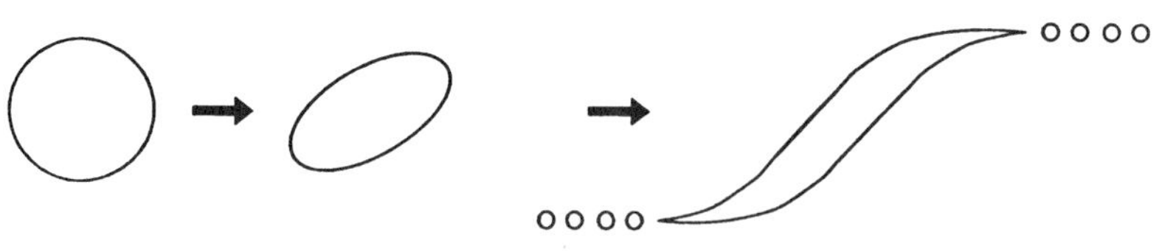

Figure 3. Deformation and breakup of drops by tip streaming in simple shear. (Adapted from Refs. 4 and 7.)

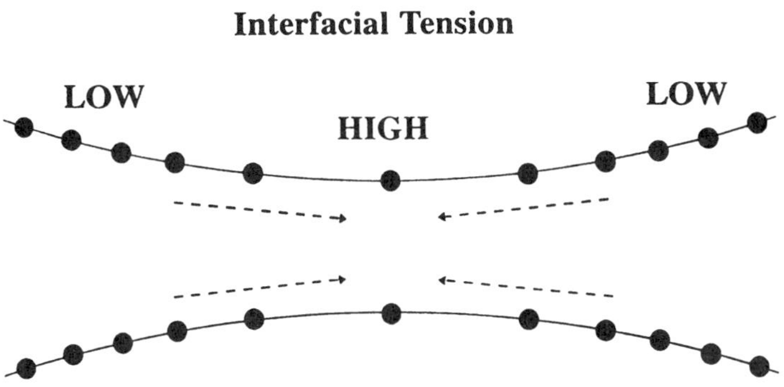

Figure 4. Marangoni effect in the liquid layer separating two emulsion drops. ●: adsorbed emulsifier molecules; arrows: flow of interface and film liquid.

dynamic stabilization, without which the drops would be liable to fast recoalescence during emulsification. Figure 4 schematically depicts two approaching drops. Without emulsifier, the liquid layer separating the two would have a uniform interfacial tension and would be insufficiently stable against drainage and rupture. In the presence of emulsifier, however, a deficit of adsorbed molecules in the film caused by local extension of the film results in a momentary gradient in adsorption, and hence a gradient in interfacial tension. In such conditions, the interface behaves like an elastic membrane, in which regions of high tension tend to contract more strongly than do low-tension regions. High-tension areas, therefore, exert a pull on adjacent lower-tension areas, which results in a flow of surface with adhering liquid toward the thin spot in the film. In this manner, emulsifiers can make a surface resist any deformation that would create a gradient in surface tension, and so they provide the film with a self-healing mechanism.

Qualitatively, the effect of a surface's resistance against tension gradients was described more than a century ago by Marangoni [15] and Gibbs [16]. Quantitatively, the measurement of this resistance took place in the last few decades, as did the interpretation of the results in terms of emulsifier properties (to be reviewed in Sec. III.A).

In sum, emulsifiers can have a significant effect on emulsification by (1) their direct effect on the dynamic interfacial tension, (2) changing the breakup conditions through interfacial viscoelasticity, and (3) preventing recoalescence through film elasticity. These three effects are closely interrelated, as Sec. III will demonstrate in more detail.

III. Interfacial Dynamics of Emulsifiers

A. Elasticity of Thin Films Separating Drops

In its simplest form, the stabilizing effect of emulsifiers on thin films is expressed by the Gibbs elasticity (E) of the film, which is defined by the increase in interfacial tension caused by a small relative increase in surface area:

$$E = 2E_S; \qquad E_S = \frac{d\sigma}{d(\ln A)} \tag{5}$$

where E_S is the elasticity of a single surface of the two-sided film. This definition, originally proposed by Gibbs for the surface elasticity of a soap-stabilized liquid film, has since been found to be of general applicability to express surface tension gradients on any liquid interface.

Figure 5 illustrates the concept of Gibbs elasticity for a plane-parallel film with adsorption Γ, surfactant concentration c, and thickness h. After rapid expansion of a film element with area A, the elasticity of each surface will have an *initial* value (i.e., before the depleted surface element is resupplied by diffusion from the bulk of the liquid layer) given by the limiting value for an insoluble monolayer, ϵ_0:

$$\epsilon_0 = \frac{-d\sigma}{d(\ln \Gamma)} \tag{6}$$

After equilibration by diffusion has run its course, the final concentration and adsorption are slightly lower because the constant total amount of material is now distributed over a larger interfacial area and the same volume. Gibbs elasticity is defined in terms of the new surface tension at the end of this equilibration. Equilibration parallel to the surface, i.e., with neighboring film elements, must be so slow as to be negligible: Direct measurements, therefore, can be performed on large films drawn up in a frame. Theoretically, for single-surfac-

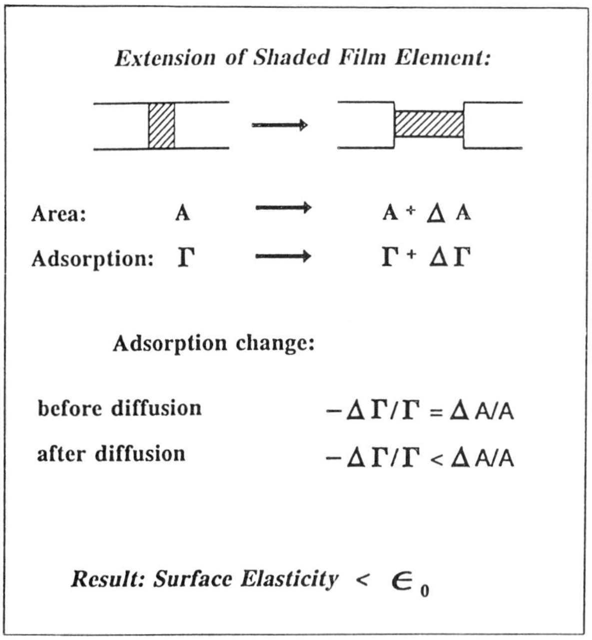

Figure 5. Elastic resistance of thin film against area extension. Note that the change in adsorption ($\Delta\Gamma$) is negative for positive ΔA.

tant solutions, the elasticity E_S of each film surface is found by a simple mass-balance argument, following the derivation by van den Tempel et al. [17]:

$$E_S = \frac{\epsilon_0}{1 + (h/2)(dc/d\Gamma)} \tag{7}$$

(Note that c here is the real concentration, not to be replaced by the activity in case of nonideal behavior.) Numerical values of E_S depend on the film thickness, h, and on the surfactant-related quantities ϵ_0 and $d\Gamma/dc$, the slope of the equilibrium adsorption isotherm. For the latter parameters we need the analytical relationships between surface tension, adsorption, and concentration. In the simplest case of ideal surface behavior of an ideally dilute solution, we have a Langmuir adsorption isotherm:

$$\frac{\Gamma}{\Gamma^\infty} = \frac{c}{c + a} \tag{8}$$

where the surface activity of the surfactant is expressed in two parameters: Γ^∞ is the limiting adsorption at saturation of the interface, and a is the concentration needed to reach half this saturation value. In this case the "surface equation of state" is given by

$$\sigma^0 - \sigma = -RT\Gamma^\infty \ln\left(1 - \frac{\Gamma}{\Gamma^\infty}\right) \tag{9}$$

This means that the surfactant-related quantities in Eq. (7) depend on concentration according to:

$$\epsilon_0 = RT\Gamma^\infty \left(\frac{c}{a}\right); \qquad \frac{dc}{d\Gamma} = \left(\frac{a}{\Gamma^\infty}\right)\left(1 + \frac{c}{a}\right)^2 \tag{10}$$

The final result for the elasticity of each film surface in Fig. 5 is obtained by substitution into Eq. (7):

$$E_S = RT\Gamma^\infty \frac{c/a}{1 + (ah/2\Gamma^\infty)(1 + c/a)^2} \tag{11}$$

Gibbs elasticity always depends on surfactant concentration in the manner shown in Fig. 6, at a given constant film thickness. In the low-concentration range, a linear increase of the elasticity is observed, followed by a maximum reached at a concentration ($c/a \approx 1$) where the surface is approximately half saturated. At still higher concentrations, the elasticity falls off due to diffusion increasingly resupplying the surface and leveling the initial gradient in interfacial tension. It must be emphasized that the characteristic maximum in the elasticity is not caused by any molecular interactions in the surface, but purely by the effect of surfactant supply to the surface from the solution at higher concentrations. Molecular interactions could not play a part in this, in any case, because Langmuir adsorption presupposes ideal surface behavior.

Apart from the surfactant's concentration and its characteristic parameters, a and Γ^∞, the elasticity depends on the thickness of the film, because the film acts as a reservoir from which molecules can be resupplied to the expanded surface: The elasticity of a thick film is always lower than that of a thinner

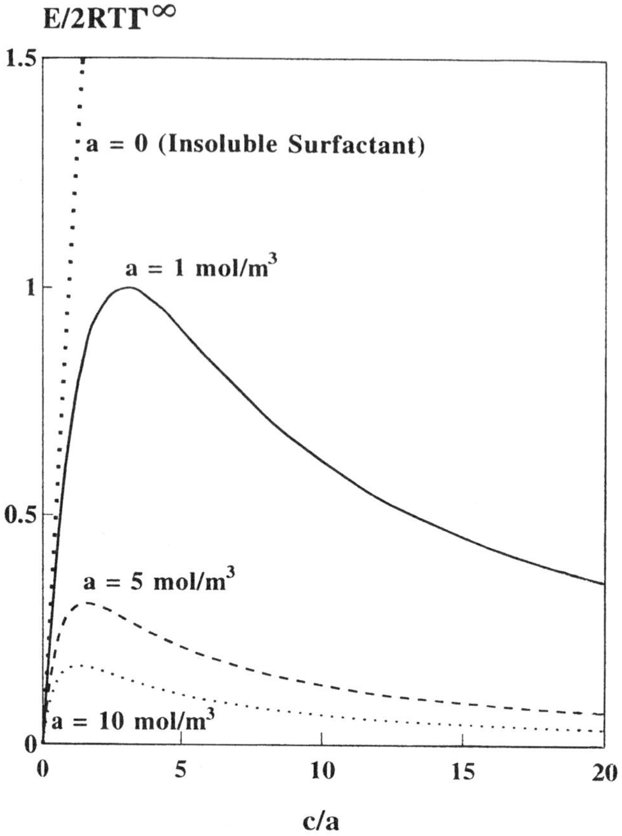

Figure 6. Numerical examples of Gibbs elasticity (E) as a function of surfactant concentration (c), calculated from Eq. (7) for Langmuir adsorption with $RT\Gamma^{\infty} = 20\,\mathrm{mN/m}$; film thickness $h = 1\,\mu\mathrm{m}$. (Reproduced from *Food Structure, 12*:1 (1993) by courtesy of Scanning Microscopy International.)

one, at the same concentration. For a very thin film from which no resupply at all can take place, the limiting value equals ϵ_0:

$$E_S = \epsilon_0 \qquad \text{for} \qquad h \ll 2\,\frac{d\Gamma}{dc} \tag{12}$$

In very thick films, on the other hand, the surface tension gradient is much reduced by the ample resupply of surfactant, and the elasticity drops to very low values:

$$E_S = \frac{2\epsilon_0}{h(dc/d\Gamma)} \qquad \text{for} \qquad h \gg 2\,\frac{d\Gamma}{dc} \tag{13}$$

Gibbs elasticity, as evaluated by Eq. (7), is regarded as a crucial parameter in stabilizing thin films against rupture [16–20]. It should be pointed out, however, that its relation to film stability is nowhere near a simple proportionality. If it were, films would be stabilized best by insoluble surfactants, since these result in the highest film elasticities: See Fig. 6 for $a = 0$. The effect of Gibbs elasticity is indirect: Required for film stability is not so much that the Gibbs elasticity be high, but that it be *higher in thin parts of the film than in its thicker parts*, as pointed out by Gibbs himself [16]. This condition is always fulfilled if the concentration of the surfactant can be kept constant, according to Eq. (7).

However, in dynamic situations where the film surfaces are continuously expanded, as they are during emulsification, the concentration is not constant: It decreases because the film liquid is depleted by adsorption to the newly created surface area. Gibbs' requirement then can be formulated as

$$\frac{dE}{dh} = \left(\frac{\partial E}{\partial h}\right)_c + \left(\frac{\partial E}{\partial c}\right)_h \cdot \frac{dc}{dh} < 0 \tag{14}$$

The second term in this expression contributes to stability only when the Gibbs elasticity *increases* with *decreasing* concentration [21]. During emulsification, therefore, the Gibbs criterion can only be fulfilled at concentrations above that of the maximum in Fig. 6. This means that any single surfactant can only slow down film drainage by its Gibbs elasticity over a limited range of film thicknesses, as illustrated in Fig. 7 for a numerical example. As a rule, emulsifiers are fairly complex mixtures of different surfactants. An interesting extension, therefore, is the evaluation of the elasticity of mixed surfactants in the light of the Gibbs criterion expressed in Eq. (14). In a continuously expanding film, the theory for mixed-surfactant systems by van den Tempel et al. [17] should be applied, not at constant concentration (c) but at constant total amount $(c + 2\Gamma/h)$ of each surfactant in the expanding film element. Figure 7 presents an example of a surfactant mixture where the main surfactant is contaminated

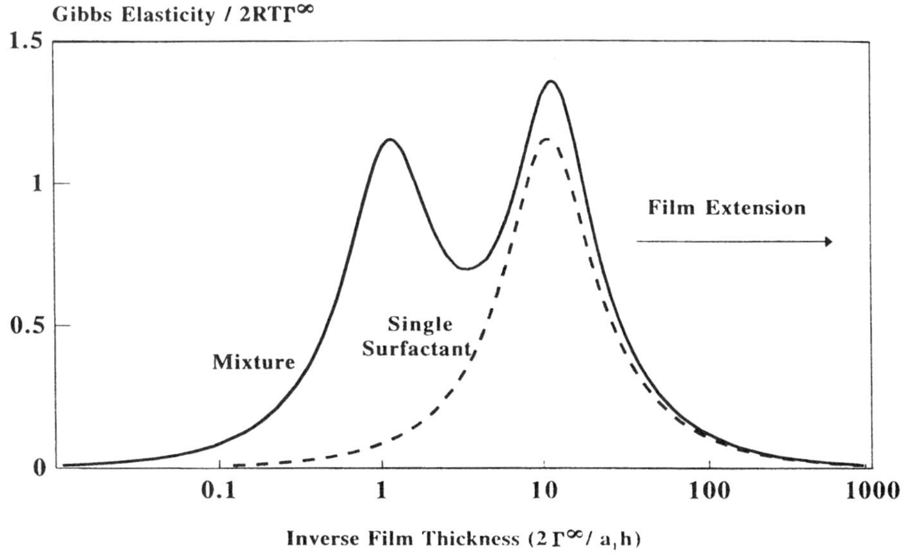

Figure 7. Gibbs elasticity during continuous extension of film. Dashed line: single pure surfactant (1); drawn line: mixture containing 10% of surfactant (2) with $a_2 = 0.01a_1$. (After Ref. 21.)

by 10% of another surfactant, which is much more surface active in terms of its a_2 value [21]. It is seen that such a mixture results in an extra barrier against fast drainage, operating over a much larger range of film thicknesses than is covered by the main surfactant on its own. This effect contributes to the much better functionality of mixed emulsifiers as compared to single pure ones.

Gibbs elasticity of liquid films is a simplified example of surface rheological behavior where surfactant transfer to an expanding interface is supposed to take place from a limited reservoir but is given unlimited time. The limited reservoir is the liquid film separating the emulsion drops; therefore, Gibbs elasticity can be significant only if the emulsifier is present in the continuous phase. This provides a physical explanation for the well-known Bancroft rule, according to which emulsifiers should be soluble in what is to become the continuous phase rather than in the drop phase.

B. Viscoelastic Behavior of the Interface

Interfacial rheology, in a broader sense, can manifest itself in at least three different types of deformation, schematically shown in Fig. 8, each of which

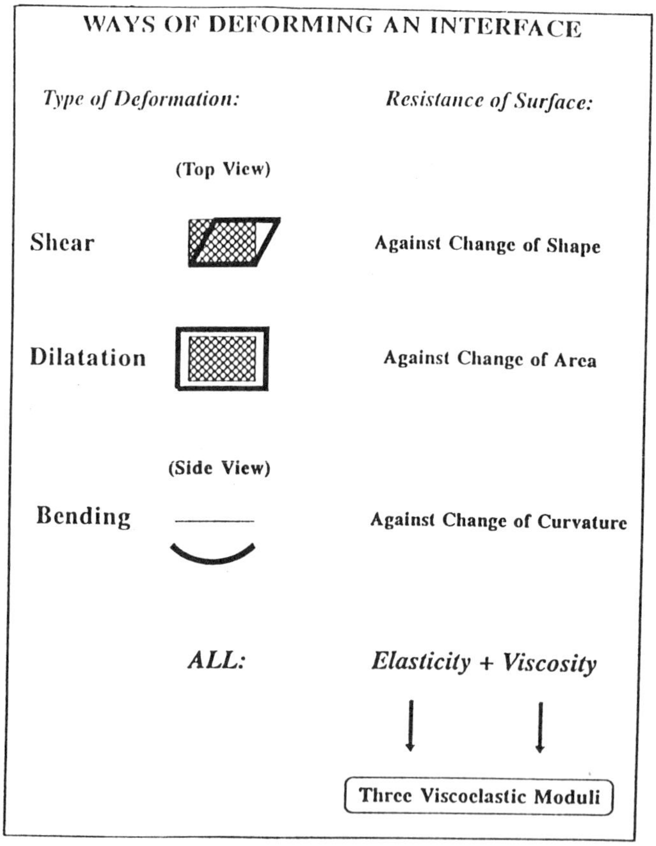

Figure 8. Different types of interfacial viscoelasticity.

can provoke its own elastic and viscous response. Emulsification involves mostly expansion and, to a lesser degree, compression of interfaces, and we shall only concern ourselves here with deformation by compression/expansion. Shearing deformation of the interface is not only less pronounced, but the concomitant shear viscosity and elasticity also are generally much smaller than their dilational counterparts for representative surfactants [22,23]. As to interfacial bending, we believe this may be relevant in high-curvature systems, such as microemulsions, but not in the fairly coarse low-curvature emulsions considered here.

In compression/expansion, the presence of an adsorbed layer of emulsifier results in a gradient in interfacial tension that can slow down movement of the interface and the adjoining bulk liquid. The breakup of a cylindrical thread into droplets can be highly retarded by such a gradient [14]. The measurable parameter expressing the interfacial tension gradients causing this effect is the surface dilational modulus ϵ defined by

$$\epsilon = \frac{\mathrm{d}\sigma}{\mathrm{d}(\ln \mathrm{A})} \tag{15}$$

This modulus has been extensively investigated both theoretically and experimentally by means of low-amplitude compression/expansion waves at a range of frequencies [24–27].

Numerically, the viscoelasticity modulus is found to depend on concentration and equation of state parameters of the adsorbed agent, and on the frequency of the wave experiment. In cases where the measurement of ϵ is performed sufficiently slowly (at low frequencies) for relaxation processes to have an effect on σ, the resulting modulus ϵ is also rate dependent and expressed in a complex number, the imaginary part of which represents the viscous component. One relaxation process of obvious importance for emulsification is the transport of surfactant to/from the interface. Surfactant transport by diffusion, illustrated in Fig. 9, has received most of the attention in measurements of its effects on small-amplitude surface waves, with characteristic time scales down to 10^{-3} sec.

Quantitatively, the viscoelasticity modulus for this relaxation mechanism is given by

$$|\epsilon| = \frac{\epsilon_0}{\sqrt{1 + 2(\omega\tau_\mathrm{d})^{-0.5} + 2(\omega\tau_\mathrm{d})^{-1}}} \tag{16}$$

where ϵ_0 is the purely elastic modulus defined in Eq. (6) and measured in the limiting case of very fast deformation, i.e., at high wave frequency ω ($\omega\tau_\mathrm{d} \gg 1$). The diffusional time scale is defined by

$$\tau_\mathrm{d} = \frac{2}{\mathrm{D}}\left(\frac{\mathrm{d}\Gamma}{\mathrm{dc}}\right)^2 \tag{17}$$

Figure 10 illustrates the frequency dependence of the viscoelasticity modulus at any given concentration in terms of the limiting modulus, ϵ_0, and the diffusional time scale, τ_d, as evaluated from Eq. (16) by Lucassen and van den Tempel [25]. This dimensionless frequency spectrum is a unique function

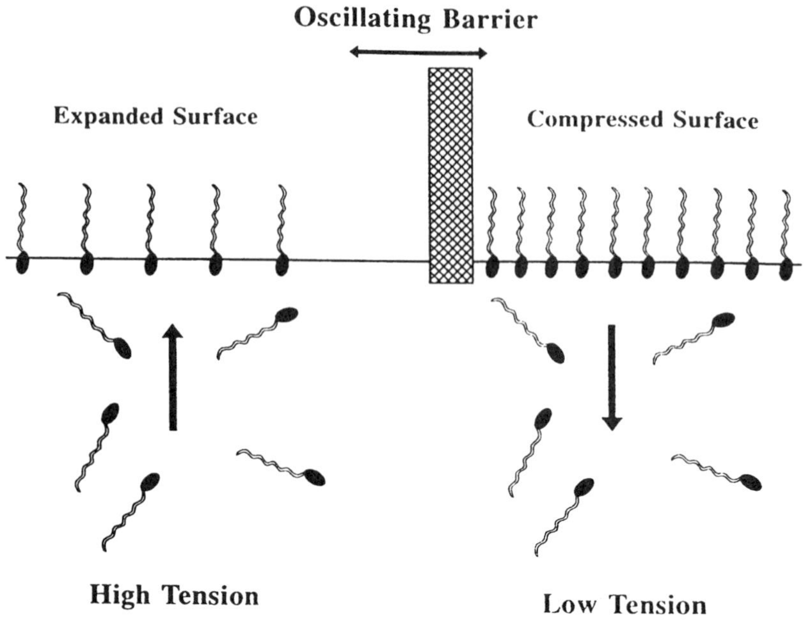

Figure 9. Diffusional relaxation. Transfer of surfactant molecules—to expanded areas and from compressed areas—partly offsets gradients in tension and introduces a viscous component into the modulus.

characteristic of the diffusion process, and independent of the surfactant's concentration and of its equation-of-state parameters. Surfactant parameters are reflected in the numerical values of limiting modulus and diffusional time scale, but not in the shape of the curve in Fig. 10, which can be used as a master curve for diffusional relaxation. An important aspect of this modulus spectrum is that there is always a range of high frequencies where diffusional exchange with the adjoining bulk solution is negligible and the limiting modulus ϵ_0 is found: Here the adsorbed surfactant behaves as if it is insoluble even though it may be very soluble over longer time spans. The transition between "soluble" and "insoluble" can be defined as indicated in Fig. 10. For rates of surface deformation above this transition frequency, the modulus retains more than half of its limiting value ϵ_0, and the viscous contribution to it is small; for rates below the transition frequency, the modulus progressively decreases until vanishingly small at $\omega\tau_d \ll 1$:

$$|\epsilon| = \epsilon_0 \left(\frac{dc}{d\Gamma}\right) \sqrt{\omega/D} \qquad \text{for} \quad \omega\tau_d \ll 1 \tag{18}$$

At such low rates of deformation, diffusion virtually eliminates all surface tension gradients: Surface behavior approaches that of a pure solvent without any adsorbed surfactant, although with a much lower tension than with the really pure solvent. In this limiting range of very slow deformation, the elastic contribution to the modulus $|\epsilon|$ is still larger than the viscous part; surface behavior for diffusional relaxation should be classed as non-Newtonian [28] because of the

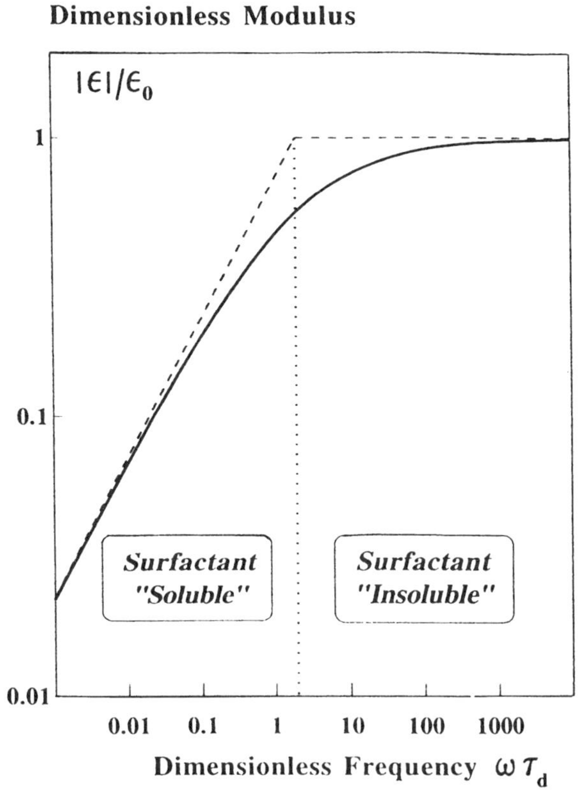

Figure 10. Modulus spectrum for diffusional relaxation. Dashed line: limiting behavior for fast and slow diffusion. (After Ref. 25.)

compositional changes inherent to mass transfer between interface and bulk solution.

Numerical values of ϵ_0 and τ_d are needed to evaluate the concentration dependence at a given frequency of the experiment. In cases where Eq. (8) is applicable, the surfactant-related quantities depend on concentration and the characteristic parameters of the surfactant according to Eq. (10). Figure 11 gives an example of the concentration dependence of the modulus. As was found for the Gibbs elasticity, the modulus also passes through a maximum caused by increasing diffusional interchange between interface and solution at the higher concentrations: At both very low and very high concentrations the behavior of a "clean" interface is observed, with interfacial tensions of σ^0 and σ_{eq}, respectively.

We may conclude that interfaces relaxing purely through diffusional exchange of molecules with a single-surfactant solution can be described quantitatively by Eqs. (15)–(17). In addition, other relaxation mechanisms may also affect tension and the viscoelastic behavior of an interface, depending on the time scale considered. Examples of relaxation processes that may occur in time scales ranging from 10^{-2} to 10^{+3} sec were reviewed by van den Tempel and

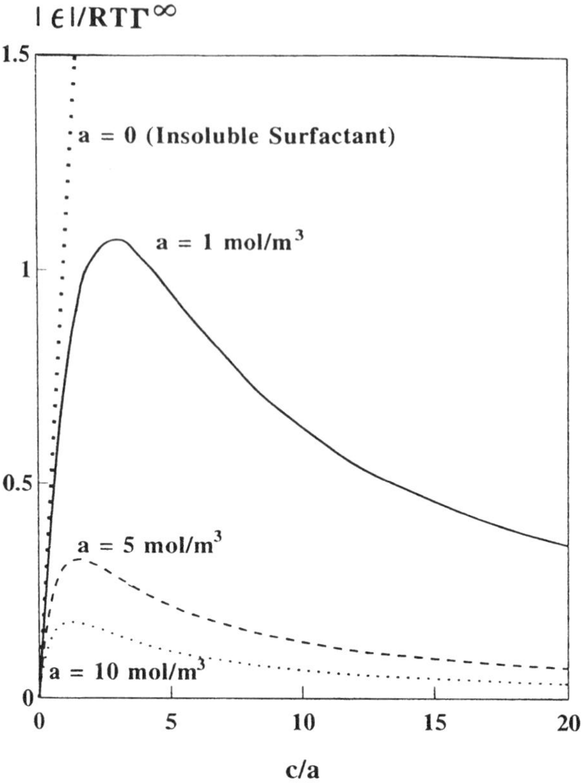

Figure 11. Numerical examples of surface dilational modulus as a function of surfactant concentration (c), calculated from Eq. (16) for Langmuir adsorption with $RT\Gamma^{\infty} = 20\,mN/m$; frequency $\omega = 400\,sec^{-1}$; $D = 10^{-10}\,m^2/s$. Half-saturation concentrations a as in Fig. 6. (Reproduced from *Food Structure*, *12*:1 (1993), by courtesy of Scanning Microscopy International.)

Lucassen-Reynders [29] and include (1) retardation of adsorption by an adsorption "barrier"; (2) slow reorientation of molecules after adsorption; (3) complex formation and phase transitions in the surface; (4) formation or destruction of 3-D structures. This last can occur either in the surface (e.g., in collapsed monolayers) or in the solution (e.g., micelles). In most technological and biological applications, these relaxation pehnomena occur in combination with diffusion. The theory for such "mixed" relaxation is fairly complex, and beyond the scope of this chapter.

For any relaxation mechanism, the modulus $|\epsilon|$ represents the viscoelastic resistance against movement of the interface, and corresponds to an apparent interfacial viscosity coefficient of $|\epsilon|/\omega$.

C. Dynamic Interfacial Tension

The dynamic (nonequilibrium) interfacial tension during emulsification depends on the rate at which surfactant can be transported to the newly created interface. Flow conditions in practice are such that this transport is likely to be determined by convection rather than by molecular diffusion [30]. Convection is most dominant when the interface is still surfactant-free and, therefore, freely moving. A diffusional boundary layer will be built up as soon as the adsorbed layer resists interfacial movement, in which case the rate of further adsorption is governed by convective diffusion. A treatment in terms of molecular rather than convective diffusion of surfactant, therefore, can give only the upper limit of the dynamic interfacial tension.

The effects of diffusional transfer depend on the rate of diffusion as compared to the relative rate of surface extension, $\delta \ln A/\delta t$, and to the reciprocal value of the breakup time defined in Eq. (4). For large-area expansions, relevant for emulsification, the time scale of convective diffusion, τ_d, is described better by

$$\tau_d \approx \frac{2}{D}\left(\frac{\Gamma}{c}\right)^2 \tag{19}$$

than by Eq. (17). This parameter was shown to be relevant to convective diffusion during large, continuous, surface expansions in a Langmuir trough [31]. At relatively high extension rates, convective diffusion is too slow to resupply the expanding interface, which will behave as an emulsifier-free interface:

$$\frac{\delta(\ln A)}{\delta t} \gg (\tau_d)^{-1} \quad \rightarrow \quad \sigma \approx \sigma^0 \tag{20}$$

At sufficiently low rates of surface extension, on the other hand, the convective diffusion of emulsifier is fast enough for the interfacial tension to have its equilibrium value:

$$\frac{\delta(\ln A)}{\delta t} \ll (\tau_d)^{-1} \quad \rightarrow \quad \sigma \approx \sigma_{eq} \tag{21}$$

The numerical value of τ_d depends very strongly on the concentration of the emulsifier and on its surface equation of state parameters. In the simple case of Szyszkowski–Langmuir adsorption, where σ_{eq} depends on concentration c according to

$$\sigma^0 - \sigma_{eq} = RT\Gamma^\infty \ln\left(1 + \frac{c}{a}\right) \tag{22}$$

the adsorption time is given by

$$\tau_d = \frac{\pi}{2D}\frac{(\Gamma^\infty/a)^2}{(1 + c/a)^2} \tag{23}$$

The saturation adsorption Γ^∞ and the half-saturation concentration a in these equations are the characteristic adsorption parameters of the emulsifier, which can be derived from the equilibrium interfacial tension vs concentration curve. A numerical example for a practically relevant emulsifier (a monoacylglycerol at the interface between water and sunflowerseed oil) is given in Fig. 12 [32]. Figure 13 shows the dependence of τ_d on concentration resulting from Eq. (23). These values for the diffusional time scale have to be compared not only to the time scale of the extension imposed by the applied flow, as in Eqs. (20) and (21), but also to the breakup time defined in Eq. (4).

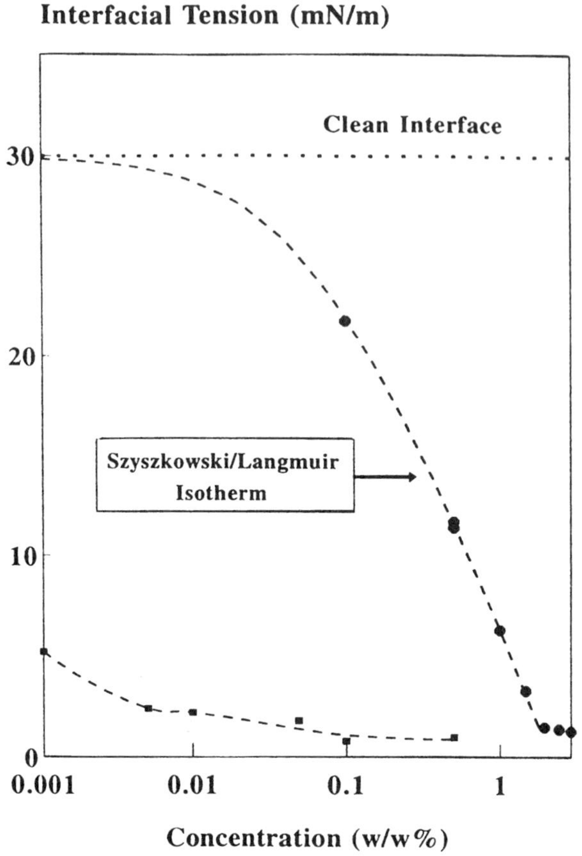

Figure 12. Equilibrium interfacial tensions of sunflowerseed oil/water interface as a function of emulsifier concentration. Emulsifiers: (●) unsaturated monoacylglycerol; (■) lecithin mixture. Line: Eq. (21) with parameter values $a = 2\,mol/m^3$, $\Gamma^{\infty} = 3.3\,\mu mol/m^2$, $RT = 2.5\,kJ/mol$. (Reproduced from *Colloids Surfaces 65*:175 (1992), by courtesy of Elsevier Science Publishers B.V.)

The interfacial tension values in the two limiting cases covered by Eqs. (20) and (21) can differ by more than one order of magnitude in practical oil/water systems. For example, using the parameter values of Fig. 12 for the sunflower seed oil/water interface, with $\sigma^0 = 30\,mN/m$, we find $\sigma_{eq} = 1.5\,mN/m$ for $c = 30a$. Substitution of these tension values into Eqs. (2) and (4) illustrates the large effects on drop size and breakup time to be expected. The effect of emulsifier transport on the interfacial tension, therefore, must be considered a major factor in emulsification under practical conditions.

D. Interrelations Between Dynamic Parameters

In the previous sections, three dynamic effects of emulsifiers were considered, viz., dynamic stabilization of the liquid films separating the emulsion drops

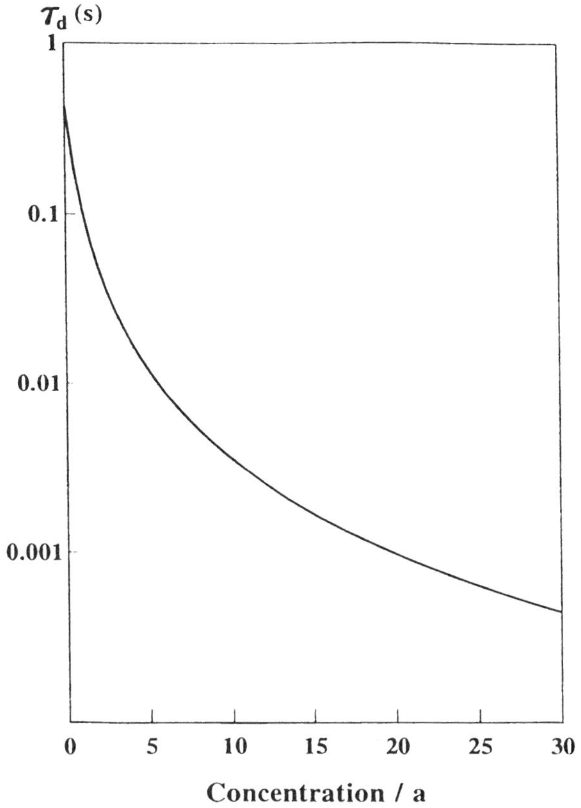

Figure 13. Diffusional time scale (τ) calculated from Eq. (23) as a function of concentration for emulsifier with parameter values $a = 2\,mol/m^3$, $\Gamma^\infty = 3.3\,\mu mol/m^2$, $D = 10^{-11}\,m^2/s$, $RT = 2.5\,kJ/mol$.

(III.A), interfacial rheology (III.B), and the effects of bulk/interface interchange of emulsifier on the overall interfacial tension (III.C). All three effects concern the dynamic tension as affected by relaxation mechanisms after a change in interfacial area, and they are in fact closely related to each other. For example, the similarity of Fig. 6 for the Gibbs elasticity (at constant film thickness) and Fig. 11 for the dilational modulus (at constant frequency) is striking and almost quantitative. The near-quantitative analogy between the two parameters was first noted by Lucassen and Hansen [33]: the *limited time* ($1/\omega$) available for diffusion during periodic area expansion/compression implies that diffusion can be effective only from/to a *limited layer* (h/2) of solution. The analogy is illustrated in Fig. 14; the wave frequency in Fig. 11 was actually chosen so as to produce a modulus isotherm virtually coincident with the Gibbs elasticity in Fig. 6.

The same relaxation processes determine the dynamic surface tension of a freshly created interface when the new area is kept constant. The kinetics of adsorption during the equilibration of such an interface by diffusional transport was evaluated by Ward and Tordai [34]. In an important development, Loglio et

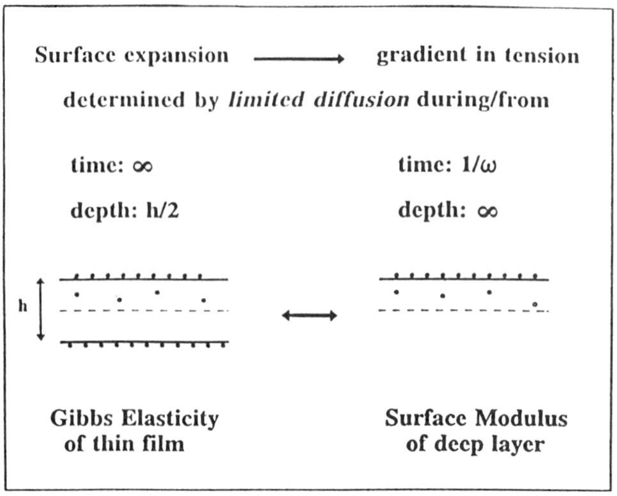

Figure 14. Analogy between the surface modulus of bulk solution and Gibbs elasticity of thin film. Modulus measured at frequency ω means Gibbs elasticity measured at thickness $h/2 = (D/\omega)^{1/2}$.

al. [35] demonstrated theoretically that the same functions that determine the frequency dependence of the modulus ϵ can also be used to describe the time dependence of the dynamic surface tension at constant area. For instance, the dynamic surface tension near the end of the equilibration, i.e., at long equilibration time t, is known to vary according to $t^{-1/2}$: this corresponds to the $\omega^{+1/2}$ dependence of the modulus at the low-frequency end of the spectrum in Eq. (18). Thus, the surface tension versus time curve of an equilibrating surfactant solution can be translated into a modulus spectrum as given in Fig. 9, and vice versa. However, this argument cannot be applied to the dynamic interfacial tension under expansion of the interface (Wantke and Lunkenheimer [36]). In emulsification, therefore, a better analogy is with the dynamic interfacial tensions studied at constant relative rate of surface extension by van Voorst Vader et al. [31]. The relaxation mechanism here is convective diffusion, and the resulting apparent dilational viscosities are fairly closely related to the values of $|\epsilon|/\omega$ in a surface wave experiment [37].

IV. Experimental Evidence for the Role of Emulsifier Dynamics

A. Single Drops: Flow-Visualization Experiments

Flow visualization of drop breakup necessitates slowing down the flow by using liquids with higher viscosity (of the order of 1 Pa) and larger drops than would occur in many practical emulsions. Its advantage is that breakup can be studied without interference by coalescence.

Phillips et al. [38] measured the effects of surfactant on drop dynamics in the simple-shear fields generated in a Couette apparatus. Drop deformation, orientation, and internal circulation were measured at low values of the capillary number (Ca), i.e., for drops extended into ellipsoids with axial ratios up to 1.5. Interfacial dynamic effects were found to be significant, particularly in low-viscosity systems, as predicted by Flumerfelt's theory [13]. Independently measured values of the interfacial elasticity and viscosity were not available for the systems under investigation; it was suggested that the measured drop dynamics could serve as a basis for evaluation of the dynamic interfacial parameters.

De Bruijn [7] demonstrated experimentally that the presence of surfactant can lead to tip streaming (see Fig. 3), caused by the adsorbed surfactant's being swept to the ends of elongated drops. This accumulation of surfactant at the tips was confirmed by collecting the tiny drops ejected from the ends and finding the interfacial tension to be much lower than that of the original drop. Interestingly, the phenomenon was also shown to be restricted to the intermediate concentration range, where deviations of the interfacial tension from its equilibrium value are most pronounced. During emulsification such tip streaming obviously has a significant effect on the final size distribution of an emulsion, but the exact mechanism for the formation of the tiny drops remains unknown.

Janssen et al. [39] studied binary breakup conditions as a function of emulsifier concentration (in the drop phase), and were able to link the dynamics of breakup to the interfacial viscoelasticity discussed in Section III.B, e.g., in Fig. 11. They determined the value of the critical capillary number (Ca_{cr} in Eq. 2) based on the equilibrium value of the interfacial tension, and found equal values for it at very low and very high surfactant concentrations. Clearly, at the high-concentration end, convective diffusion of surfactant was fast enough for immediate re-equilibration of the extended surface. At intermediate concentrations, however, the surfactant impeded breakup: Ca_{cr} passed through a maximum, as illustrated by the example in Fig. 15. In this work, an independent estimate was obtained for the interfacial viscoelasticities from Eq. (16), by using values of the surfactant parameters ϵ_0 and τ_d by applying a corrected version of Eq. (22) to the equilibrium tensions. The maximum in Ca_{cr} was found at the same concentration where the viscoelasticity of the surfactant monolayer was expected to be maximal given the rate of surface extension imposed by the simple shear flow. It was concluded that the dynamic interfacial tension (or rather its deviation from the static value of the tension) was linearly related to the viscoelasticity, with an empirical proportionality constant of approximately 0.25 over a wide range of viscosity ratios. Relevant to these experiments was not the overall expansion of the deformed drop, but rather the local expansion at the waist, while the ends of the drop were compressed. The rotational component of the simple shear flow results in a compression/expansion wave traveling around the drop, which explains the close analogy with the viscoelasticity in surface wave experiments.

B. Single Films: Rate of Drainage

Studies of surfactant effects on the thinning of liquid films between emulsion drops are rare in comparison to those on films enclosed by air. The latter are expected to be relevant for emulsions with a very low viscosity ratio η_d/η_c. Recent experimental and theoretical work by Ivanov and Dimitrov [9], Tambe

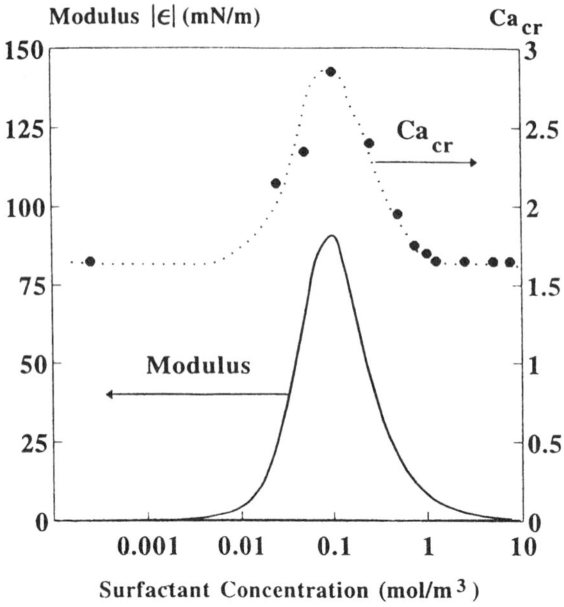

Figure 15. Effect of surfactant (Dobanol) on the breakup of water drops in simple shear as a function of surfactant concentration, at viscosity ratio = 0.01. ●: critical capillary number based on equilibrium tension (Ca_{cr}); drawn line: viscoelasticity modulus ($|\epsilon|$). (After Ref. 39.)

and Sharma [40], and Sonin et al. [10] deals with the effect of surface elasticity on the drainage of circular horizontal soap films. The analysis is simple only in cases where the film surfaces are either very mobile or immobile, as illustrated in Fig. 16. Immobility is observed when a rigid monolayer of surfactant is present, i.e., when the modulus ϵ is extremely high. Such films thin out according to the Reynolds equation:

$$-\frac{dh}{dt} = \frac{h^3}{3\eta_c R_f^2}\Delta P_f \tag{24}$$

where R_f is the radius of the film and ΔP_f is the pressure difference between the film center and the border. Correcting the Reynolds equation for finite elasticity is not easy and requires a number of simplifying assumptions [9,40]. Drainage measured by Sonin et al. [10] was much faster than predicted by Eq. (24). The authors sought to correlate the moduli required for such fast drainage by current theory [9] with viscoelasticity moduli measured independently by surface waves on thick layers of solution. However, the modulus values necessary to describe the film drainage were found to be much higher than the moduli measured at the frequency corresponding to the measured rate of drainage. The authors attributed the discrepancy to the effect of surface convection: As surfactant was carried away by the draining film liquid, the adsorption was not constant, as required by the assumptions of the theory. Such surface convection

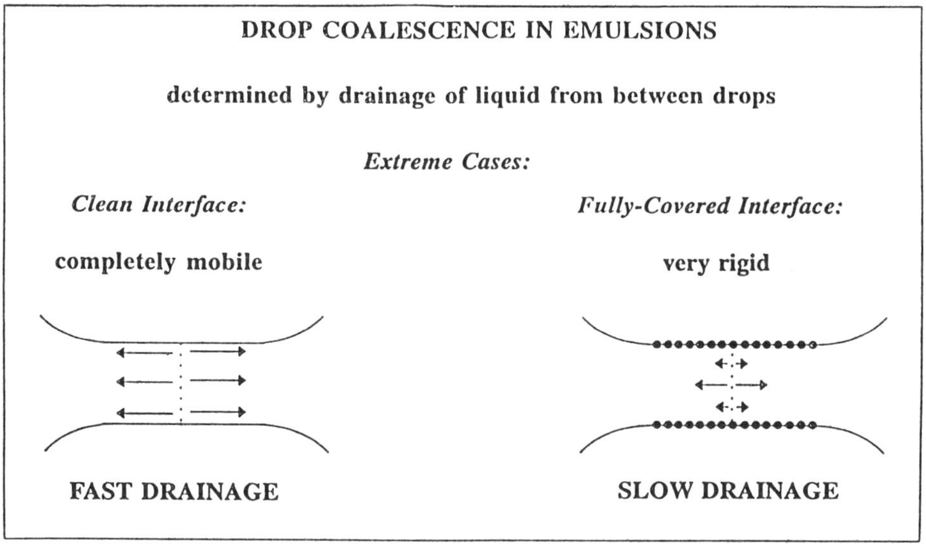

Figure 16. Drainage of plane-parallel liquid films between drops or bubbles. ●: adsorbed emulsifier molecules.

was estimated to be a far more important factor than exchange with the film liquid during film drainage. Their analysis demonstrates the crucial role of surface elasticity, and points to gaps in the current treatment of film drainage.

C. Complete Emulsions: Drop Size and Dynamic Stability

Evidence for dynamic emulsifier effects on drop size of emulsions has been reported for a number of systems.

Koshy and Das [41] measured the effect of surfactants on drop breakage for dilute oil-in-water emulsions in turbulent flow. The volume fraction of dispersed phase was kept at less than 0.02 to minimize coalescence. The presence of surfactant at intermediate concentrations was found to decrease drop size, even if the decrease in interfacial tension was accounted for. The authors interpreted their data by adding the local interfacial tension gradient to the turbulent stress applied. Such promotion of drop breakup by surfactant is contrary to the results in single-drop experiments discussed in Section IV.A. Once again, clean-interface behavior was found in the limits of very low and very high surfactant concentration.

Armbruster et al. [42] prepared more concentrated oil-in-water emulsions (volume fractions from 0.05 to 0.2) in laminar flow, generated in a colloid mill, for a range of viscosities of the two phases. They found deviations from Fig. 1 that depended on the adsorption kinetics of the emulsifier: Fast kinetics (i.e., low time scale in terms of Eq. 17) produced reasonable agreement with surfactant-free behavior. Slow kinetics resulted in positive deviations from Fig. 1 at low viscosity ratio, while deviations at high ratio tended to be negative. These deviations were at least in part attributed to coalescence.

Mean drop diameters ($D_{3.3}$) of 40% water-in-oil emulsions with different tensions, σ, of the macroscopic oil–water interface, resulting from different concentrations of added emulsifiers were measured by Lucassen-Reynders and Kuijpers [32]. The equilibrium interfacial tensions of two emulsifiers are given in Fig. 12. Long-term stability of these emulsions derived from the presence of solid particles (triacylglycerol crystals), which have no effect on the interfacial tension. Figure 17 compares the experimental dependence of drop size on interfacial tension to the theoretical slope expected from Eq. (2). No significance is to be attached to the position of this line, which is arbitrary because quantitative knowledge of flow conditions was lacking. Nevertheless, two conclusions are clear from Fig. 17:

1. Interfacial tension is not the only interfacial parameter determining drop size of complete emulsions. If it were, all measured points would have coincided on a single line. In fact, the emulsions containing

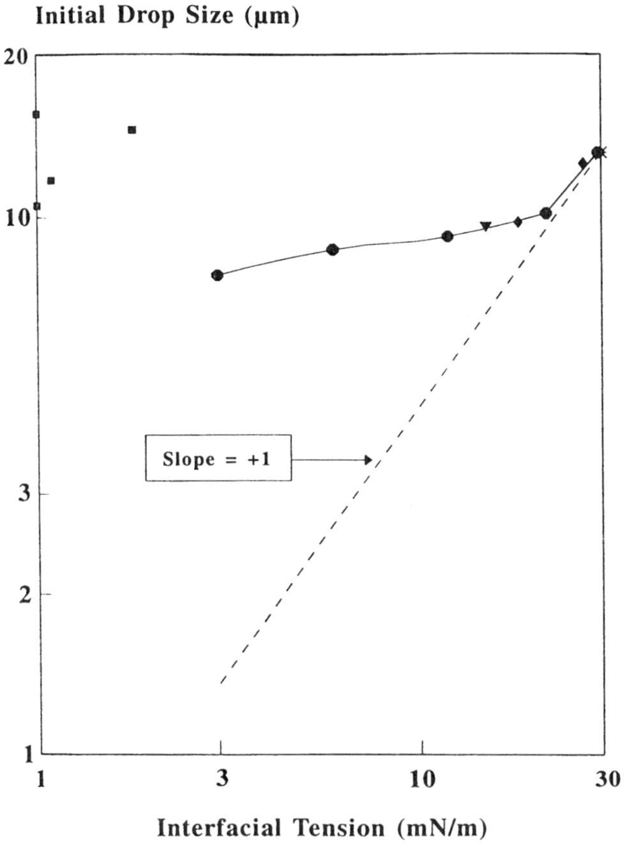

Figure 17. Initial drop size of emulsion as a function of interfacial tension at equilibrium. Emulsifiers: (*) none; (●) unsaturated monoacylglycerol; (◆) saturated monoacylglycerol; (■) lecithin; (▼) protein. (Reproduced from *Colloids Surfaces 65*:175 (1992), by courtesy of Elsevier Science Publishers B.V.)

lecithin were significantly coarser—even though their equilibrium tensions were lower—than those containing the other emulsifiers.
2. Most of the emulsifiers (all except the lecithin) did produce points coinciding on a single line, but the slope of this line is mostly of the order of 0.1 rather than the order of 1 expected from Eq. (2): Drop size varied much less with interfacial tension than one would expect on the basis of Eq. (2).

These results were attributed to the effects of interfacial viscoelasticity during breakup and to recoalescence. The authors assumed that the interfacial viscosity represents an additional resistance against movement necessary for the breakup of the disperse phase. In a crude approximation, the total resistance of a layer of disperse phase with thickness r containing the interface then would be determined by the sum of the bulk viscosity η_d contributing over a distance r and the interfacial viscosity. This would mean that the interface effectively increases the viscosity of the disperse phase to

$$\eta_d^{eff} = \eta_d + \frac{|\epsilon|}{r\alpha} \tag{25}$$

where the relative rate of interfacial expansion, α, replaces the wave frequency, ω, in Eq. (16). Apart from a numerical factor, this expression is equivalent to the equation for the retarding effect of surfactants on gravity-induced drop movement [43,44]. It is also in line with the drop breakup treatments of Flumerfelt [13] and Milliken et al. [14] and with Oldroyd's considerations on the effect of surfactant in emulsion rheology [45].

The effect of an emulsifier at any given values of the relative extension rate α and layer thickness r can be evaluated by substituting $|\epsilon|$ from Eq. (16). In a numerical example, a practically relevant rate would be $\alpha = 10^4\ \text{sec}^{-1}$, estimated for an emulsifying apparatus with a time of passage of 0.1 to 1 msec. For the characteristic parameter values approximating the properties of the unsaturated monoacylglycerol used, combined with values of $\eta_d = 1\,\text{mPa}$ and $r = 100\,\mu\text{m}$, representing water drops of 0.1 mm, the effect of the interfacial viscoelasticity would be to increase the effective drop viscosity by more than an order of magnitude, as illustrated in Fig. 18. The maximum in effective viscosity in this case is approximately

$$(\eta_d^{eff})_{max} \approx 30\eta_d$$

This maximum reflects the maximum that is generally found in the surface viscoelasticity modulus for bulk/surface interchange.

Generally speaking, the effect of interfacial rheology on effective drop viscosity will increase with decreasing radius r, with increasing limiting elasticity ϵ_0, and increasing time scale τ_d. In this model, interfacial viscoelasticity can raise the effective viscosity of the drop phase to such high levels that the viscosity ratio η_d/η_c becomes unfavorable for breakup in simple shear. As a result, emulsions formed can be relatively coarser than they would have been for zero interfacial viscoelasticity at the same value of the interfacial tension. This is in qualitative agreement with the experimental results in Fig. 17. A more quantitative comparison of theory with experiment cannot yet be given. This would

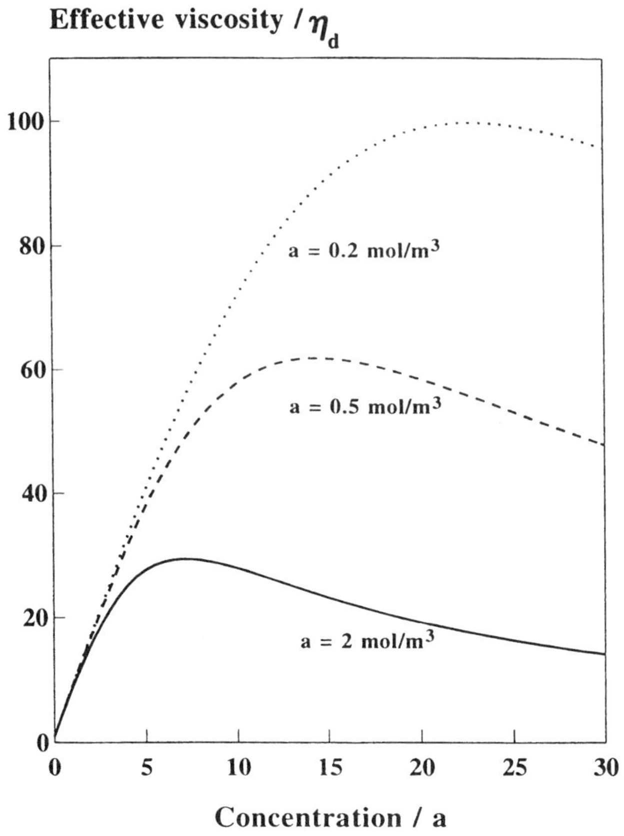

Figure 18. Effective drop viscosity according to Eq. (25) for emulsifiers with different a, for $\Gamma^{\infty} = 3.3\,\mu mol/m^2$, $RT = 2.5\,kJ/mol$, $r = 100\,\mu m$, and relative rate of surface extension $\alpha = 10^4\,sec^{-1}$. (Reproduced from *Colloids Surfaces 65*:175 (1992), by courtesy of Elsevier Science Publishers B.V.)

require a reasonable estimate of the rate of interface extension (α) during breakup and an experimental setup with better-defined flow conditions.

Recoalescence of drops during emulsification, on the other hand, is also likely to be a major factor in the present results, particularly in view of the high volume fraction of the dispersed phase. Such recoalescence is almost certainly the cause of the deviating behavior of the lecithin-containing emulsions. This would imply that the commercial lecithin mixture used confers less dynamic stability to droplets immediately after breakup and, as a result, produces coarser emulsions than do the other emulsifiers. Possibly, this lack of dynamic stabilization is related to the known tendency of lecithins to aggregate into "structured" solutions [46], from which monomers may not be released quickly enough for fast adsorption.

Finally, Groeneweg et al. [47] found that the presence of emulsifier, in water-in-oil emulsions with volume fractions up to 70%, resulted in a bimodal size distribution with a very large number of tiny droplets. The average drop size of the full size distribution could not be described with simple breakup relations such as Eq. (2). The average drop size of the "coarse" part of the distribution, however, was reported to agree fairly well with Eq. (2) for viscosity-dominated breakup in terms of the equilibrium interfacial tension. This indicates that, at the concentrations employed, emulsifier transport was fast enough for the tension to be close to its equilibrium value. The tiny droplets at the "fine" end of the distribution were attributed to tip streaming; they induced a considerable decrease in average drop size of the total distribution.

Summarizing, most of the work on full emulsions tallies with the results of single-drop experiments if coalescence can be prevented. On the one hand, the breakup of drops under critical conditions ($Ca \approx Ca_{cr}$) is hindered by interfacial tension gradients at the waist of a drop; but on the other hand, breakup by tip streaming is helped by the same gradient under high-shear conditions ($Ca \gg Ca_{cr}$). These effects are found only at intermediate concentrations of emulsifier, where the viscoelasticity modulus ϵ has finite values, as illustrated in Figs. 11 and 15. At both very low and very high concentrations the interface behaves as if clean, but with the different interfacial tensions, σ^0 and σ_{eq}, respectively. If coalescence does play a significant part, it is not easy to separate emulsifier effects on coalescence from those on drop breakup in the analysis of drop size.

VI. Concluding Remarks

The essential functions of emulsifiers in emulsification and stabilization against coalescence are related to the viscoelastic properties of the emulsifier-covered interface. First-order models exist for the effect of such properties on drop breakup in terms of the emulsifier's characteristic adsorption parameters. The interfacial viscoelasticity modulus not only adds to the effective viscosity of the drop phase, but is also the parameter in which to express interfacial tension gradients on any liquid interface. Depending on the strength of the flow field, drop breakup in simple shear can be either hampered or promoted significantly by the effect of a finite modulus.

No quantitative treatment is available as yet for the dynamic stabilization of the liquid films separating newly formed droplets against rupture, but the viscoelasticity and the Gibbs film elasticity, which is closely related to it, are generally recognized as important parameters. Therefore, a quantitative model for the function of emulsifiers in recoalescence may be achieved in terms of the same parameters found in the model for drop breakup. The development of a comprehensive model will be no easy task, since the effects of interfacial dynamics depend on the local values of the rates of liquid flow and the rate of surface deformation. Application of our knowledge of interfacial dynamics, therefore, is hampered by our lack of a quantitative model for such local processes during emulsification.

Finally, the simplified picture used here for the hydrodynamic conditions is laminar flow, ignoring effects of inertial forces and turbulence. This is more likely to be approximately correct for emulsification of water in fairly viscous oil than for oil in water. Turbulent flow results in viscosity-dominated breakup if the size of the turbulent eddies is much larger than that of the drops to be broken up. As the size of the eddies increases with the viscosity of the continuous phase, this condition is also more easily obeyed during emulsification in viscous oil than in water. In any case, even if inertia and turbulence are not negligible, emulsifier-induced dynamic interfacial properties should be accounted for, and this can be achieved by considerations similar to those reviewed earlier.

Symbols

a	half-saturation concentration of emulsifier
A	interfacial area
c	emulsifier concentration
Ca	capillary number defined in Eq. (2)
D	diffusion coefficient of emulsifier
f_1	function of viscosity ratio in Eq. (3)
f_2	function of viscosity ratio in Eq. (4)
L	length of cylinder into which drop is deformed
N	number of droplets into which cylinder breaks up
r	radius of cylinder into which drop is deformed
R	radius of drop to be broken up
R_f	radius of film separating drops
t	time

Greek

α	relative rate of interface extension ($= d(\ln A)/dt$)
$\dot{\gamma}$	rate of shear
Γ	emulsifier adsorption
Γ^{∞}	saturation value of Γ
ϵ	interfacial modulus of dilational viscoelasticity
ϵ_0	limiting value of ϵ for large α
η_c	viscosity of continuous phase
η_d	viscosity of drop phase
σ	interfacial tension
σ^0	interfacial tension in the absence of emulsifier
σ_{eq}	interfacial tension at equilibrium adsorption of emulsifier
τ_b	time scale of drop breakup defined in Eq. (4)
τ_d	time scale of diffusion defined in Eqs. (17) and (19)

References

1. M. van den Tempel, in *IIIrd International Congress on Surface-Active Agents*, Vol. II, Cologne, Verlag der Universitätsdruckerei Mainz, Germany, 1960, p. 573.
2. P. Walstra, in *Encyclopedia of Emulsion Technology*, Vol. 1 (P. Becher, ed.), Dekker, New York, 1983, p. 57.
3. S. Tomotika, *Proc. Roy. Soc. A150*:322 (1935).
4. F. D. Rumscheidt and S. G. Mason, *J. Colloid Sci. 16*:238 (1961).
5. F. D. Rumscheidt and S. G. Mason, *J. Colloid Sci. 17*:260 (1962).
6. H. P. Grace, *Chem. Eng. Commun. 14*:2257 (1982).
7. R. A. de Bruijn, *Chem. Eng. Sci. 48*:277 (1993).
8. D. T. Wasan, J. J. McNamara, S. M. Shah, K. Sampath, and M. Aderangi, *J. Rheology 23*:181 (1979).

9. I. B. Ivanov and D. S. Dimitrov, in *Thin Liquid Films: Fundamentals and Applications* (I. B. Ivanov, ed.), Dekker, New York, 1988, p. 379.
10. A. A. Sonin, A. Bonfillon, and D. Langevin, *J. Colloid Interface Sci. 162*:323 (1994).
11. A. C. Juriaanse, D. W. de Bruijne, and D. J. P. Out, in *Engineering and Food*, Vol. 1 (W. E. L. Spiess and H. Schubert, eds.), Elsevier Science Publ., Barking, UK, 1990, p. 207.
12. J. M. H. Janssen and H. E. H. Meijer, *J. Rheology 37*:597 (1993).
13. R. W. Flumerfelt, *J. Colloid Interface Sci. 76*:350 (1980).
14. W. J. Milliken, H. A. Stone, and L. G. Leal, *Phys. Fluids A5*:69 (1993).
15. C. Marangoni, *Nuovo Cimento, Ser. 2, 5/6*:239 (1872).
16. J. W. Gibbs, *Trans. Connecticut Acad. III*:108, 343 (1878). Reprinted in *The Scientific Papers of J. Willard Gibbs*, Vol. I. Dover, New York, 1961, p. 55.
17. M. van den Tempel, J. Lucassen, and E. H. Lucassen-Reynders, *J. Phys. Chem. 69*:1798 (1965).
18. J. A. Kitchener, in *Recent Progress in Surface Science*, Vol. 1 (J. F. Danielli, K. G. A. Pankhurst, and A. C. Riddiford, eds.), Academic Press, New York, 1964, p. 51.
19. A. Prins, C. Arcuri, and M. van den Tempel, *J. Colloid Interface Sci. 24*:84 (1967).
20. K. Malysa, R. Miller, and K. Lunkenheimer, *Colloids Surfaces 53*:47 (1991).
21. J. Lucassen, in *Anionic Surfactants: Physical Chemistry of Surfactant Action* (E. H. Lucassen-Reynders, ed.), Dekker, New York, 1981, p. 217.
22. E. H. Lucassen-Reynders, in *Anionic Surfactants: Physical Chemistry of Surfactant Action* (E. H. Lucassen-Reynders, ed.), Dekker, New York, 1981, p. 173.
23. J. Benjamins and F. van Voorst Vader, *Colloids Surfaces 65*:161 (1992).
24. J. Lucassen and M. van den Tempel, *J. Colloid Interface Sci. 41*:491 (1972).
25. J. Lucassen and M. van den Tempel, *Chem. Eng. Sci. 27*:1283 (1972).
26. J. Lucassen and D. Giles, *J. Chem. Soc., Faraday Trans. I 171*:217 (1975).
27. L. Ting, D. T. Wasan, and K. Miyano, *J. Colloid Interface Sci. 107*:345 (1985).
28. M. van den Tempel, *J. Non-Newtonian Fluid Mech. 2*:205 (1977).
29. M. van den Tempel and E. H. Lucassen-Reynders, *Adv. Colloid Interface Sci. 18*:281 (1983).
30. P. Walstra, *Chem. Eng. Sci. 48*:333 (1993).
31. F. van Voorst Vader, T. F. Erkens, and M. van den Tempel, *Trans. Faraday Soc. 60*:1170 (1964).
32. E. H. Lucassen-Reynders and K. A. Kuijpers, *Colloids Surfaces 65*:175 (1992).
33. J. Lucassen and R. S. Hansen, *J. Colloid Interface Sci. 23*:319 (1967).
34. A. F. H. Ward and L. Tordai, *J. Chem. Phys. 14*:453 (1946).
35. G. Loglio, U. Tesei, R. Miller, and R. Cini, *Colloids Surfaces 61*:219 (1991).
36. K. D. Wantke and K. Lunkenheimer, private communication (1993).
37. E. H. Lucassen-Reynders and J. Lucassen, *Colloids Surfaces 85*:211 (1994).
38. W. J. Phillips, R. W. Graves, and R. W. Flumerfelt, *J. Colloid Interface Sci. 76*:350 (1980).
39. J. J. M. Janssen, A. Boon, and W. G. M. Agterof, *AIChE J. 40*:1929 (1994).
40. D. E. Tambe and M. M. Sharma, *J. Colloid Interface Sci. 147*:137 (1991).
41. A. Koshy and T. R. Das, *Chem. Eng. Sci. 43*:649 (1988).
42. H. Armbruster, H. Karbstein, and H. Schubert, *Chem.-Ing.-Tech. 63*:266 (1991).
43. J. Boussinesq, *Ann. Chim. Phys. 29*:364 (1913).
44. V. G. Levich, *Physicochemical Hydrodynamics*, Prentice-Hall, Englewood Cliffs, N.J., 1962, p. 420.
45. J. G. Oldroyd, *Proc. Roy. Soc. A232*:567 (1955).
46. K. Shinoda, *J. Phys. Chem. 89*:2429 (1985).
47. F. Groeneweg, F. van Dieren, and W. G. M. Agterof, *Colloids Surfaces 91*:207 (1994).

3
Rheology of Emulsions Containing Polymeric Liquids

RAJINDER PAL / Department of Chemical Engineering, University of Waterloo, Waterloo, Ontario, Canada

I. Introduction

The applications of emulsions are numerous. Many products of commercial importance are sold or handled in the form of emulsions. The industries where emulsions play an important role include petroleum, foods, medical and pharmaceuticals, cosmetics, agriculture, explosives, polishes, leather, textiles, paints, lubricants, and polymers [1–8]. Some examples of petroleum and nonpetroleum emulsions are: well-head emulsions (generally water-in-oil type), emulsion drilling fluids (both oil-in-water and water-in-oil), oil spill mousse emulsions (water-in-oil), milk, mayonnaise, butter, margarine, cake batters, meat emulsions, salad dressings, several antiallergic and antibacterial creams, sunscreen lotions and creams, foundations and vanishing creams, hand lotions and creams, cold creams, shaving creams, hair creams, emulsified explosives, polishes, antistatic and waterproofing agents employed in the treatment of leather, bitumen emulsions for the construction of roads and highways, emulsion paints, lubricants for drilling and cutting tools, and emulsified polymer blends.

In many commercial emulsion formulations, high-molecular-weight polymers (macromolecules) are often incorporated into the emulsion system, although the reasons for adding the polymers vary from one application to another. In many applications, polymers are used as *thickeners* to modify the rheology of the continuous phase and to confer viscosity and yield value on that phase [9–25]. The thickening effect causes the emulsion droplets to remain suspended in the continuous medium, and thus sedimentation/creaming of emulsions is eliminated; for example, pharmaceutical mineral oil emulsions containing 0.25% xanthan gum are generally stable with no sign of phase separation for a period of more than six months [14–26]. Likewise, salad dressings (which are vegetable-oil-in-water emulsions) containing xanthan gum at a level of 0.25–0.4% have storage stability for at least one year with little or no phase separation [14,15,27]. In food emulsions, there is an additional purpose for adding polymeric thickeners (mainly polysaccharides) to the continuous phase (other than simply to extend the shelf life of the product), that is, to impart the required mouth-feel properties, pleasing appearance, and texture to the food products and to improve moisture retention. Some examples of polymeric thickening agents used widely in emulsion formulations are: sodium carboxymethyl cellulose (CMC), carboxymethyl cellulose, methyl cellulose, methylethyl cellulose, hydroxyethyl cellulose, methyl hydroxyethyl cellulose, hydroxypropyl cellulose, methyl hydroxypropyl cellulose, xanthan gum, Rhamsan gum, gum agar, gum tragacanth, gum karaya, gum arabic, locust bean gum, poly(acrylic acid), polyacrylamides [9–28].

However, it is important to note that polymeric thickeners may not necessarily exhibit the tendency to adsorb at the oil–water interface and, therefore, they may not, by themselves, prevent coalescence of droplets. Thus it may be necessary to add low-molecular-weight surfactants (emulsifiers) for emulsification and stabilization of droplets against coalescence. Figure 1 shows the schematic diagram of polymer-thickened surfactant-stabilized emulsions.

In many applications, polymers are used as *surfactants* [16,20,25,29–34]. High-molecular-weight polymeric surfactants have the advantage that their

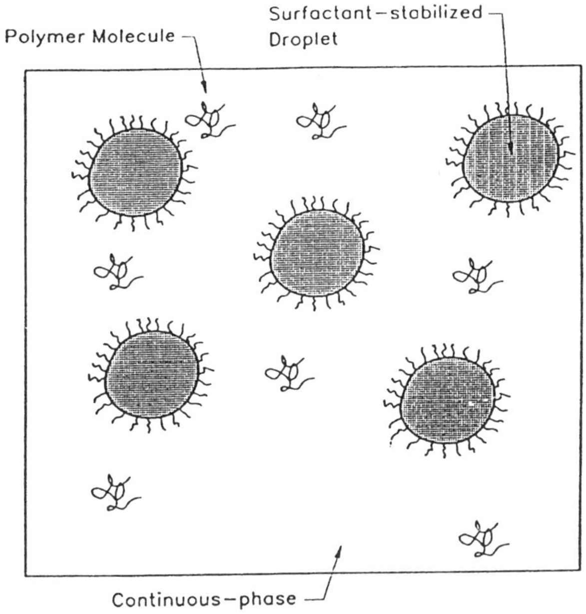

Figure 1. Polymer-thickened surfactant-stabilized emulsion.

adsorption onto the droplet surface is much firmer than with low-molecular-weight surfactants. This occurs because polymeric surfactants attach to the droplet surface via several segments (see Fig. 2). Furthermore, polymeric surfactants provide a thick steric barrier against flocculation and coalescence at the droplet surface. Consequently, they can be used in low concentrations and much more efficiently. Naturally occurring polymers such as gums and proteins are widely used for steric stabilization of food and pharmaceutical emulsions [4,35,36] (fat globules in milk are sterically stabilized against coalescence by adsorbed proteins). Asphaltenes, which are high-molecular-weight polyaromatic components of some crude oils, are known to play a major role in steric stabilization of some crude oil emulsions [37]. Polymeric surfactant/stabilizers are also used in other industrial preparations, such as cosmetics, paints, dyestuffs, and pesticides. Many amphiphilic graft (comb) copolymers composed of a hydrophobic backbone and hydrophilic branches have been utilized. Some examples of synthetic polymeric surfactants are: polymethacrylic esters-graft-poly(ethylene glycol) monomethyl ethers, poly(*p*-methylstyrene)-graft-poly(ethylene glycol), polysiloxane-graft-poly(ethylene glycol), poly(methacrylic acid), polymethylmethacrylate, carboxymethyl cellulose, poly(vinyl alcohol), poly(ethylene oxide), block copolymers of poly(ethylene oxide) and poly(vinyl acetate), poly(-acrylamide) and poly(vinyl acetate), poly(acrylic acid) and poly(vinyl acetate), poly(lauryl methacrylate) and poly(acrylic acid), polystyrene and poly(ethylene oxide) [30,31,35].

High-molecular-weight polymers are also used as *flocculants* to flocculate, and hence demulsify, emulsions [37–40]. Indeed, polymer-induced flocculation is

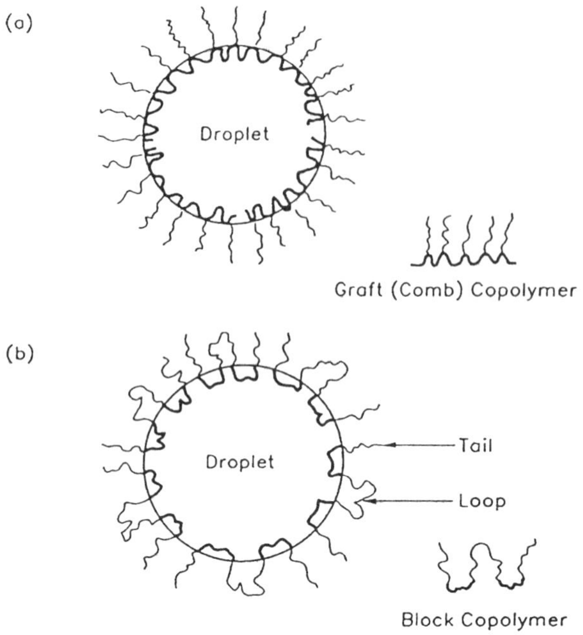

Figure 2. Schematic diagram of possible configurations of polymeric surfactant at the droplet surface.

the most common method for breaking water-in-oil emulsions in the oil industry [37,40]. Examples of some commercial polymeric flocculants are: ionic and nonionic acrylamide polymers, sodium polystyrene sulphonate, polyethyleneimine, polydiallyl dimethyl ammonium chloride ("Catfloc"), poly(2-vinylimidazoline), alkyd resins with polyalkylene glycol comprising part of the polyhydric alcohol component of the resin, polyethylene glycol tetrapropenyl succinate, polycondensates of reaction of alkanol amines with poly 2-alkenals, polyoxyalkylene ether, neutralized polyester, acrylated poly(oxyalkylene) polyamines, block copolymers of ethylene and propylene oxides, and thiazine quaternary ammonium salts of polyepihalohydrin [40,41].

The flocculation of emulsion droplets by added polymers is caused by at least three different mechanisms, namely: polymer bridging, charge neutralization, and depletion flocculation [34,41]. *Bridging flocculation* occurs when segments of the same polymer molecule adsorb onto different droplets, thus linking the droplets together (see Fig. 3). However, if the droplets of an emulsion are stabilized by ionic surfactants—that is, the droplets carry a charge on their surface—the emulsion could be flocculated by the addition of a polyelectrolyte of opposite charge. This mechanism is referred to as *charge neutralization*. It is generally believed that polyelectrolytes adsorb onto the droplet surface in a flat configuration leading to the formation of patches of uneven charge distribution (see Fig. 4). Flocculation of droplets then occurs mainly due to attraction between positive and negative patches of the neighboring

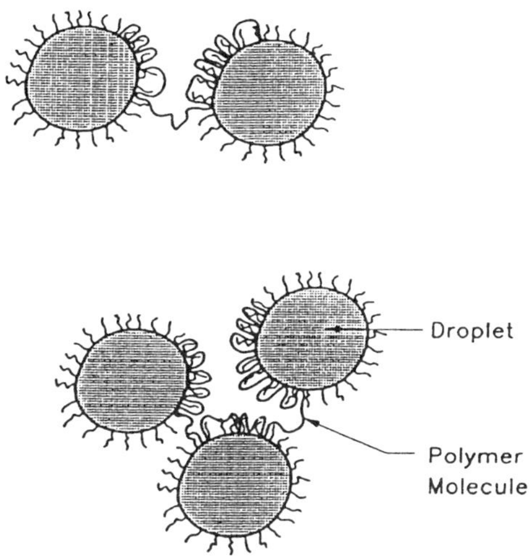

Figure 3. Bridging flocculation of emulsion droplets.

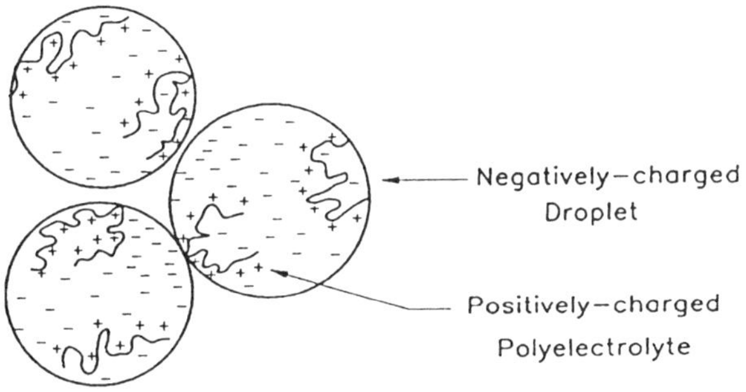

Figure 4. Flocculation of emulsion droplets due to charge neutralization.

droplets. Flocculation of emulsion droplets could also be induced by the addition of a nonadsorbing free polymer. The mechanism of flocculation in this case is called *depletion flocculation*. The depletion flocculation mechanism can be explained, albeit briefly, as follows: In a system containing free nonadsorbing polymer, when the emulsion droplets approach each other to within a distance that is smaller than the size of the polymer coil, exclusion of polymer molecules from the interstices between the droplets takes place, leading to the formation of a polymer-free zone. This phenomenon results in an attractive force between the droplets due to the lowering of osmotic pressure in the region between the droplets, and consequently flocculation of droplets occurs (see Fig. 5).

Another important application of emulsions containing polymeric liquids relates to immiscible polymer blends [42–63]. Physical blending of two immiscible polymer melts, to generate new performance materials, is of growing

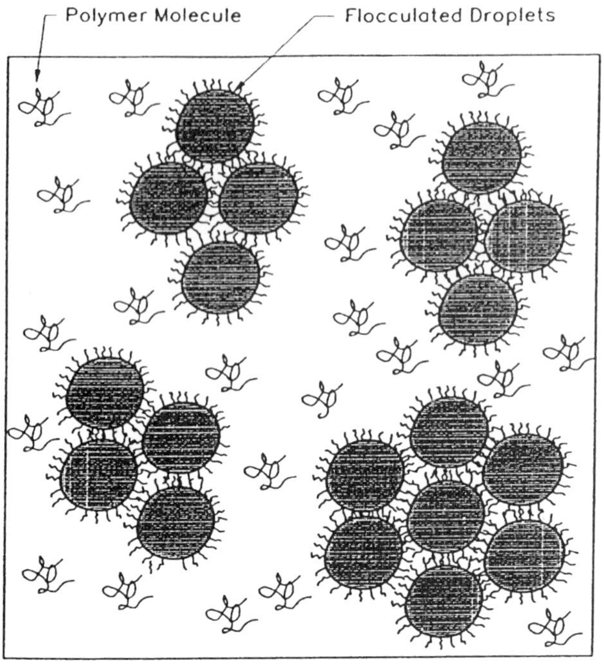

Figure 5. Depletion flocculation of emulsion droplets.

industrial importance. The concept of physical blending of immiscible polymeric melts is attractive, for one can create a large variety of new products with desirable properties without the need of synthesizing new chemical structures. In general, the blending of immiscible polymer melts gives a product whose properties reflect a combination of the useful properties of the individual components. For example, many plastics lack sufficient toughness, and hence they are unsuitable for many applications. However, this deficiency can be easily eliminated by dispersing small quantities of a suitable rubbery polymer into the plastic matrix. As an example, Fig. 6 shows a typical photomicrograph of emulsified polymeric melts (polyoxyethylene-diol droplets dispersed in polydimethylsiloxane [58]).

In most or all of the applications just discussed, knowledge of the rheological properties of emulsions of polymeric liquids is required. The term *polymeric liquids* includes very dilute polymer solutions, concentrated polymer solutions, and polymeric melts. For example, in the formulation, handling, mixing, processing, storage, and pipeline transportation of emulsions, the rheological data are required for the design, selection, and operation of the equipment involved. Furthermore, rheological studies shed light on droplet–polymer, droplet–droplet, polymer–surfactant, and polymer–solvent interactions that govern the stability, flocculation characteristics, and hence internal microstructure of emulsions.

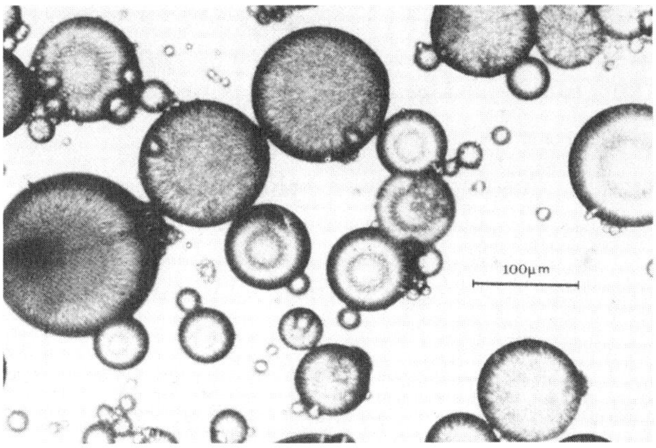

Figure 6. Photomicrograph of emulsified polymeric melts. (From Ref. 58.)

II. Basic Concepts

A. Polymer Solutions

1. Criteria for Polymer Solubility

The solubility of a polymer in any given solvent depends on the sign of the Gibbs free energy of mixing (ΔG_m) of polymer with solvent [64–66]. The Gibbs free energy of mixing is given as

$$\Delta G_m = \Delta H_m - T\,\Delta S_m \tag{1}$$

where ΔH_m is enthalpy of mixing, T is temperature, and ΔS_m is entropy of mixing. The polymer is soluble if ΔG_m is *negative*. In general, when the attractive forces of the solvent molecules for the polymer molecules exceed those between pairs of polymer molecules and between pairs of solvent molecules, ΔG_m is negative. According to the Flory–Huggins theory, the entropy of mixing for polymer solutions is given as

$$\Delta S_m = -R_0[n_1 \ln \nu_1 + n_2 \ln \nu_2] \tag{2}$$

where R_0 is the universal gas constant, n_1 and n_2 are moles of solvent and polymer, respectively, and ν_1 and ν_2 are volume fractions of solvent and polymer, respectively. The volume fractions of solvent and polymer are given as

$$\nu_1 = \frac{n_1}{n_1 + xn_2} \quad \text{and} \quad \nu_2 = \frac{n_2 x}{n_1 + n_2 x} \tag{3}$$

where $x = \bar{V}_2/\bar{V}_1$, in which $\bar{V}_1$ is the molar volume of pure solvent and $\bar{V}_2$ is the molar volume of pure polymer.

Since ΔS_m is always positive, the sign of ΔG_m is determined by the sign and magnitude of enthalpy of mixing, ΔH_m. The enthalpy of mixing of polymer solutions, according to Flory–Huggins theory, is given by

$$\Delta H_m = N_1 \nu_2 (\chi_1 kT) \tag{4}$$

where N_1 is the number of solvent molecules, $\chi_1 kT$ is the difference in energy of a solvent molecule immersed in pure polymer compared with that in pure solvent, k is the Boltzmann constant, and χ_1 is the interaction parameter. Thus,

$$\Delta G_m = R_0 T[(n_1 \nu_2 \chi_1) + (n_1 \ln \nu_1 + n_2 \ln \nu_2)] \tag{5}$$

This is the well-known Flory–Huggins expression for the free energy of mixing of polymer with solvent. For theta-solvents, the interaction parameter χ_1 has a numerical value of 1/2. If $\chi_1 < 1/2$, the solvent is better than theta-solvent; that is, as χ_1 decreases in numerical value, the solvency of the polymer improves.

2. *The Osmotic Pressure*

If a polymer solution and its pure solvent are placed on opposite sides of a semipermeable membrane that allows only the solvent to pass through (see Fig. 7), the solvent tends to migrate into the solution compartment [67] because the chemical potential of pure solvent (μ_1^*) is greater than the chemical potential of solvent in the solution (μ_1) at any given temperature (T) and pressure (P). However, the exertion of additional pressure Π on the polymer solution results in μ_1 (chemical potential of solvent in solution) being rendered equal to μ_1^* (chemical potential of pure solvent); that is, at equilibrium:

$$\mu_1(T, P + \Pi) = \mu_1^*(T, P) \tag{6}$$

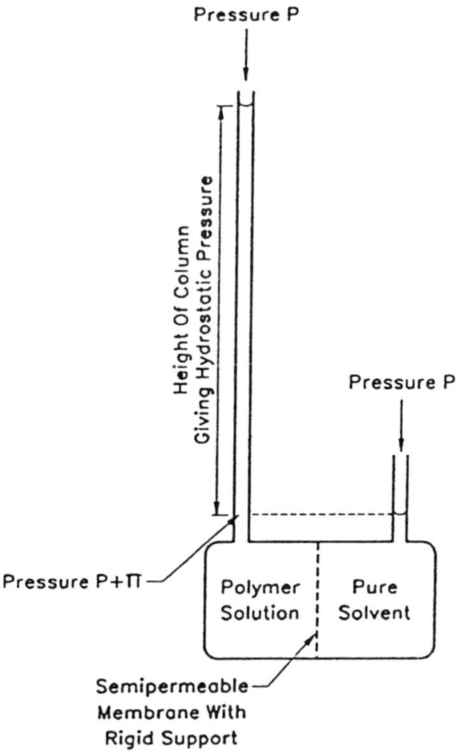

Figure 7. Schematic diagram of an osmometer. (From Ref. 67.)

$$\mu_1(T, P + \Pi) \approx \mu_1(T, P) + \int_P^{P+\Pi} \bar{V}_1 \, dp \tag{7}$$

$$\mu_1(T, P + \Pi) = \mu_1(T, P) + \bar{V}_1 \Pi \tag{8}$$

$$\rightarrow \quad \mu_1^* - \mu_1 = \bar{V}_1 \Pi \tag{9}$$

or

$$\Pi = -\frac{\mu_1 - \mu_1^*}{\bar{V}_1} \tag{10}$$

Since

$$\Delta G_{\text{mixing}} = G_{\text{mix}} - G_{\text{solvent}} - G_{\text{polymer}} \tag{11}$$

$$\begin{aligned} \frac{\partial \Delta G_{\text{mixing}}}{\partial n_1} &= \frac{\partial G_{\text{mix}}}{\partial n_1} - \frac{\partial G_{\text{solvent}}}{\partial n_1} \\ &= \mu_1 - \mu_1^* \end{aligned} \tag{12}$$

From the Flory–Huggins expression for ΔG_{mixing}, we obtain

$$\frac{\partial \Delta G_{\text{mixing}}}{\partial n_1} = R_0 T \left[\ln \nu_1 + \left(1 - \frac{1}{x}\right) \nu_2 + \chi_1 \nu_2^2 \right] \tag{13}$$

$$\rightarrow \quad \mu_1 - \mu_1^* = R_0 T \left[\ln \nu_1 + \left(1 - \frac{1}{x}\right) \nu_2 + \chi_1 \nu_2^2 \right] \tag{14}$$

Therefore,

$$\begin{aligned} \Pi &= \left(-\frac{R_0 T}{\bar{V}_1}\right) \left[\ln\,(1 - \nu_2) + \left(1 - \frac{1}{x}\right) \nu_2 + \chi_1 \nu_2^2 \right] \\ \Pi &= \left(-\frac{R_0 T}{\bar{V}_1}\right) \left[\left(-\nu_2 - \frac{\nu_2^2}{2} + \ldots\right) + \left(1 - \frac{1}{x}\right) \nu_2 + \chi_1 \nu_2^2 \right] \end{aligned} \tag{15}$$

$$\rightarrow \quad \Pi = \left(\frac{R_0 T}{\bar{V}_1}\right) \left[\frac{\nu_2}{x} + \left(\frac{1}{2} - \chi_1\right) \nu_2^2 + \ldots \right] \tag{16}$$

Since the density of polymer is equal to $(\bar{M}_n/\bar{V}_2)$, where $\bar{M}_n$ is the number-average molecular weight of polymer, Eq. (16) can be written in terms of concentration of polymer (C, g/L):

$$\nu_2 = \frac{C\bar{V}_2}{\bar{M}_n} \tag{17}$$

$$\rightarrow \quad \frac{\Pi}{C} = R_0 T \left[\frac{1}{\bar{M}_n} + \left(\frac{\bar{V}_2}{\bar{M}_n}\right)^2 \left(\frac{1}{\bar{V}_1}\right) \left(\frac{1}{2} - \chi_1\right) C + \ldots \right] \tag{18}$$

or

$$\frac{\Pi}{C} = R_0 T [B_1 + B_2 C + \ldots] \tag{19}$$

where the first virial coefficient (B_1) is equal to $1/\bar{M}_n$ and the second virial coefficient (B_2) is equal to

$$B_2 = \left(\frac{\bar{V}_2}{\bar{M}_n}\right)^2 \left(\frac{1}{\bar{V}_1}\right)\left(\frac{1}{2} - \chi_1\right) \tag{20}$$

The second virial coefficient is a measure of the quality of solvent for a given polymer. For theta-solvents, $\chi_1 = 1/2$ and the second virial coefficient vanishes. For poor solvents, $\chi_1 > 1/2$ and the second virial coefficient is negative, but for good solvents, $\chi_1 < 1/2$, and the second virial coefficient is positive. Figure 8 shows schematically the osmotic pressure variation for theta-solvent, good solvent, and poor solvent. From the plot of Π/C versus C, one can determine $\bar{M}_n$ and the interaction parameter χ_1. The intercept of Π/C versus C plot gives the first virial coefficient (B_1), and the slope at $C \rightarrow 0$ gives the second virial coefficient (B_2). From B_1 and B_2, one can determine $\bar{M}_n$ and χ_1. Note that the quality of the solvent can be changed in either of two ways, i.e., by adding a nonsolvent to the polymer solution or by changing the temperature. Thus, for a given polymer–solvent system, the solvent could be a good solvent, a poor solvent, or a theta-solvent, depending on the temperature. The temperature at which the second virial coefficient (B_2) vanishes (that is, $\chi_1 = 1/2$) is referred to as theta-temperature. Under theta-conditions, polymer–solvent interaction is zero, and the molecules interpenetrate one another freely with no net interactions. Also, the polymer molecules adopt their unperturbed dimensions at the theta-point.

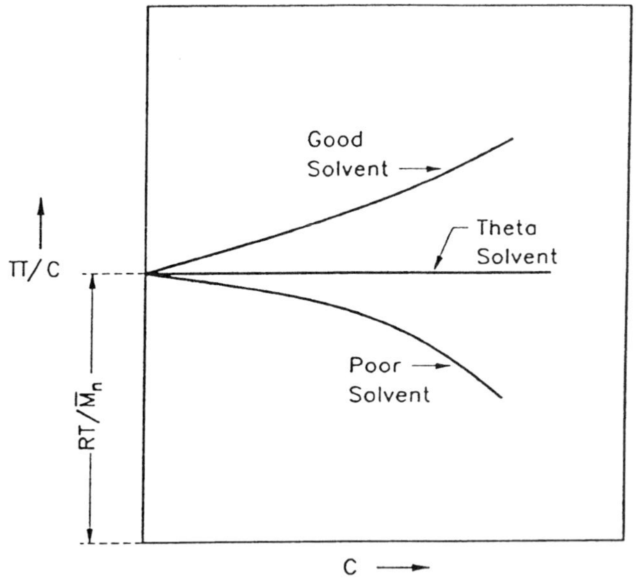

Figure 8. Schematic variation of osmotic pressure for good solvent, theta-solvent, and poor solvent.

3. Configuration of Polymer Molecules

The polymer conformation in a solution has an enormous influence on the rheology of the solution. For example, when polymer is dissolved in a good solvent, the viscosity of the resultant solution is relatively high because in such a solvent the polymer–solvent contacts are favored and, consequently, the polymer coil undergoes significant expansion.

The average dimension of a polymer molecule is generally described in terms of the root-mean-square (rms) end-to-end length of the polymer chain. The rms end-to-end length of the polymer chain depends on the nature of the polymer and the nature of the solvent.

The simplest model to visualize the polymer conformation in a solution (or in a melt) is the "*random flight model*" for a *freely jointed chain* (completely unrestricted chain). Here the polymer molecule is considered to be a long chain consisting of a series of N segments of length ℓ joined in a linear sequence with no restrictions on the angles between successive segments. Such a freely jointed chain has a tendency to coil in on itself to form a spherical cloud of segments having radial symmetry [65]. The schematic diagram of the random coil conformation of a polymer molecule is shown in Fig. 9 [34]. Note that the polymer molecule in a solution (or in a melt) is not stationary and is constantly changing its conformation. Therefore, the end-to-end distance (r_i) of the freely joined chain is ever changing, and the root-mean-square (rms) end-to-end length of a macromolecule is given as

$$(\overline{r_0^2})^{1/2} = \left(\frac{\Sigma\, n_i r_i^2}{\Sigma\, n_i}\right)^{1/2} \tag{21}$$

where n_i = the number of polymer configurations with the end-to-end length r_i. For the random flight chain, the rms end-to-end distance turns out to be

$$(\overline{r_0^2})^{1/2} = \ell N^{1/2} \tag{22}$$

where N and ℓ are as defined earlier. However, in actual polymer chains, there are restrictions on bond angles and on rotation of bonds; consequently, the actual rms end-to-end distance $(\overline{r^2})^{1/2}$ is significantly larger than the value

Figure 9. Schematic diagram of the random coil conformation of a polymer molecule. (From Ref. 34.)

$(\overline{r_0^2})^{1/2}$ given by Eq. (22). To account for such restrictions, the real polymer chain can be represented by an equivalent random flight chain composed of a smaller number of hypothetical larger segments chosen so the contour length (fully extended length of the chain) remains unchanged. Let the number of segments and length of each segment in an equivalent chain be N^* ($N^* < N$) and ℓ^* ($\ell^* > \ell$), respectively. Thus, the contour length (L) can be written as:

$$L = N\ell = N^*\ell^* \tag{23}$$

and the rms end-to-end length $(\overline{r_0^{*2}})^{1/2}$ can be written as

$$(\overline{r_0^{*2}})^{1/2} = \ell^*(N^*)^{1/2} \tag{24}$$

The ratio of $\overline{r_0^{*2}}$ to $\overline{r_0^2}$ is called the characteristic ratio (β), which reflects the intrinsic flexibility of the polymer chain; that is, β depends only on the nature of the polymer (such as bond angle or the degree of restriction of rotation):

$$\beta = \frac{\overline{r_0^{*2}}}{\overline{r_0^2}} = \frac{\overline{r_0^{*2}}}{N\ell^2} = \frac{(\overline{r_0^{*2}})}{(M/m)\ell^2} \tag{25}$$

where M is the molecular weight and m is the molecular weight of the unit represented by the segment in the freely jointed chain. From Eq. (25),

$$(\overline{r_0^{*2}})^{1/2} = \left(\frac{\beta}{m}\ell^2\right)^{1/2} M^{1/2} \tag{26}$$

Thus, the rms end-to-end length of a polymer chain is proportional to the square root of the molecular weight of the polymer.

However, it should be noted that Eqs. (22)–(26) neglect the interaction between the polymer molecules and the solvent molecules; that is, they are valid under theta-conditions only. Under nontheta-conditions, the size of the molecular coil is very much influenced by the polymer–solvent interaction. In a "good" solvent, the polymer cloud (coil) expands considerably relative to the unperturbed (theta-condition) dimension. But in a "poor" solvent, the polymer coil undergoes contraction relative to the unperturbed dimension. The expansion of a polymer coil under nontheta-conditions is specified in terms of an expansion factor (α), defined as

$$\alpha = \left(\frac{\overline{r^2}}{\overline{r_0^{*2}}}\right)^{1/2} \tag{27}$$

where $(\overline{r^2})^{1/2}$ is the actual rms end-to-end distance of a polymer coil. Generally, α lies in the range from 1 to 1.5 for good solvents; for bad solvents, $\alpha < 1$ ($\alpha = 1$ for theta-solvent).

Combining Eqs. (26) and (27) gives

$$(\overline{r^2})^{1/2} = \alpha\left(\frac{\beta}{m}\ell^2\right)^{1/2} M^{1/2} \tag{28}$$

Thus, the average size of a polymer molecule in a solvent depends on the solvent–polymer interaction (α), the polymer chain structure (bond angle, bond length, degree of restriction of rotation, molecular weight of the structural unit), and the molecular weight of the polymer.

III. Rheology of Polymeric Liquids

A knowledge of the flow properties of polymeric liquids (dilute polymer solutions, concentrated polymer solutions, and polymer melts) is essential for understanding the rheological behavior of emulsions prepared from such polymer liquids. This is particularly the case when the continuous phase (as opposed to the dispersed phase) of an emulsion is a polymeric liquid; the rheology of emulsions is strongly influenced by the continuous phase rheology.

In very dilute polymer solutions, little or no interaction occurs between polymer molecules, and therefore the solution viscosity varies in direct proportion to the polymer concentration. However, in concentrated polymer solutions, the rate of change of viscosity with polymer concentration is much faster because of the interaction between polymer molecules. Also, in concentrated polymer solutions and polymer melts, formation of entanglements between polymer molecules is quite significant, resulting in highly viscoelastic behavior.

A. Dilute Polymer Solutions

The polymer molecule in a solution could be viewed as a spherical (deformable) particle consisting of the polymer molecule and the solvent immobilized with the polymer coil. Thus, it is instructive to discuss some of the viscosity equations originally developed for suspensions of spherical particles.

For very dilute suspensions of spherical (rigid) particles, Einstein [68,69] developed the following relation:

$$\eta_r = 1 + 2.5\Phi \tag{29}$$

where η_r is the relative viscosity (defined as the ratio of suspension viscosity to suspending medium viscosity) and Φ is the volume fraction of the dispersed phase.

The Eq. (29) could be applied to very dilute polymer solutions. The volume fraction of solvated polymer particles is related to polymer concentration (C, g/L), molecular weight of polymer, and hydrodynamic volume (V_h) of solvated polymer molecules as follows:

$$\Phi = \frac{N_A C V_h}{M} \tag{30}$$

where N_A is Avogadro's number. Thus Einstein's relation becomes:

$$\eta_r = 1 + 2.5\left(\frac{N_A C V_h}{M}\right) \tag{31}$$

From Eq. (31), the intrinsic viscosity $[\eta]$ of a polymer solution, defined as

$$[\eta] = \lim_{C \to 0} \frac{\eta_r - 1}{C} \tag{32}$$

is given by

$$[\eta] = \frac{2.5N_A V_h}{M} \tag{33}$$

For flexible (nonrigid) polymer molecules, the radius of an equivalent hydrodynamic spherical particle is given [64] by

$$R_e = \varepsilon R_G \tag{34}$$

where R_G is the radius of gyration of polymer molecule and ε is some constant. Thus,

$$[\eta] = \frac{2.5N_A}{M}\,\frac{4}{3}\,\pi\varepsilon^3 R_G^3 = \frac{10\pi N_A}{3M}\,\varepsilon^3 R_G^3 \tag{35}$$

The radius of gyration for *linear* polymers is related to the rms end-to-end distance of the polymer coil as follows:

$$R_G = \left(\frac{\overline{r^2}}{6}\right)^{1/2} \tag{36}$$

Combining Eqs. (35) and (36) gives

$$[\eta] = \left(\frac{10\pi N_A \varepsilon^3}{3(6)^{3/2}}\right)\frac{(\overline{r^2})^{3/2}}{M} = \Phi'\,\frac{(\overline{r^2})^{3/2}}{M} \tag{37}$$

where Φ' is believed to be a universal constant [$\approx 2.1 \times 10^{21}$ dL/(gmole) (cm^3)]. Since $(\overline{r^2})^{1/2}$ is $\alpha(\overline{r_0^{*2}})^{1/2}$, Eq. (37) can be rewritten as

$$[\eta] = \Phi'\left(\frac{\overline{r_0^{*2}}}{M}\right)^{3/2}\alpha^3 M^{1/2} \tag{38}$$

The ratio $(\overline{r_0^{*2}}/M)$ is a function of chain structure only (bond angle, bond length, degree of restriction of rotation), independent of solvent or molecular weight of polymer (refer to Eq. 26). Thus, for a given polymer,

$$[\eta] = K\alpha^3 M^{1/2} \tag{39}$$

where K is a constant (dependent on chain structure only). Under theta-conditions, $\alpha = 1$ and therefore,

$$[\eta]_\theta = KM^{1/2} \tag{40}$$

Equation (40) is generally found to be valid for a number of polymer systems [64,65]. As an example, Fig. 10 shows the intrinsic viscosities of polymer solutions at a theta-temperature [64]. The data confirm the square-root dependence of $[\eta]$ on molecular weight.

Equation (39) could be modified further if we take into account the fact that $\alpha = AM^z$, where A and z are constants whose values depend on the nature of the polymer and solvent and on the temperature:

$$[\eta] = KM^a \tag{41}$$

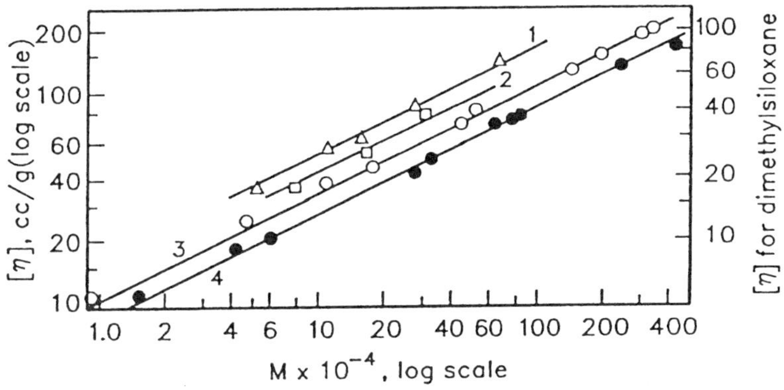

Figure 10. Intrinsic viscosities of polymer solutions at the theta-temperature. Curve 1 represents poly(dimethylsiloxane) in methylethyl ketone at 20°C (with $[\eta]$ given by the right-hand ordinate); curve 2 is cellulose tricaprylate in γ-phenylpropanol at 48°C; curve 3 is polyisobutylene in benzene at 24°C; curve 4 is polystyrene in cyclohexane at 34°C. (From Ref. 64.)

where K and a are new constants (dependent on the nature of polymer and solvent and on the temperature). For flexible randomly coiled chains, the exponent a varies from 0.5 (in a theta-solvent) to about 0.8. For rod-shaped molecules, the exponent a is approximately 1.8. However, Eq. (41) is valid only for linear polymers with uniform molecular weight. For linear polymers with molecular weight distribution, Eq. (41) becomes

$$[\eta] = K(\bar{M}_v)^a \tag{42}$$

where $\bar{M}_v$ is the viscosity-average molecular weight.

1. Effect of Polymer Concentration on the Viscosity of Dilute Polymer Solutions

From Einstein's equation:

$$\eta_r = 1 + [\eta]C \tag{43}$$

and, therefore, the specific viscosity η_{sp} (defined as $\eta_r - 1$) becomes

$$\eta_{sp} = [\eta]C \tag{44}$$

or

$$\frac{\eta_{sp}}{C} = [\eta] \tag{45}$$

This equation is valid in the limit $C \rightarrow 0$. For finite concentration, yet dilute polymer solutions, the following relationship, proposed by Huggins [70], is generally found to be valid:

$$\frac{\eta_{sp}}{C} = [\eta] + k'[\eta]^2C \tag{46}$$

where k′ is a constant. Another relation, originally proposed by Kraemer [71], is also found to be valid for dilute polymer solutions:

$$\frac{\ln \eta_r}{C} = [\eta] - k''[\eta]^2 C \tag{47}$$

where k″ is another constant. This equation reduces to the Arrhenius relationship in the limit $C \to 0$:

$$\ln \eta_r = [\eta]C \tag{48}$$

Figure 11 shows the application of the Huggins and Kraemer relations to a dilute polymer solution of polystyrene in benzene. The relationship between η_{sp}/C [or $(\ln \eta_r)/C$] and C is linear, as expected [65].

2. *Effect of Shear Rate on the Viscosity of Dilute Polymer Solutions*

Even in the case of dilute polymer solutions, the effect of shear rate is important for molecules of high intrinsic viscosity (high molecular weight). With the increase in shear rate, the polymer coils deform and orient with applied shear and, consequently, offer less resistance to flow; that is, the viscosity decreases. Figure 12 shows an example of the effect of shear rate [64]. Notice that the effect of shear rate exists even at infinite dilution ($C \to 0$). According to the theories [72] of the shear-rate dependence, the ratio of $[\eta]$ to its value $[\eta]_0$ at zero shear rate is a function of reduced shear rate ($\dot{\gamma}_r$) given as:

$$\dot{\gamma}_r = \left(\frac{[\eta]_0 \eta_S M}{R_0 T}\right)\dot{\gamma} \tag{49}$$

where η_S is solvent viscosity, R_0 is the gas constant, and $\dot{\gamma}$ is the shear rate. As an example, Fig. 13 shows the ratio $[\eta]/[\eta]_0$ as a function of $\dot{\gamma}_r$ for polystyrene solutions with various solvents [72,73]. All solutions follow the same curve.

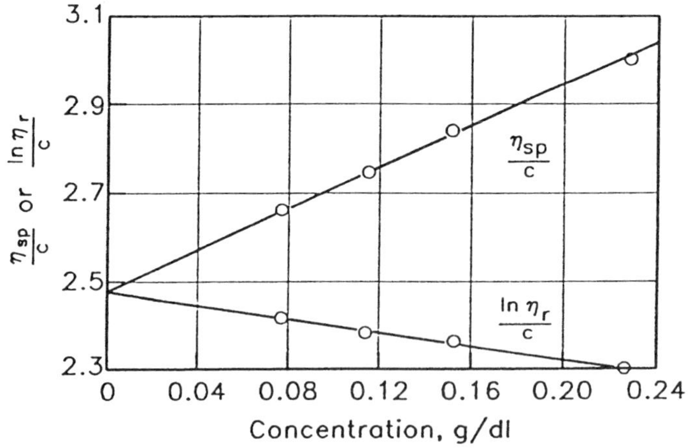

Figure 11. Application of the Huggins and Kraemer relations to a dilute solution of polystyrene in benzene. (From Refs. 70, 71, and 65.)

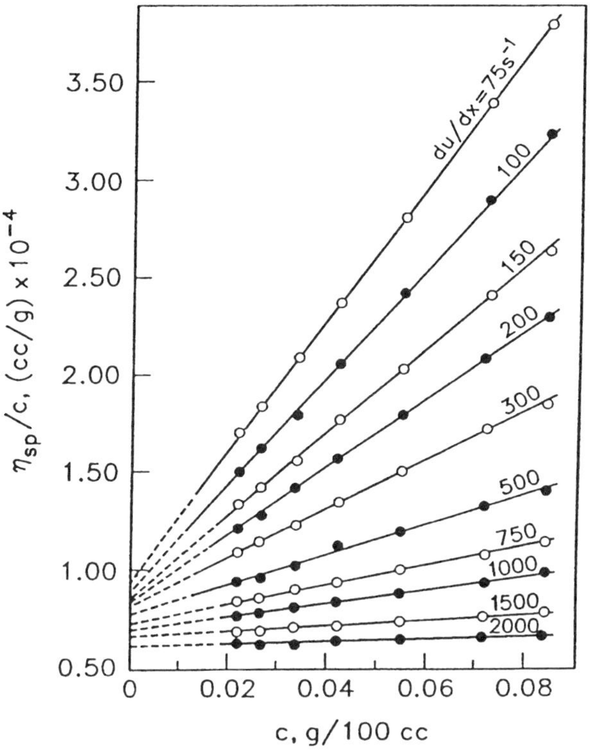

Figure 12. Effect of shear rate on the viscosity of a dilute solution of cellulose nitrate. (From Ref. 64.)

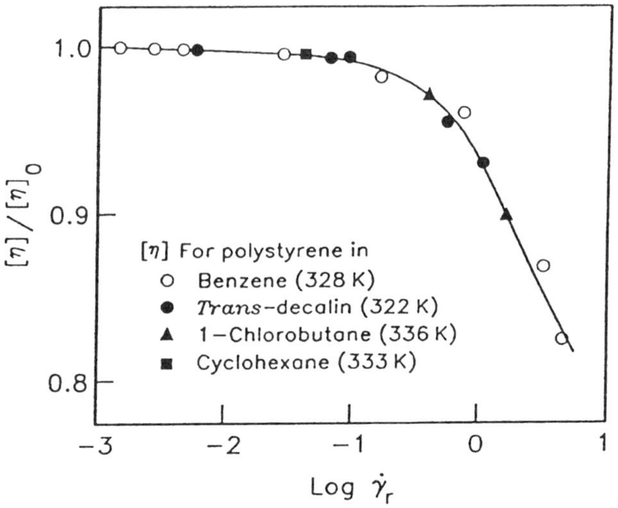

Figure 13. The ratio $[\eta]/[\eta]_0$ as a function of $\log \dot{\gamma}_r$ for polystyrene solutions in various solvents. (From Refs. 72 and 73.)

B. Concentrated Polymer Solutions

Concentrated polymer solutions generally exhibit non-Newtonian behavior. Figure 14 shows a typical flow curve for concentrated polymer solution [74]. One can observe three different regions: (1) *lower Newtonian region* at low shear rates, where the viscosity of the polymer solution is constant; (2) *shear-thinning region* (power-law region) at intermediate shear rates, where the viscosity of the polymer solution decreases with an increase in the shear rate; and (3) *upper Newtonian region* at high shear rates, where the viscosity of the polymer solution again becomes constant.

The non-Newtonian character of polymer solutions is due to at least three factors:

1. a change in the shape of flexible molecules with an increase in the shear rate
2. an increase in the orientation of nonsymmetrical molecules with an increase in the shear rate
3. a reduction in the degree of entanglements between the polymer molecules

In the case of dilute polymer solutions, the non-Newtonian behavior is due to factors 1 and 2, whereas in the case of concentrated polymer solutions, the main cause of non-Newtonian behavior is physical entanglements between the polymer molecules. The shape of the flow curve in Fig. 14 could easily be explained in terms of these factors.

In the lower Newtonian region, the rate of deformation (shear stress) is so small that there are no significant changes in the shape and orientation of polymer chains or in the degree of entanglements between the polymer chains; consequently, the viscosity remains constant. In the shear-thinning region, the rate of deformation (shear stress) is high enough to cause changes in the shape and orientation of polymer chains and to reduce the degree of entanglements;

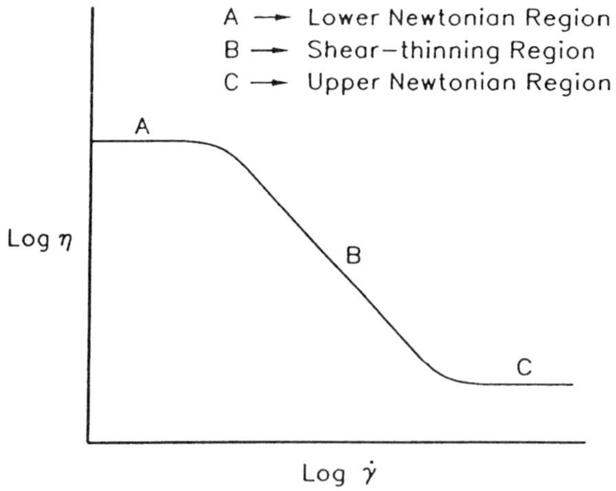

Figure 14. Typical flow curve for a concentrated polymer solution.

therefore, the viscosity decreases with an increase in the shear rate. In the upper Newtonian region, the rate of deformation (shear stress) is so high that any further increase in the shear rate or shear stress produces little or no further change either in the degree of entanglements or in molecular shape and orientation; consequently, the viscosity remains constant.

Figure 15 shows the actual flow curves for aqueous solutions of poly(ethylene oxide) resin (Polyox WSR-301) at different concentrations [75]. At low polymer concentrations, only lower Newtonian region is observed in the range of shear stresses investigated. At high polymer concentrations, lower Newtonian region and shear-thinning region both are observed. Because of the experimental limitations, it was not possible to reach sufficiently high shear rates (shear stresses) to check on the existence of the upper Newtonian region. Note that the flow curves shift upwards toward higher viscosities with an increase in polymer concentration.

Figure 16 shows the flow curves for aqueous solutions of xanthan gum at different concentrations [76]. Here, we do observe three different regions: lower Newtonian region, shear-thinning region, and upper Newtonian region. Xanthan gum solutions are well known to form a complex network of molecules in solutions due to physical entanglements and hydrogen bonding; therefore, we observe non-Newtonian behavior in these solutions.

The entanglement of polymer chains is a function of both the size (radius of gyration) and the concentration of polymer in the solution. Since the size of the polymer molecule is related to the intrinsic viscosity $[\eta]$, a useful dimensionless

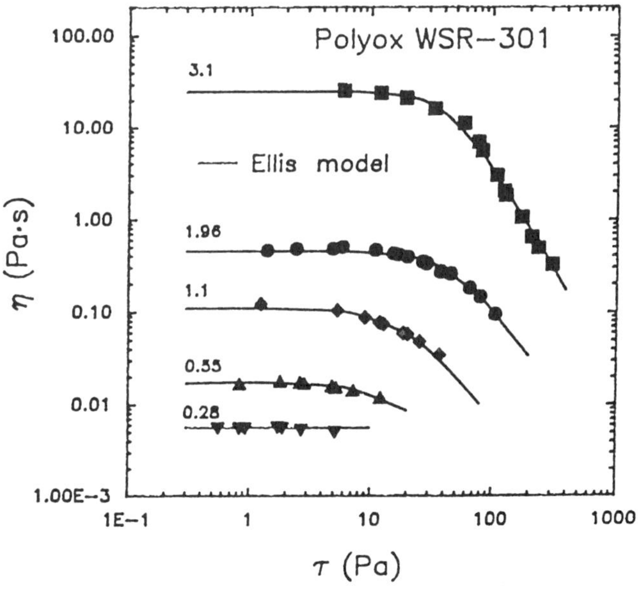

Figure 15. Flow curves for aqueous solutions of Polyox WSR-301 at various concentrations (the numbers in the graph indicate weight % polymer concentration). (From Ref. 75.)

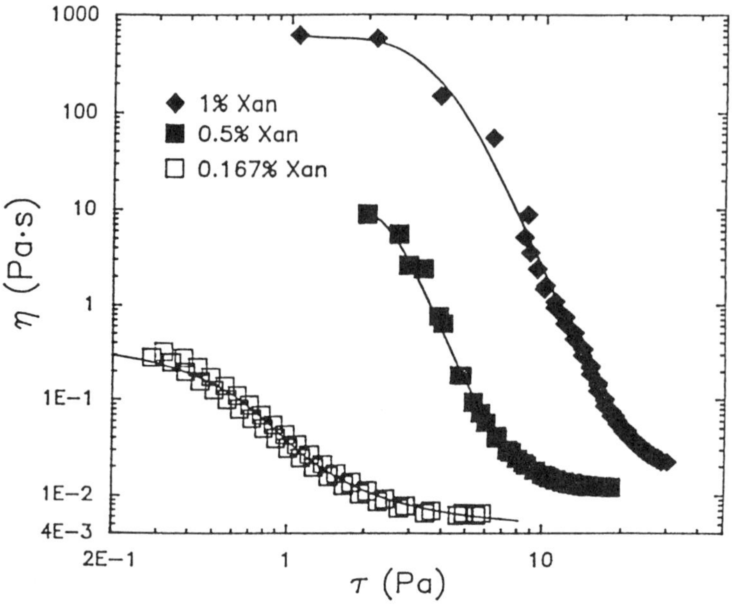

Figure 16. Apparent viscosity as a function of shear stress for xanthan gum solutions. (From Ref. 76.)

parameter that can be taken as a measure of the extent of entanglement is $C[\eta]_0$, where $[\eta]_0$ is intrinsic viscosity at zero shear rate. The parameter $C[\eta]_0$ is often called "coil overlap parameter." The entanglements of polymer chains govern the flow properties of polymer solutions when $C[\eta]_0$ attains a value in the range from 5 to 10. In the presence of entanglements of polymer chains, the polymer solutions exhibit not only shear-dependent viscosity, but also strongly viscoelastic behavior. It should be further noted that one generally observes a sudden increase in the slope of $(\eta_0 - \eta_S)$ versus C plot when the onset of entanglements occur, as can be seen in Fig. 17 [77].

1. Models for Viscosity/Shear-Rate Behavior of Polymer Solutions

Several models have been proposed in the literature to describe the apparent viscosity versus shear-rate (shear-stress) behavior of polymer solutions. The prominent ones are summarized in Table 1.

2. Viscoelastic Properties of Concentrated Polymer Solutions

Concentrated polymer solutions are generally viscoelastic in nature; that is, they exhibit both viscous and elastic properties (fluids exhibiting elastic properties are also referred to as memory fluids).

In *steady shearing* flow, the presence of elastic effects is manifested as finite nonzero "normal stresses." If we have steady simple shear flow given by

$$V_x = \dot{\gamma} y \qquad V_y = V_z = 0 \tag{68}$$

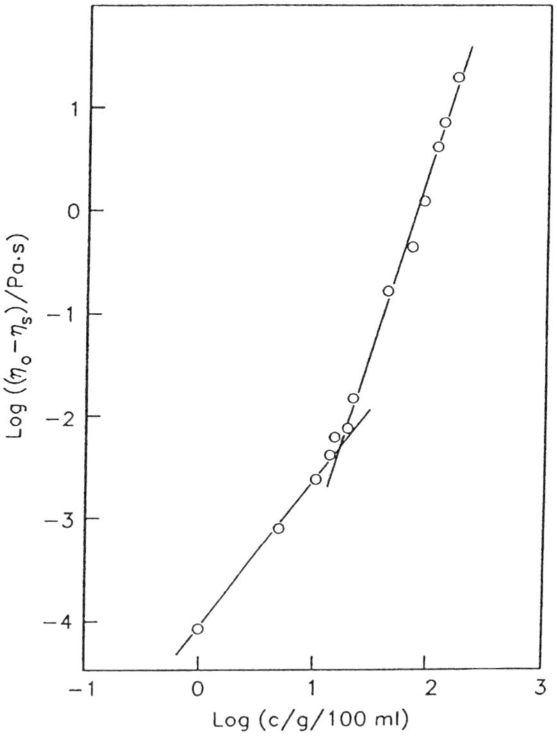

Figure 17. Plot of $(\eta_0 - \eta_s)$ versus C for solutions of polystyrene in benzene. Note that a sudden change in slope from 1.5 to 4 occurs at the onset of entanglements. (From Ref. 77.)

Table 1. Rheological Models for Polymer Solutions

Model name	Brief description	Equation number	Reference number
Power-law	$\tau = K\dot{\gamma}^n$	(50)	78,79
	or		
	$\eta = K\dot{\gamma}^{n-1}$	(51)	
Cross	$\frac{\eta - \eta_\infty}{\eta_o - \eta_\infty} = \frac{1}{1 + (K\dot{\gamma})^{1-n}}$	(52)	80
	when $\eta \gg \eta_\infty$ and $\eta_o \gg \eta_\infty$		
	$\eta = \frac{\eta_o}{K}\dot{\gamma}^{n-1} = K_1\dot{\gamma}^{n-1}$	(53)	
	when $\eta_\infty \simeq 0$		
	$\frac{\eta}{\eta_o} = \frac{1}{1 + (K\dot{\gamma})^{1-n}}$	(54)	

(continued)

Table 1. (continued)

Model name	Brief description	Equation number	Reference number
Moore	$\dfrac{\eta-\eta_\infty}{\eta_0-\eta_\infty}=\dfrac{1}{1+K\dot{\gamma}}$	(55)	81
Carreau	$\dfrac{\eta-\eta_\infty}{\eta_0-\eta_\infty}=\dfrac{1}{[1+(K\dot{\gamma})^2]^{(1-n)/2}}$	(56)	82
	when $\eta_\infty \simeq 0$		
	$\dfrac{\eta}{\eta_0}=\dfrac{1}{[1+(K\dot{\gamma})^2]^{(1-n)/2}}$	(57)	
Carreau–Yasuda	$\dfrac{\eta-\eta_\infty}{\eta_0-\eta_\infty}=\dfrac{1}{[1+(K\dot{\gamma})^m]^{(1-n)/m}}$	(58)	72
Williamson	$\dfrac{\eta-\eta_\infty}{\eta_0-\eta_\infty}=\dfrac{1}{1+K\tau}$	(59)	83
Ellis	$\dot{\gamma}=(K_1+K_2\tau^{\alpha-1})\tau$	(60)	84
	The above can be rewritten as:		
	$\dfrac{\eta}{\eta_0}=\dfrac{1}{1+K_2\eta_0\tau^{\alpha-1}}$	(61)	
Sisko	$\tau=\eta_\infty\dot{\gamma}+K\dot{\gamma}^n$	(62)	85
	or		
	$\eta=\eta_\infty+K\dot{\gamma}^{n-1}$	(63)	
Bingham-plastic	$\tau=\tau_0+K\dot{\gamma}$	(64)	86
Herschel–Bulkley	$\tau=\tau_0+K\dot{\gamma}^n$	(65)	87
Vocadlo	$\tau=(\tau_0^{1/n}+K\dot{\gamma})^n$	(66)	88
Casson	$\tau^{1/2}=\tau_0^{1/2}+K\dot{\gamma}^{1/2}$	(67)	89

the stress distributions for viscoelastic fluids are:

$$\tau_{xy}=\eta\dot{\gamma};\qquad \tau_{xz}=\tau_{yz}=0 \tag{69}$$

$$\tau_{xx}-\tau_{yy}=\Psi_1\dot{\gamma}^2;\qquad \tau_{yy}-\tau_{zz}=\Psi_2\dot{\gamma}^2 \tag{70}$$

where V_x, V_y, V_z are the three velocity components; η, Ψ_1, Ψ_2 are the viscometric functions, which are functions of shear rate $\dot{\gamma}$; $(\tau_{xx}-\tau_{yy})$ is called the "first normal stress difference"; and $(\tau_{yy}-\tau_{zz})$ is called the "second normal stress difference." In the case of Newtonian fluids, η is independent of shear rate and the two normal stress differences are zero; that is, $\tau_{xx} - \tau_{yy} = 0$ and $\tau_{yy} - \tau_{zz} = 0$.

Figure 18 shows the plots of two normal stress differences $(\tau_{xx} - \tau_{yy})$ and $(\tau_{yy} - \tau_{zz})$ as functions of shear rate [77]. The first normal stress difference is positive for all shear rates, and it increases rapidly with the increase in shear rate. The second normal stress difference is relatively very small and is negative for all shear rates. The elasticity of the fluid is generally measured in terms of the stress ratio $(\tau_{xx} - \tau_{yy})/\tau_{yx}$; this stress ratio is zero for Newtonian fluids, but for elastic fluids it is an increasing function of shear rate (see Fig. 19 [72]).

In *oscillatory* flow, the elastic effects in polymer solutions are manifested as high storage modulus, high storage compliance, and a phase lag angle much less than 90° (strain lags behind the stress).

If we apply an oscillatory stress with controlled amplitude τ_0 and angular frequency ω to a concentrated polymer solution such that

$$\tau = \tau_0 \cos \omega t \tag{71}$$

the resulting strain γ will be out of phase with the stress (stress is in advance of strain):

$$\gamma = \gamma_0 \cos(\omega t - \delta) = (\gamma_0 \cos \delta) \cos \omega t + (\gamma_0 \sin \delta) \sin \omega t \tag{72}$$

where γ_0 is the amplitude of the strain response and δ is the phase lag angle. For Newtonian fluids, $\delta = 90°$; for ideally elastic solids, $\delta = 0°$. However, for viscoelastic fluids, such as concentrated polymer solutions, the phase lag δ varies from 0 to 90°, depending on the nature of the material and the frequency of oscillation.

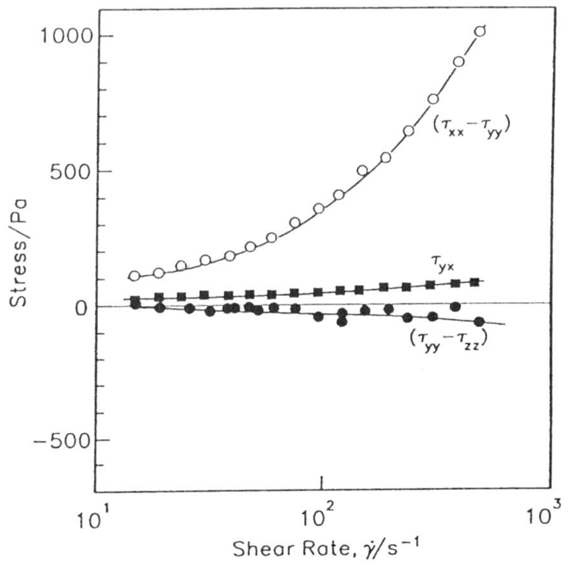

Figure 18. The plots of shear stress and two normal stress differences as functions of shear rate for 2% by wt. polyisobutylene solution in decalin. (From Ref. 77.)

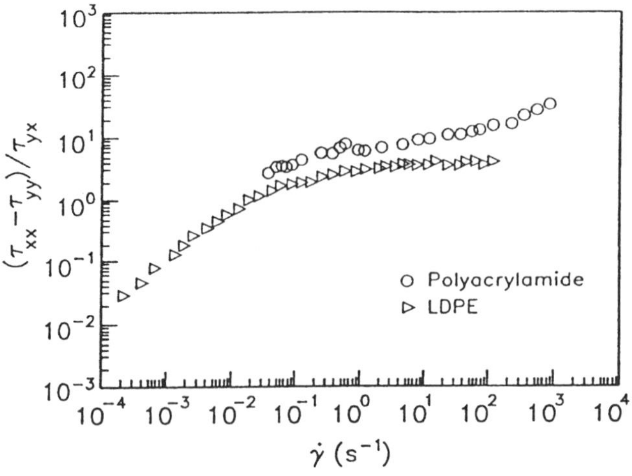

Figure 19. The stress ratio $(\tau_{xx} - \tau_{yy})/\tau_{yx}$ as a function of shear rate for the low-density polyethylene melt and the polyacrylamide solution. (From Ref. 72.)

From the known (measured) values of τ_0, ω, δ, and γ_0, several other useful rheological parameters characteristic of viscoelastic material can be evaluated. Some of the important parameters are defined in Table 2.

Table 2. Definition of Rheological Parameters

Parameter	Symbol	Definition	Equation number
Storage compliance	J'	$J' = \dfrac{\text{in-phase strain amplitude}}{\text{stress amplitude}} = \dfrac{\gamma_0 \cos\delta}{\tau_0}$	(73)
Loss compliance	J''	$J'' = \dfrac{90^\circ\text{-out-of-phase strain amplitude}}{\text{stress amplitude}} = \dfrac{\gamma_0 \sin\delta}{\tau_0}$	(74)
Complex compliance	J^*	$J^* = J' - jJ''$	(75)
Magnitude of complex compliance	$\lvert J^* \rvert$	$\lvert J^* \rvert = \sqrt{(J')^2 + (J'')^2}$	(76)
Complex modulus	G^*	$G^* = G' + jG'' = \dfrac{1}{J^*}$	(77)
Magnitude of complex modulus	$\lvert G^* \rvert$	$\lvert G^* \rvert = \dfrac{1}{\lvert J^* \rvert} = \dfrac{\tau_0}{\gamma_0}$	(78)

Table 2. (continued)

Parameter	Symbol	Definition	Equation number
Storage modulus	G'	$G' = \|G^*\| \cos\delta = \frac{\tau_0}{\gamma_0}\cos\delta = \frac{J'}{(J')^2 + (J'')^2}$	(79)
Loss modulus	G''	$G'' = \|G^*\| \sin\delta = \frac{\tau_0}{\gamma_0}\sin\delta = \frac{J''}{(J')^2 + (J'')^2}$	(80)
Complex viscosity	η^*	$\eta^* = \eta' - j\eta'' = \frac{G^*}{j\omega} = \frac{1}{j\omega J^*}$	(81)
Magnitude of complex viscosity	$\|\eta^*\|$	$\|\eta^*\| = \sqrt{(\eta')^2 + (\eta'')^2} = \frac{\|G^*\|}{\omega} = \frac{\tau_0}{\gamma_0\omega}$	(82)
Dynamic viscosity	η'	$\eta' = \|\eta^*\| \sin\delta = \frac{G''}{\omega} = \frac{\tau_0 \sin\delta}{\gamma_0\omega}$	(83)
Imaginary part of complex viscosity	η''	$\eta'' = \|\eta^*\| \cos\delta = \frac{G'}{\omega} = \frac{\tau_0 \cos\delta}{\gamma_0\omega}$	(84)
Loss tangent	$\tan\delta$	$\tan\delta = \frac{J''}{J'} = \frac{G''}{G'} = \frac{\eta'}{\eta''}$	(85)

In a typical oscillatory experiment, one generally measures the rheological parameters $|G^*|$, G', G'', η', etc. as a function of stress amplitude (at a fixed frequency) to determine the *linear visoelastic* region; in the linear viscoelastic region, all these parameters are independent of the applied stress. Once the linear viscoelastic region is established, measurements are then made at a fixed stress amplitude (within the linear viscoelastic region) as a function of frequency. As an example, Fig. 20 shows the results of stress sweep experiments done on three differently concentrated xanthan gum solutions. The storage modulus (G') of xanthan solutions is plotted as a function of shear stress, at a fixed frequency of 0.05 Hz. The linear viscoelastic region is characterized by constant G' at low shear stresses. Note that the linear viscoelastic region covers a wider range of shear stress when the polymer concentration is increased.

Figure 21 shows the plots of phase lag (δ) as a function of frequency for three differently concentrated xanthan gum solutions (in the linear region). At any

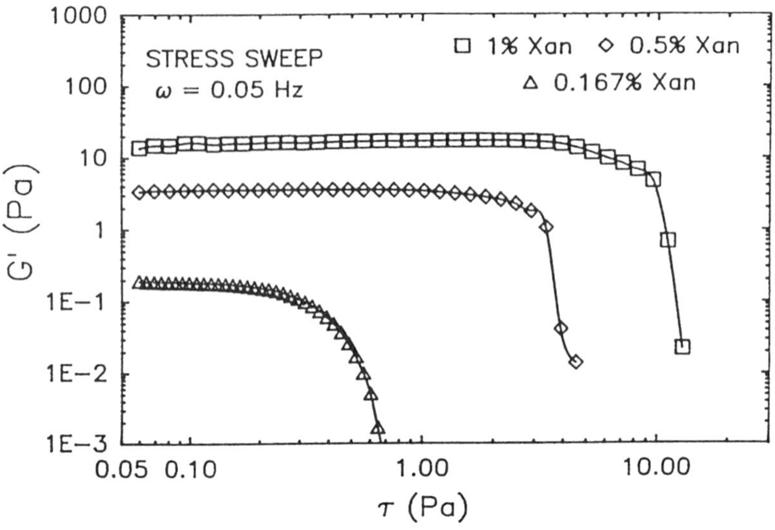

Figure 20. Storage modulus as a function of shear stress for xanthan gum solutions.

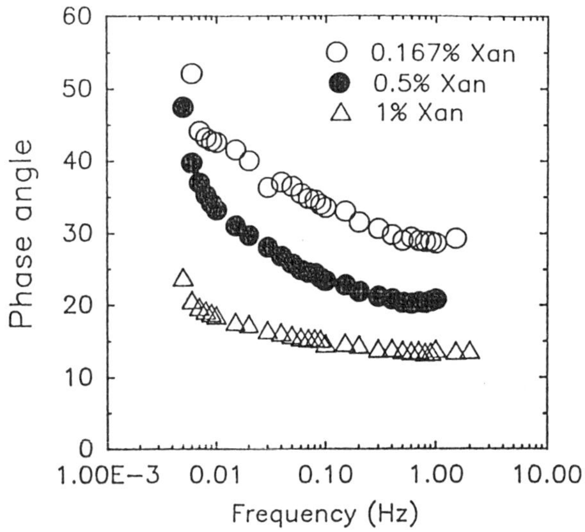

Figure 21. The plots of phase lag (δ) as a function of frequency for three differently concentrated xanthan gum solutions (in the linear region).

given frequency, the phase lag decreases with the increase in polymer concentration, indicating that xanthan solutions become more viscoelastic with the increase in polymer concentration.

Thus far we have discussed steady shearing and oscillatory flows for determining the viscoelastic properties of concentrated polymer solutions. The third approach to characterize viscoelastic polymer solutions involves *creep tests*. In creep tests, a constant shear stress of low amplitude is applied on the polymer solution and the creep compliance (J), defined as the ratio of strain to stress, is measured as a function of time. Not only do viscoelastic fluids exhibit instantaneous elastic compliance upon the application of stress, but they also exhibit a significant recovery in compliance upon removal of the stress. Figure 22 shows creep/recovery behavior of 1 wt% xanthan gum solution (with 0.4 wt% Triton X-100) in the linear region. This fluid exhibits an instantaneous elastic compliance as well as a recoverable compliance, indicating viscoelastic behavior.

3. Relationships Between Linear Viscoelastic Properties and Steady-Shear Properties

In many situations, it is easier to carry out oscillatory measurements (linear viscoelastic measurements) than steady-shear-flow measurements; therefore, it is important to determine if steady-shear properties (apparent viscosity, normal stresses) can be predicted from oscillatory measurements.

In the lower limits of frequency and shear rate, the exact relationships between oscillatory and steady-shear properties are as follows:

$$\eta'(\omega)|_{\omega\to 0} = \eta(\dot{\gamma})|_{\dot{\gamma}\to 0} \tag{86}$$

$$\left.\frac{G'(\omega)}{\omega^2}\right|_{\omega\to 0} = \left.\frac{\Psi_1(\dot{\gamma})}{2}\right|_{\dot{\gamma}\to 0} \tag{87}$$

For larger shear rates and frequencies, several empirical rules have been proposed in the literature. According to the Cox–Merz rule [90], η is the same function of $\dot{\gamma}$ as $|\eta^*|$ is of ω

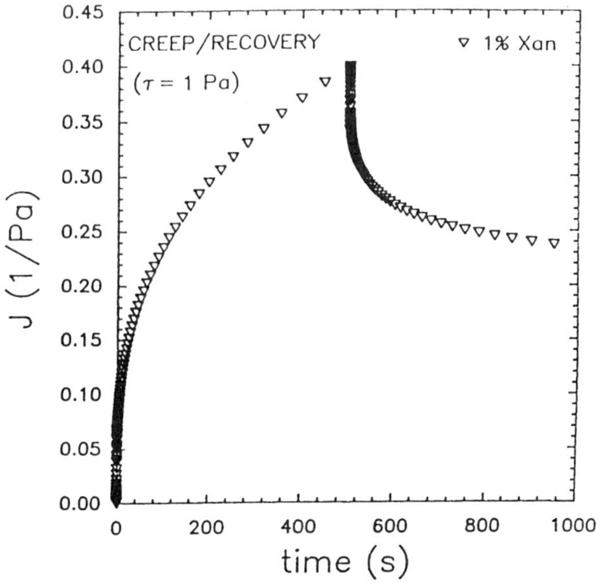

Figure 22. Creep/recovery behavior for 1% xanthan gum solution.

$$\eta(\dot{\gamma}) = |\eta^*(\omega)|\big|_{\omega=\dot{\gamma}} \tag{88}$$

$$\eta(\dot{\gamma}) = \left[(\eta')^2 + \left(\frac{G'}{\omega} \right)^2 \right]^{1/2} \Bigg|_{\omega=\dot{\gamma}} \tag{89}$$

Laun [91] proposed the following empirical rule relating the first normal stress difference and dynamic properties:

$$\frac{\Psi_1(\dot{\gamma})}{2} = \frac{\eta''(\omega)}{\omega} \left[1 + \left(\frac{\eta''}{\eta'} \right)^2 \right]^{0.7} \Bigg|_{\omega=\dot{\gamma}} \tag{90}$$

In the limit $\omega \to 0$ and $\dot{\gamma} \to 0$, both the Cox–Merz rule and the Laun rule reduce to the limiting relationships (Eqs. 86–87). Note that one can compute η'' from η' data or vice-versa using Kramers–Kronig relations [92,93]:

$$\eta''(\omega) = \frac{2\omega}{\pi} \int_0^\infty \frac{\eta'(x) - \eta'(\omega)}{\omega^2 - x^2} dx \tag{91}$$

$$\eta'(\omega) - \eta'(\infty) = \frac{2}{\pi} \int_0^\infty \frac{x\eta''(x) - \omega\eta''(\omega)}{x^2 - \omega^2} dx \tag{92}$$

As an example of the application of some of the rules just discussed, Fig. 23 compares the steady-shear viscosity $\eta(\dot{\gamma})$ with complex viscosity $|\eta^*(\omega)|$ for a polymer solution on the same plot [77]. The experimental data clearly follow the Cox–Merz rule. Figure 24 shows another example, one in which the Cox–Merz rule is not very accurate; the experimental data are shown for 1 wt% xanthan gum solution (containing 0.4 wt % Triton X-100). Clearly, the complex viscosities fall consistently above the steady-shear viscosities.

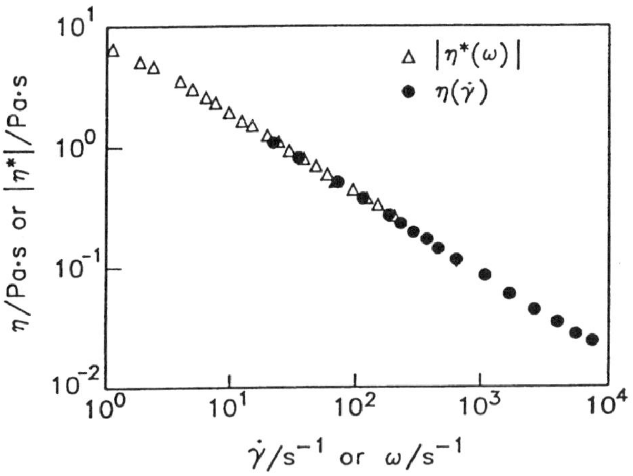

Figure 23. Rheological data for 2 wt % polyisobutylene solution in decalin. The Cox–Merz rule is obeyed by the data. (From Ref. 77.)

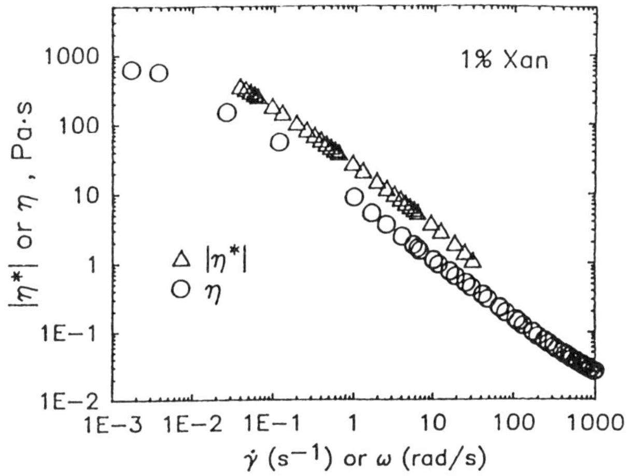

Figure 24. Rheological data for 1 wt % xanthan gum solution. The Cox–Merz rule is not obeyed.

C. Polymeric Melts

The rheological properties of polymer melts are closely related to the molecular weight of the polymer. Figure 25 shows a typical relationship between zero-shear-rate viscosity (η_0) and molecular weight (M) for a polymer melt [74]. As can be seen, η_0 of the polymer melt varies linearly with molecular weight (M) below a critical value of molecular weight (M_c):

$$\eta_0 \propto M \qquad (M < M_c) \tag{93}$$

but above M_c, η_0 varies with M as follows:

$$\eta_0 \propto M^{3.5} \qquad (M > M_c) \tag{94}$$

These relationships are found to apply to a vast majority of linear polymers. It is generally believed that the critical molecular weight M_c corresponds to a value beyond which the *entanglements* between the polymer molecules dominate the flow properties of a polymer melt. Thus, above M_c, non-Newtonian flow is generally observed and the polymer melts are highly elastic in nature.

The preceding relationships as such are valid for polymers with narrow molecular weight distributions. These relations apply to polymers with broad molecular weight distribution if M is replaced by the weight-average molecular weight $\bar{M}_w$; that is,

$$\eta_0 \propto \bar{M}_w \qquad (\bar{M}_w < \bar{M}_c) \tag{95}$$

$$\eta_0 \propto \bar{M}_w^{3.4} \qquad (\bar{M}_w > \bar{M}_c) \tag{96}$$

Table 3 gives values of $\bar{M}_c$ for some polymers [74].

Figure 26 shows the rheological properties of molten polypropylene ($\bar{M}_w = 4.4 \times 10^5$) at 200°C [62]. As can be seen, this polymer melt exhibits

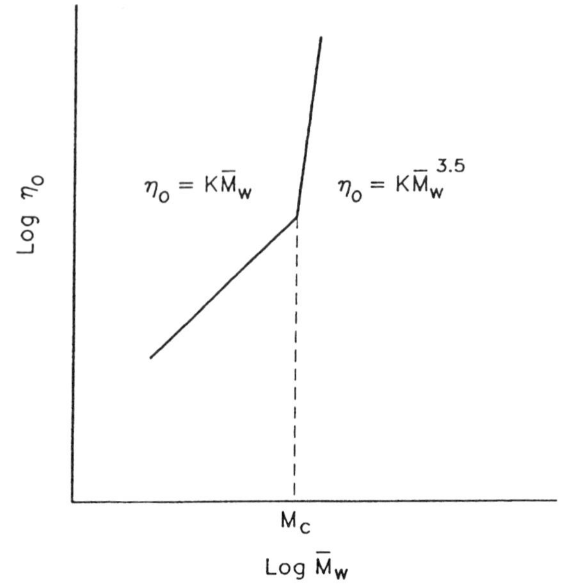

Figure 25. Typical relationship between zero-shear viscosity and molecular weight for a polymer melt. (From Ref. 74.)

Table 3. Values of $\bar{M}_c$ for Some Polymers

Polymer	$\bar{M}_c$
Polyethylene	4,000
Polybutadiene	5,600
Polybutylene	17,000
Polyvinylacetate	22,500
Polymethylsulphoxide	29,000
Polystyrene	35,000

Source: Ref. 74.

non-Newtonian viscoelastic behavior. The steady-shear viscosity versus shear rate plot shows the lower Newtonian plateau followed by a shear-thinning region. In the limit $\omega \rightarrow 0$ and $\dot{\gamma} \rightarrow 0$, the dynamic viscosity and steady-shear viscosity agree with each other. The storage modulus (G') and the first normal stress difference ($\tau_{xx} - \tau_{yy}$) are quite high, and they increase with the increase in frequency or shear rate (as expected). These characteristics are typical of many polymer melts. It should be further noted that the rheological properties of polymer melts are generally similar to those of concentrated polymer solutions discussed earlier. Also, the empirical rules relating linear viscoelastic properties and steady-shear properties (discussed earlier) are applicable to polymeric melts as well.

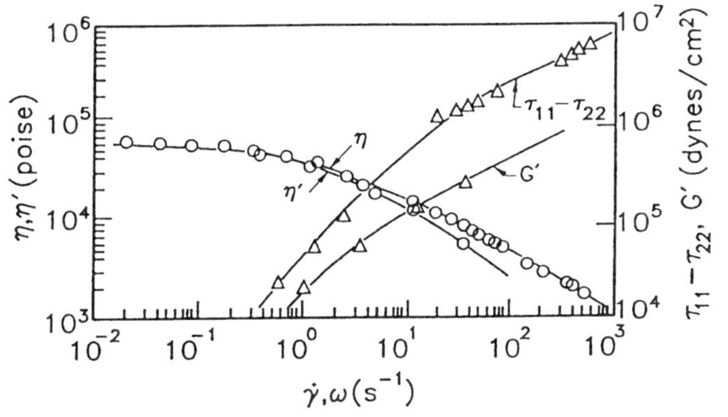

Figure 26. Rheological properties of molten polypropylene ($M_w = 4.4 \times 10^5$) at 200°C. (From Ref. 62.)

Figure 27 compares the plots of viscosity versus shear rate for a highly branched low-density polyethylene melt and a linear high-density polyethylene melt [62]. The molecular weights and molecular weight distributions of these polymers are similar: For low-density polyethylene, $\bar{M}_w = 4 \times 10^5$ and $\bar{M}_w/\bar{M}_n = 20$; for high-density polyethylene, $\bar{M}_w = 2.2 \times 10^5$ and $\bar{M}_w/\bar{M}_n = 16$. The branched polymer melt has much lower viscosity than the linear polymer melt. Thus, the presence of branching results in lower viscosities.

Figure 28 compares the plots of first normal stress difference ($\tau_{xx} - \tau_{yy}$) versus shear stress τ_{xy} for a highly branched low-density polyethylene melt and a linear high-density polyethylene melt [62]. These polymers are the same as those of Fig. 27. Again, one observes that the presence of branching has a significant effect on the elastic properties of polymeric melts; according to the plots shown in Fig. 28, the elasticity of a polymeric melt increases due to the presence of branching.

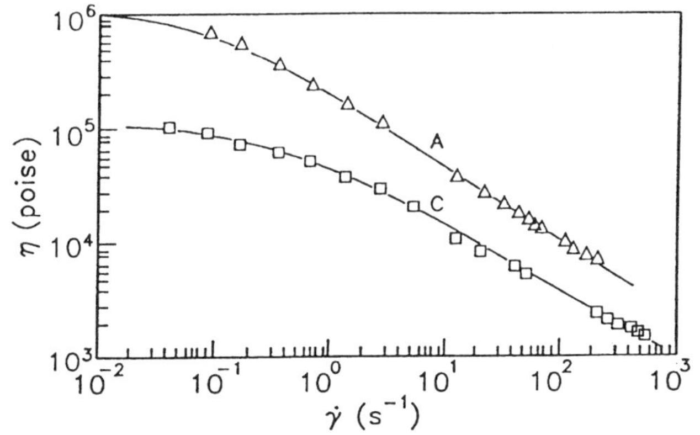

Figure 27. Comparison of flow curves for a highly branched low-density polyethylene melt (denoted by C) and a linear high-density polyethylene melt (denoted by A). (From Ref. 62.)

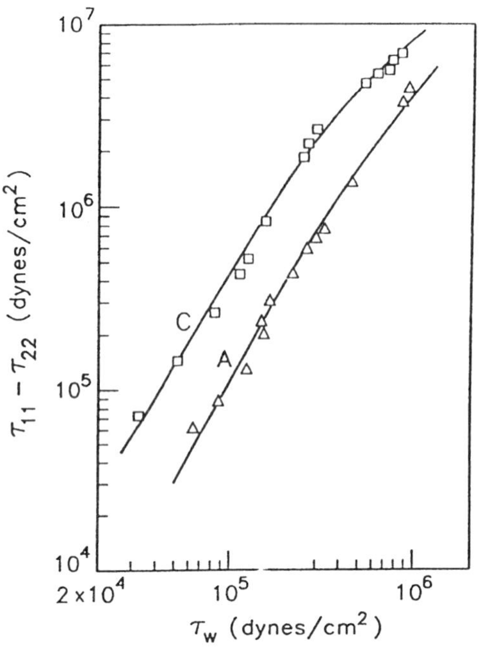

Figure 28. Comparison of first normal stress difference versus shear stress plots for a highly branched low-density polyethylene melt (denoted by C) and a linear high-density polyethylene melt (denoted by A). (From Ref. 62.)

IV. Emulsions Containing Polymeric Liquids: Basic Concepts

A. Adsorption of Polymers

The adsorption behavior of polymers onto the droplet surface is quite complex. Several factors influence the adsorption process, such as:

1. The affinities of the polymer molecule for oil and water phases
2. Polymer/surfactant interaction (if a low-molecular-weight surfactant is present in addition to the high-molecular-weight polymer)
3. The molecular weight of the polymer
4. Temperature

In the case of a *surfactant-free* system, the adsorption behavior of polymer molecules onto the droplet surface is governed largely by the affinities of the polymer molecule for oil and water phases. If the polymer molecule has a strong affinity for the continuous phase relative to its affinity for the dispersed (droplet) phase, it may not adsorb at all and may remain dissolved in the continuous phase. For the polymer molecule to adsorb firmly at the oil/water interface and to provide good steric stabilization to emulsion droplets, the polymer molecule must be amphipathic in nature; that is, it should possess both hydrophilic and hydrophobic parts. An amphipathic polymer molecule has the ability to be strongly anchored onto the droplet surface as well as to provide a highly extended steric layer protruding into the continuous phase to protect the droplets from flocculation and coalescence. A large number of block and graft (comb) copolymers possessing amphipathic properties have been synthesized for this purpose; the likely configuration of such amphipathic polymers at the droplet surface was shown in Fig. 2.

The *homopolymers* generally do not adsorb as strongly at the oil/water interface as do the block and graft copolymers possessing amphipathic properties. Figure 29 presents some likely configurations of a homopolymer at the droplet surface. Figure 29a shows a loop-and-train type of configuration. In this configuration, some segments (loops and tails) of the polymer chain protrude into the continuous phase liquid, and some segments (trains) are adsorbed onto the droplet surface. Figure 29b shows another type of configuration, one in which the polymer molecule is adsorbed almost flat onto the droplet surface, with a large number of segments being at the surface of the droplet. In the configuration in Fig. 29c, the polymer molecule adsorbs onto the droplet surface in nearly the same configuration as it exists in the continuous medium, that is, in random coil form, with minimum points of attachment. Figure 29d shows a type of

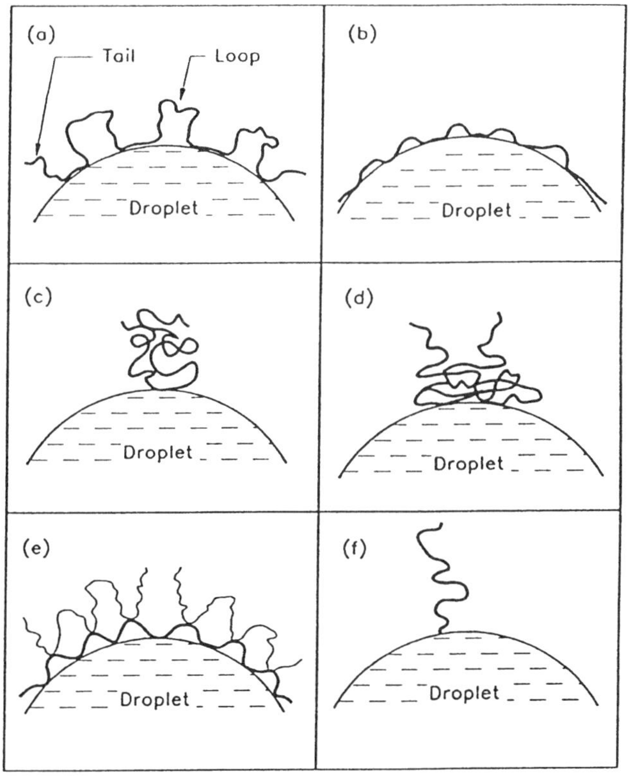

Figure 29. Some likely configurations of a homopolymer at a droplet surface.

adsorption in which the density of the polymer chain segments have a finite distribution, being maximum at the droplet surface and decreasing with distance away from the droplet surface. Figure 29e shows the case in which the polymer molecules are adsorbed in two steps: In the first step, the polymer molecules are adsorbed almost flat onto the droplet surface; in the second step, another layer of polymer molecules is adsorbed onto the first layer, with few points of attachment. Figure 29f shows yet another possibility; in this case the polymer molecule is attached onto the droplet surface at only one point, with the whole chain extending into the continuous phase.

For an emulsion system containing both low-molecular-weight surfactant and high-molecular-weight polymer, the adsorption of polymer onto the droplet surface is governed by polymer/surfactant, polymer/droplet, and surfactant/droplet interactions. The surfactant and the polymer not only compete for adsorption sites at the interfaces, but they may also interact in the bulk phase to form a range of polymer–surfactant complexes [94–107]. These complexes may themselves be surface-active. The exact nature of polymer/surfactant interactions—and, therefore, the adsorption behavior of polymers in the presence of surfactants—is not well understood at present.

1. *Effect of Molecular Weight on Adsorption*

The molecular weight of the polymer is an important parameter affecting both the adsorption kinetics as well as the amount of adsorption. It is expected that the higher the molecular weight of the polymer, the slower would be the rate of adsorption, because of the slower rate of diffusion of large molecules. The saturated amount of adsorption m_s (m_s is the mass of polymer adsorbed per unit mass of adsorbent at saturation) generally varies with the molecular weight (M) of the polymer according to [108] the relation

$$m_s = KM^a \tag{97}$$

where K and a are constants for a given system. According to Perkel and Ullman [109], the constant a is directly linked to the conformation of the polymer molecule at the interface. They suggest the following classification:

1. When $a \approx 1$, the adsorption is directly proportional to the molecular weight. It is thought that in this case, the polymer molecules attach to the surface at a single point (Fig. 29f), with the polymer chain extending into the continuous phase. According to Sato and Ruch [108], the loop-and-tail configuration (Fig. 29a) also may give $a \approx 1$.
2. When $a \approx 0.5$, the adsorption of a polymer is proportional to the square root of the molecular weight of the polymer. It is believed that in this case, the polymer molecules adsorb according to the configuration shown in Fig. 29d; that is, the polymer chain is entangled and intertwined at the interface.
3. When a is in the approximate range of $0 < a < 0.1$, the polymer molecule is thought to adsorb according to the configuration shown in Fig. 29c; that is, the polymer molecule adsorbs onto the droplet surface in

nearly the same configuration as it exists in the solution (that is, random coil), with minimum points of attachment.

4. When $a \approx 0$, the adsorption of polymer is independent of its molecular weight. It is believed that in this case, the polymer molecule is adsorbed flat onto the droplet surface, with multiple points of attachment (see Fig. 29b).

It should be noted that Eq. (97) is generally restricted to adsorption of homopolymers. Also, this equation needs to be modified for situations in which polymer molecules are adsorbed according to the configuration shown in fig. 29e [108]. In such situations, the polymer molecules adsorb in two steps: In the first step, the polymer adsorbs flat onto the droplet surface; in the second step, the polymer molecules adsorb on the first layer, with few points of attachment (thus forming loops and tails). According to Sato and Ruch [108], the following modified equation is applicable under these conditions:

$$m_s = K_1 + K_2 M^a \tag{98}$$

where K_1 takes into account the adsorption of polymer in the first layer, and the term $K_2 M^a$ arises from the adsorption of the second layer.

2. *Effect of Temperature*

The extent of polymer adsorption onto the droplet surface is a strong function of the quality of the solvent (continuous phase) in which the polymer is dissolved. The quality of the solvent depends strongly on the temperature. Quite often, the quality of the solvent improves with an increase in temperature. Thus, the extent of polymer adsorption is directly linked to temperature.

In a good solvent, the polymer molecules are more solvated and, consequently, have a larger radius of gyration as compared with a poor solvent, where the polymer molecules are more tightly coiled. If one assumes that the polymer molecule adsorbs onto the droplet surface in almost the same configuration as the molecule exists in the continuous medium (solvent), the saturated amount of adsorption per unit mass of adsorbent (m_s) is given as follows:

$$m_s = \frac{A_s M}{\pi (R_G^2) N_A} \tag{99}$$

where A_s is the droplet surface area available for adsorption per unit mass of adsorbent (dispersed droplets), M is the molecular weight of the polymer, N_A is Avogadro's number, and R_G is the rms radius of gyration of the polymer molecule. From Eq. (99), the amount of adsorption varies inversely proportional to the square of the rms radius of gyration. From Eq. (35), the radius of gyration is given as

$$R_G = \left(\frac{3M}{10\pi N_A \varepsilon^3} \right)^{1/3} [\eta]^{1/3} \tag{100}$$

Therefore,

$$m_s = \frac{(\text{constant})A_s M^{1/3}}{[\eta]^{2/3}} \tag{101}$$

Equation (101) indicates that for a given polymer, as the quality of the solvent decreases ($[\eta]$ decreases), the amount of adsorption of polymer molecules increases. This is an expected behavior because, in a poor solvent, the polymer molecule adsorbed onto the droplet surface occupies a smaller surface area and, therefore, more polymer molecules can adsorb onto the available droplet surface area. In summary, one would generally observe a decrease in the amount of polymer adsorption with an increase in temperature because the quality of the solvent improves with an increase in temperature.

3. Adsorption Isotherms

Little experimental work has been reported on adsorption isotherms of polymers for liquid/liquid interface, although a substantial amount of literature is available on adsorption of polymers at the solid/liquid interface [108]. At a solid/liquid interface, the adsorption isotherm for polymers has been frequently described in terms of the classical Langmuir equation [110]:

$$\theta = \frac{KC}{1 + KC} \tag{102}$$

where θ is the fraction of the surface covered with the polymer molecules, C is the equilibrium concentration of the polymer in solution, and K is a constant. The amount of polymer adsorbed m (m is the mass of polymer adsorbed per unit mass of adsorbent) is proportional to θ for a given adsorbent, so that $m = a\theta$ when a is a constant. Therefore,

$$m = \frac{aKC}{1 + KC} \tag{103}$$

which can be rewritten as

$$\frac{1}{m} = \frac{1}{a} + \frac{1}{aKC} \tag{104}$$

By plotting 1/m versus 1/C, the constants K and a can be determined from the slope and intercept of the line, and, hence, one can calculate θ (the fraction of surface covered with the polymer molecules).

4. Adsorption Kinetics

The adsorption kinetics of macromolecules from solution depends on several factors, such as the nature of the adsorbent (the droplet surface in the present case), the nature of the polymer, and the rate of stirring of the system. The rate of adsorption of the polymer from dilute solutions can be written as

$$\left(\frac{d\theta}{dt}\right)_a = K_a C(1 - \theta) \tag{105}$$

where K_a is the adsorption rate constant. According to this equation, the rate of adsorption of polymer is the first order in the concentration of polymer in the solution and is also proportional to $(1 - \theta)$, the fraction of surface sites available for adsorption.

The rate of desorption of the polymer from the surface can be written as

$$\left(\frac{d\theta}{dt}\right)_d = K_d\theta \tag{106}$$

where K_d is the desorption rate constant. Combining Eqs. (105) and (106), the net rate of adsorption of polymer from the solution can be written as

$$\frac{d\theta}{dt} = K_aC(1 - \theta) - K_d\theta \tag{107}$$

Upon integration, one obtains

$$-\ln\left[1 - \left(1 + \frac{K_d}{K_aC}\right)\theta\right] = (K_aC + K_d)t \tag{108}$$

It should be noted that at equilibrium, $(d\theta/dt)_a = (d\theta/dt)_d$ and that under such conditions, combining Eqs. (105) and (106) would lead to the Langmuir equation with $K = K_a/K_d$. However, as already noted, little experimental work has been done on adsorption kinetics of polymers at the liquid/liquid interfaces, although a substantial amount of literature is available on polymer adsorption at solid/liquid interfaces.

5. *Thickness of Adsorbed Polymer Layer*

The adsorbed polymer layer thickness (δ) is one of the most important interfacial parameters, for it determines the range of steric barrier on the droplet surface. Several methods have been developed for the measurement of the adsorbed polymer layer thickness on dispersed particles [108,111]. These include ellipsometry, viscometry, measurement of sedimentation rate, and measurement of the diffusion coefficient.

The ellipsometric method of measuring δ involves two steps. In the first step, a change in the state of polarization of light upon reflection from a bare surface (without polymer coating) is measured. In the second step, the same measurement is made for polymer-covered surface. From these two measurements, the thickness (δ) of the polymer film is determined [112].

The viscometric method for measuring δ is based on measuring the relative viscosity (η_r) of a dilute emulsion. Assuming that the relative viscosity of a dilute emulsion is governed by the Einstein equation, we can calculate the effective volume fraction of the emulsion as follows:

$$\phi_{\text{effective}} = \frac{\eta_r - 1}{2.5} \tag{109}$$

The effective volume fraction $\phi_{\text{effective}}$ is related to the true volume fraction ϕ as follows:

$$\phi_{\text{effective}} = \phi\left(1 + \frac{\delta}{R}\right)^3 \tag{110}$$

where R is the radius of the emulsion droplets. Thus, from the knowledge of R, η_r, and ϕ, we can calculate δ.

The sedimentation rate method involves measuring the sedimentation of base particles (without polymer coating) and polymer-coated particles. The sedimentation rates are measured via the ultracentrifuge. The sedimentation coefficient for the particle is defined as follows [113,114]:

$$S = \frac{1}{\omega^2 r}\left(\frac{dr}{dt}\right) \tag{111}$$

where ω is the angular velocity of ultracentrifuge rotor, and r is the distance of the particle from the axis of rotation. Upon integration of Eq. (111), one obtains

$$S = \frac{\ln(r_2/r_1)}{\omega^2(t_2 - t_1)} \tag{112}$$

Thus, S can be calculated using measurements of the positions r_1 and r_2 of the particle at times t_1 and t_2.

From Stokes law, the sedimentation coefficient for base particles is given as

$$S^0 = \frac{2R^2(\rho_1 - \rho_c)}{9\eta_c} \tag{113}$$

where R is the particle radius, η_c is the continuous phase viscosity, ρ_c is the density of continuous phase, and ρ_1 is the density of the particle. For a polymer-coated particle, the sedimentation coefficient is given [114,115] by

$$S^\delta = \frac{2R^3(\rho_1 - \rho_c) + 2(3R^2\delta + 3R\delta^2 + \delta^3)(\rho_2 - \rho_c)}{9\eta_c(R + \delta)} \tag{114}$$

where ρ_2 is the density of the adsorbed polymer layer and δ is the adsorbed layer thickness. From Eqs. (113) and (114),

$$\frac{S^0}{S^\delta} = \frac{R^2(R + \delta)(\rho_1 - \rho_c)}{[R^3(\rho_1 - \rho_c) + (3R^2\delta + 3R\delta^2 + \delta^3)(\rho_2 - \rho_c)]} \tag{115}$$

The density of the adsorbed polymer layer can be obtained from the absorption isotherm [115]. However, if one assumes that the density of the adsorbed layer is close to that of the continuous phase, Eq. (115) reduces to

$$\frac{S^0}{S^\delta} = \frac{R + \delta}{R} \tag{116}$$

Thus, by measuring S_0 and S^δ, one can calculate δ.

The diffusion coefficient of the particles is determined by measuring fluctuation in the intensity of light scattered by the particles as they undergo Brownian motion [114,116]. From the knowledge of the diffusion coefficient (D), the thickness of the adsorbed layer (δ) is calculated from the Stokes–Einstein relation:

$$D = \frac{kT}{6\pi\eta_c(R + \delta)} \tag{117}$$

$$\delta = \frac{kT}{6\pi\eta_c D} - R \tag{118}$$

where k is the Boltzmann constant and T is the absolute temperature.

B. Steric Stabilization

The theory of emulsion stability has been reviewed by Tadros and Vincent in the first volume of this encyclopedia [116a] and updated by Walstra [116b] in the current volume. However, in this and the following sections (IV.B–IV.F), a brief summary of the theory is given, emphasizing the role of polymer.

Consider the close approach of two droplets covered with adsorbed polymer layers (see Fig. 30). When the distance between the droplets b is greater than twice the thickness of the adsorbed polymer layer (i.e., $b > 2\delta$), only the van der Waals interaction is present between the droplets (assuming that the droplets have no electric charge and that electrical double layer repulsion is absent). However, when the distance of separation between the droplets becomes less than 2δ, a strong steric interaction takes place between the droplets due to the interference or overlapping of the two adsorbed polymer layers.

There are two types of steric interactions that take place between the droplets, namely, mixing (osmotic) interaction and elastic (entropic) interaction. The entropic interaction is also referred to as the volume restriction effect.

When $\delta \leq b \leq 2\delta$, the major steric interaction is the mixing (osmotic) type due to interpenetration of adsorbed polymer chains, without significant compression of chains. The interpenetration of adsorbed polymer chains, which results in *demixing* of polymer segments from molecules of the continuous medium, leads to an increase in the polymer segment density in the interaction zone. This generates a difference in the chemical potential between the molecules of the continuous medium in the interaction zone and those in the rest of the adsorbed layer. If the continuous medium is a good solvent (better than theta-solvent) for the adsorbed polymer chains, the interpenetration of polymer chains is thermodynamically unfavored, resulting in a repulsive force between the droplets. The origin of this repulsive force could be explained in terms of chemical potentials. Because the chemical potential of the solvent (continuous medium) molecules in the interaction zone is smaller than that in the rest of the adsorbed layer, the solvent molecules from the bulk region tend to diffuse into the interaction region, tending to force the droplets apart. However, if the continuous medium is a poor solvent (worse than theta–solvent) for the polymer chains, the interpenetration of adsorbed polymer chains is thermodynamically favored, resulting in an *attractive* force between the droplets.

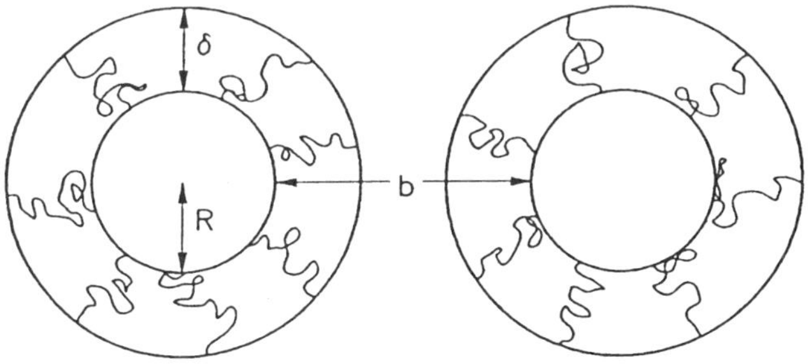

Figure 30. Schematic representation of steric interaction between two emulsion droplets covered with adsorbed polymer layers.

Once the distance of separation between the droplets becomes less than δ (i.e., $b < \delta$), the spatial extension of the polymer segments normal to the droplet surface is reduced due to compression imposed by the presence of the opposing droplet. Consequently, there occurs a loss in configurational entropy of the polymer segments. This type of steric interaction whereby the adsorbed polymer chains undergo compression, and hence loss in configuration entropy, is called elastic (entropic) interaction. Note that this entropic interaction is always repulsive, irrespective of the solvency of the continuous medium. Furthermore, when $b < \delta$, mixing interaction is also operative along with the entropic interaction.

Thus, the total free energy of interaction (G_i) between two polymer-covered droplets is given as

$$G_i = G_A + G_M + G_{EL} \tag{119}$$

where G_A is the van der Waals interaction, G_M is the mixing interaction, and G_{EL} is the elastic interaction.

1. van der Waals Attraction

It is well known that in a single-component system, individual molecules or atoms always attract each other at short distances owing to van der Waals forces. These attractive forces could be of three kinds: permanent dipole–permanent dipole interaction (Keesom force), permanent dipole–induced dipole interaction (Debye force), and interaction based on fluctuations in the electron-density distribution (London force).

The free energy corresponding to van der Waals attraction between two atoms (or molecules) varies inversely as the sixth power of the distance b between the two atoms:

$$G_A = -\frac{\beta_{11}}{b^6} \tag{120}$$

where β_{11} is a constant referring to identical atoms of material 1. The negative sign in Eq. (120) indicates that the free energy always decreases upon close approach of atoms when there exists an attraction between the atoms (molecules or particles).

From Eq. (120), it is clear that the van der Waals attraction between two atoms or molecules is of relatively short range. However, particles or droplets are composed of a large number of molecules, and each molecule or atom of one particle attracts every atom in the other particle. The net effect of adding all the possible atomic interactions among the particles is a long-range attraction (between the particles) of considerable strength. It should be noted that one has to consider only the London interactions when dealing with particles. This is because large assemblies have neither a net dipole moment nor a net polarization [117].

For two identical *parallel plates*, each of thickness δ, the van der Waals attraction at a separation distance b in vacuum is given by

$$G_A = -\frac{A_{11}}{12\pi}\left[\frac{1}{b^2} + \frac{1}{(b+2\delta)^2} - \frac{2}{(b+\delta)^2}\right] \tag{121}$$

where A_{11}, known as the Hamaker constant, is defined by

$$A_{11} = \pi^2 \beta_{11} n_1^2 \tag{122}$$

where β_{11} is the London constant (which depends on the properties of the specific atoms) and n_1 is the number of atoms (molecules) per unit volume.

For short distances ($b \ll \delta$), Eq. (121) reduces to

$$G_A = -\frac{A_{11}}{12\pi b^2} \tag{123}$$

The van der Waals interaction between two identical *spherical particles* in vacuum is given by

$$G_A = -\frac{A_{11}}{6}\left[\frac{2}{s^2-4} + \frac{2}{s^2} + \ln\left(\frac{s^2-4}{s^2}\right)\right] \tag{124}$$

where $s = (b + 2R)/R$ and R is the particle radius. For small distances ($b \ll R$), Eq. (124) reduces to

$$G_A = -\frac{RA_{11}}{12b} \tag{125}$$

Upon comparing Eqs. (123) and (125) with Eq. (120), it is clear that the range of van der Waals attractive force between macroscopic particles (droplets) is orders of magnitude greater than that for individual atoms or molecules.

The preceding treatment is restricted to particles or spheres in a vacuum. When a liquid medium is present between the particles, the Hamaker constant is modified, although the same equations are valid. The Hamaker constant for particles of material 1 dispersed in a liquid medium 2, $A_{11(2)}$, is defined as

$$A_{11(2)} = A_{11} + A_{22} - 2A_{12} \tag{126}$$

where A_{11} is the Hamaker constant referring to the interaction of material 1 with material 1, A_{22} is the Hamaker constant referring to the interaction of material 2 with material 2, and A_{12} is the Hamaker constant referring to the interaction of material 1 with material 2. As an approximation, A_{12} could be written as

$$A_{12}^2 = A_{11}A_{22} \tag{127}$$

$$\rightarrow \quad A_{11(2)} = [A_{11}^{1/2} - A_{22}^{1/2}]^2 \tag{128}$$

$A_{11(2)}$ is always a positive quantity, so two particles of the same material, in any medium, always attract each other. Table 4 gives some data on Hamaker constants in aqueous systems [117,118]. Using Eq. (125) with a Hamaker constant $A_{11(2)}$ of 10^{-19} joules, the van der Waals attraction was calculated [108,119]. The plots of G_A for different-sized particles as a function of interparticle distance are shown in Fig. 31. As can be seen, the larger the particles, the greater is the van der Waals attraction.

In the preceding discussion, no allowance was made for the presence of any adsorbed polymer layers on the particles. Thus, the equations discussed there are applicable only to the cases in which there is no adsorbed polymer layer on

Table 4. Hamaker Constants $A_{11(2)}$ for Some Aqueous Dispersions

Material	10^{20} $A_{11(2)}$ (joules)
Metals	10–20
SiO_2	0.3–0.9
Low-mol.-wt. organic compounds	0.08–1.8
Polymers	0.2–2.5
AgI	2–4

Source: Refs. 117 and 118.

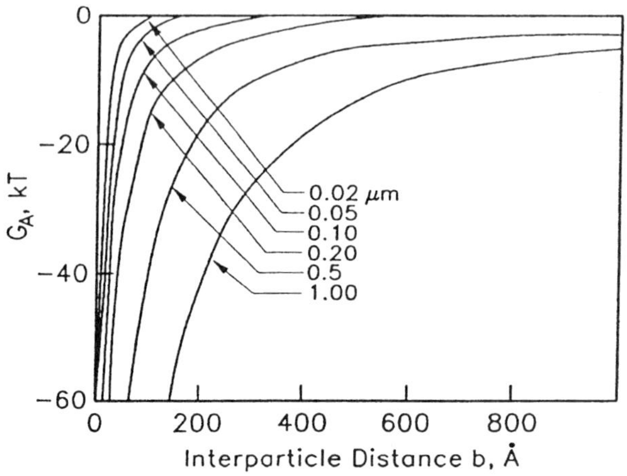

Figure 31. Van der Waals' attraction potential between particles of different sizes. (From Refs. 108 and 119.)

the particles. The presence of an adsorbed polymer layer on the particles or droplets can have a significant influence on the van der Waals attraction between the particles.

Vold [120] derived the following expression for van der Waals attraction between particles covered with an adsorbed polymer layer of thickness δ:

$$G_A = -\frac{1}{12}\left[(A_{22}^{1/2} - A_{33}^{1/2})^2\left(\frac{R+\delta}{b}\right) + (A_{33}^{1/2} - A_{11}^{1/2})^2\left(\frac{R}{b+2\delta}\right) + \frac{4R(A_{22}^{1/2} - A_{33}^{1/2})(A_{33}^{1/2} - A_{11}^{1/2})(R+\delta)}{(b+\delta)(2R+\delta)}\right] \tag{129}$$

where A_{33} is the Hamaker constant referring to the interaction of material 3 (adsorbed polymer) with material 3. When A_{33} and δ are zero (that is, the adsorbed layer is absent), Eq. (129) reduces to

$$G_A = -\frac{R}{12b}[A_{11} + A_{22} - 2(A_{11}A_{22})^{1/2}] \tag{130}$$

which is the same as Eq. (125) with the Hamaker constant $A_{11(2)}$ (see Eq. 128).

As an example, Fig. 32 shows the effect of adsorbed layer thickness (δ) on G_A for two spherical particles [35,121]. The Hamaker constants for the data shown in the figure are: $A_{11} = 2 \times 10^{-19}$ J, $A_{22} = 6 \times 10^{-20}$ J, and $A_{33} = 5.6 \times 10^{-20}$ J. The figure clearly indicates that the van der Waals attraction between the particles decreases as the thickness of the adsorbed layer increases. However, note that the presence of the adsorbed layer can also increase the van der Waals attraction, depending on the magnitude of A_{33} [35].

Becher [121a] has calculated the Vold effect for parallel plates.

2. *Osmotic Effect*

Fischer [122] was the first to consider the mixing (osmotic) interaction between the adsorbed polymer layers. As discussed earlier, the interpretation of adsorbed polymer chains leads to an increase in the polymer segment density in the interaction zone. This generates a difference in the chemical potential between the solvent (continuous medium) molecules in the interaction zone and the solvent molecules in the rest of the adsorbed layer (see Fig. 33).

If the chemical potentials of the solvent molecules in the overlap region dV is μ_1 and in the rest of adsorbed layer is μ_1^*, the excess chemical potential is given by

$$\Delta\mu_1 = \mu_1^* - \mu_1 = \pi_E \bar{V}_1 \tag{131}$$

where π_E is excess osmotic pressure and $\bar{V}_1$ is molar volume of pure solvent. From Eq. (131),

$$\pi_E = -\left(\frac{\mu_1 - \mu_1^*}{\bar{V}_1}\right) \tag{132}$$

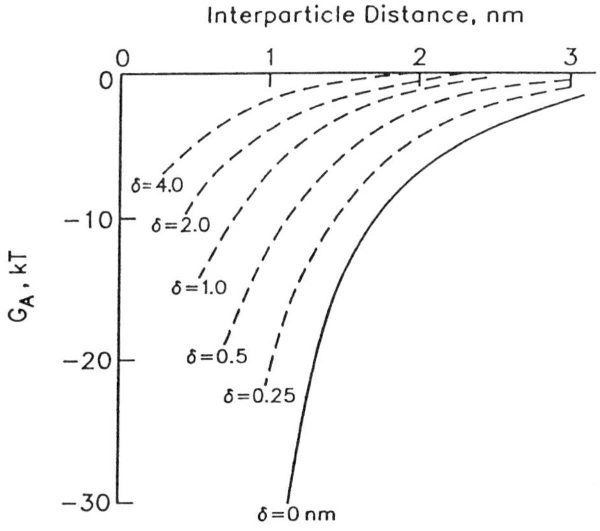

Figure 32. Variation of van der Waals' attraction potential with interparticle distance at various values of δ. (From. Refs. 35 and 121.)

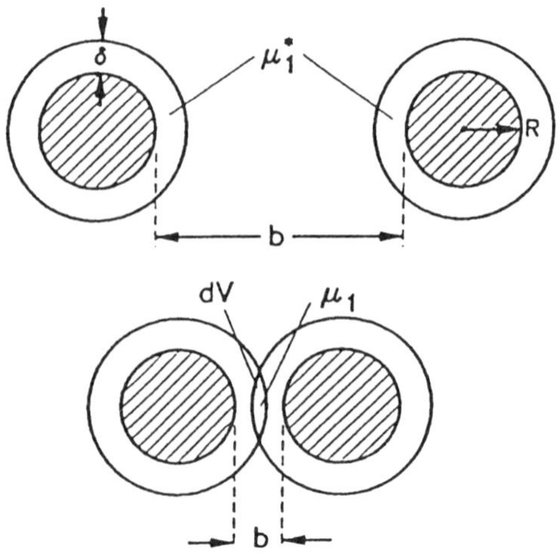

Figure 33. Overlapping of adsorbed polymer layers. (From Ref. 35.)

The excess osmotic pressure π_E can be related to the second virial coefficient B_2 of the polymer solution [108] as follows:

$$\pi_E = R_0 T B_2 C^2 \tag{133}$$

where C is the concentration of the polymer in the adsorbed layer and R_0 is the universal gas constant. Thus, the free energy of interpenetration of adsorbed polymer layers is equal to

$$\begin{aligned} G_M &= 2\int_V \pi_E \, dV = 2\int_V R_0 T B_2 C^2 \, dV \\ &= 2 R_0 T B_2 C^2 (\Delta V) \end{aligned} \tag{134}$$

where ΔV is the overlap volume (lens-shaped), which is given as

$$\Delta V = \frac{2}{3}\pi \left(\delta - \frac{b}{2}\right)^2 \left(3R + 2\delta + \frac{b}{2}\right) \tag{135}$$

where δ = the steric layer thickness, R is the particle radius, and b is the minimum distance between the surfaces of the two spheres. Note that $\Delta V = 0$ when $b = 2\delta$ (no overlap). From Eqs. (134) and (135) then:

$$G_M = \frac{4}{3}\pi R_0 T B_2 C^2 \left(\delta - \frac{b}{2}\right)^2 \left(3R + 2\delta + \frac{b}{2}\right) \tag{136}$$

Since $R_0 = kN_A$ (k is the Boltzmann constant and N_A is Avogadro's number), Eq. (136) could be rewritten as

$$\frac{G_M}{kT} = \frac{4}{3}\pi N_A B_2 C^2 \left(\delta - \frac{b}{2}\right)^2 \left(3R + 2\delta + \frac{b}{2}\right) \tag{137}$$

This equation indicates that the mixing interaction is zero when $b = 2\delta$ (no interpenetration of adsorbed layers occurs when $b \geq 2\delta$). Also, the mixing interaction could be attractive or repulsive depending on the solvency of the dispersion medium. If the dispersion medium is a good solvent (better than theta-solvent) for the polymer, the second virial coefficient B_2 is positive and, therefore, the mixing interaction is repulsive ($G_M > 0$). However, in a poor solvent (worse than theta-solvent), B_2 is negative, implying that the mixing interaction is attractive ($G_M < 0$).

As an example, Fig. 34 shows the free energy of mixing interaction as a function of the distance of separation between two particles with an adsorbed layer of thickness $\delta = 1.5 \times 10^{-5}$ cm and $C = 0.02\,\text{gm/cm}^3$ for different values of B_2 and particle size. Since B_2 is positive, the interaction is repulsive ($G_M > 0$). The repulsion between the particles increases with the increase in particle size. Also, the repulsion is zero when $b \geq 2\delta$, and it increases with a decrease in the separation distance b between the particles.

3. *Elastic Effect*

Mackor [123] was the first to consider entropic interaction between adsorbed polymer layers. As discussed earlier, the spatial extension of adsorbed polymer chains normal to the droplet surface is reduced on close approach of two droplets because of the compression of the adsorbed layers. This results in a loss of configurational entropy of the polymer chains, and hence repulsion takes place between the droplets. Mackor modeled the entropic interaction between the particles as follows: He treated the particles as flat parallel plates and assumed that the polymer molecules were inflexible rods of length δ attached to one end to the particle surface by a freely hinged joint. It was further assumed that the surface concentration of the polymer molecules is very low so that there is no lateral interaction between neighboring rod-type molecules. Under these conditions, the number of possible configurations (Ω_∞) that a single rod can

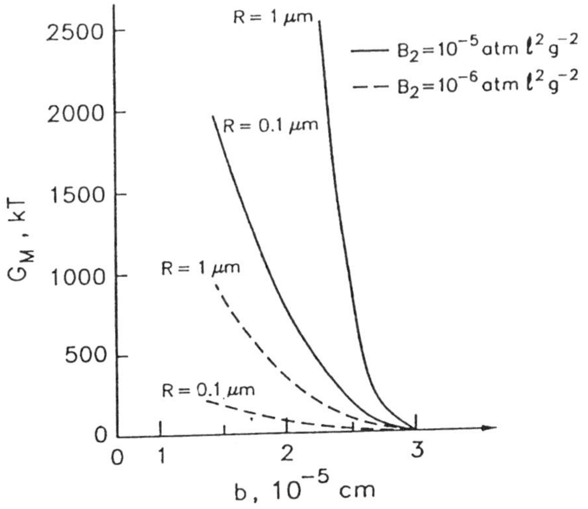

Figure 34. Variation of mixing interaction potential G_M with interparticle distance. (From Refs. 108 and 122.)

have when the particles are far apart ($b > \delta$) is proportional to the surface area swept out by the free end of the rod (see Fig. 35):

$$\Omega_\infty \propto 2\pi\delta^2 \tag{138}$$

However, when the separation distance between the particle surfaces becomes less than $\delta (b < \delta)$, the number of accessible configurations Ω of the rod is reduced and is given by

$$\Omega \propto 2\pi\delta^2 \left(\frac{b}{\delta}\right) \tag{139}$$

Thus there occurs a loss in configurational entropy, which is given by the Boltzmann relation:

$$S = k \ln \Omega \tag{140}$$

where k is the Boltzmann constant. Substituting the number of configurations into Eq. (140), we obtain

$$\Delta S = k \ln\left(\frac{\Omega}{\Omega_\infty}\right) = k \ln\left(\frac{b}{\delta}\right) \tag{141}$$

where ΔS is the loss in configurational entropy of a single polymer molecule.

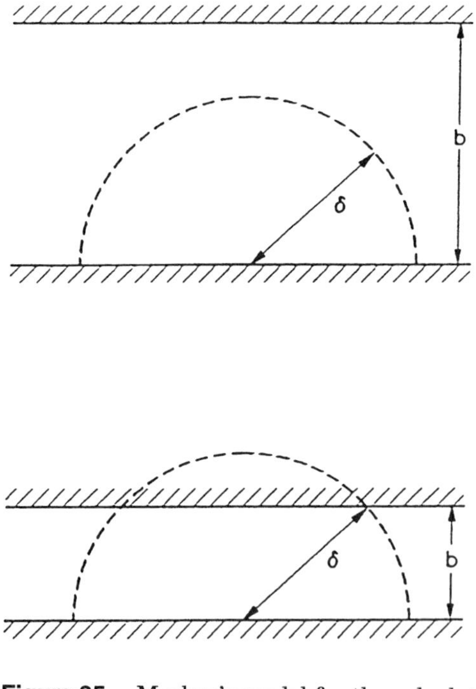

Figure 35. Mackor's model for the calculation of entropic interaction between adsorbed polymer layers.

Thus the free energy of repulsion due to the loss in configurational entropy is given as follows:

$$G_{EL} = -T\,\Delta S = -kT\ln\left(\frac{b}{\delta}\right) = kT\ln\left(\frac{\delta}{b}\right) \tag{142}$$

For values of b not too far removed from δ, Eq. (142) reduces to

$$G_{EL} = kT\left(1 - \frac{b}{\delta}\right) \tag{143}$$

This value of G_{EL} refers to just one polymer molecule. If θ_∞ is the fraction of particle surface covered by polymer molecules when the particles are far apart and N_S is the number of molecules adsorbed per unit surface area, then the total repulsive free energy per unit particle surface area is given by

$$G_{EL} = N_S kT\theta_\infty\left(1 - \frac{b}{\delta}\right) \tag{144}$$

Equation (144) clearly indicates that $G_{EL} = 0$ when $b = \delta$; i.e., entropic interaction becomes important only when $b < \delta$. Furthermore, G_{EL} is always positive (repulsive).

As an example, Fig. 36 shows the plots of G_{EL} as a function of b for the case in which $N_S = 2 \times 10^{14}\ \mathrm{cm}^{-2}$ and $\theta_\infty = 0.20$ [108,123]. The value of G_{EL} increases with a decrease in b and with an increase in the length of the polymer molecule (adsorbed polymer layer thickness).

4. *Total Interaction*

The total free energy of interaction (G_i) between the two polymer-coated droplets is given as

$$G_i = G_A + G_M + G_{EL} \tag{145}$$

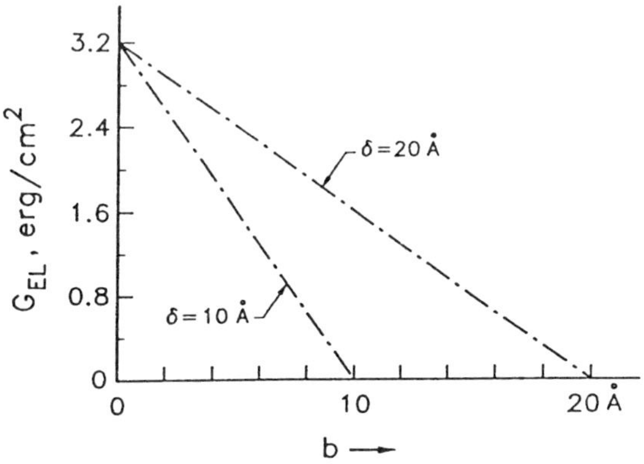

Figure 36. Variation of elastic interaction potential with interparticle distance. (From Refs. 108 and 123.)

provided that the electrical double layer interaction is absent. Here G_A is the van der Waals attraction, G_M is the mixing interaction (interpenetration of adsorbed polymer chains), and G_{EL} is the elastic (entropic) interaction between adsorbed polymer layers.

Substituting the expressions for G_A, G_M, and G_{EL} into Eq. (145) for the total free energy of interaction gives

$$G_i = -\frac{RA}{12b} + \frac{4}{3}\pi R_0 T B_2 C^2 \left(\delta - \frac{b}{2}\right)^2 \left(3R + 2\delta + \frac{b}{2}\right) + N_S k T \theta_\infty \left(1 - \frac{b}{\delta}\right) s \qquad (146)$$

where A is the effective Hamaker constant and s is the surface area of particles in the interaction zone (all other parameters have been defined earlier).

In Eq. (146), one of the key parameters controlling the stability of emulsions is the second virial coefficient B_2. Depending on the solvency of the dispersion medium, B_2 could be positive or negative. Figure 37 shows how a change in solvency (sign of B_2) of the dispersion medium can influence the emulsion

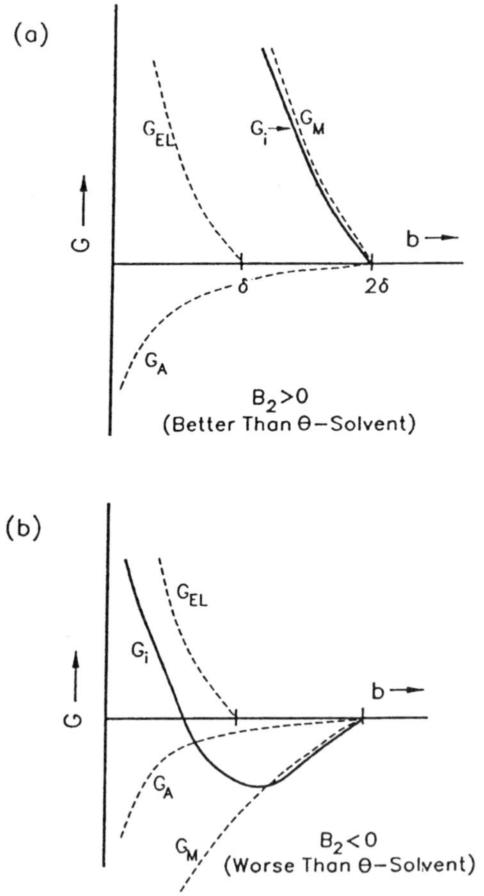

Figure 37. Effect of solvency on free energy of interaction.

stability. In Fig. 37a, the dispersion medium is a good solvent for the polymer and, therefore, the mixing term is repulsive and the emulsion is stable with respect to flocculation. However, upon changing the solvency of the dispersion medium from a good solvent to a poor solvent ($B_2 < 0$), the mixing term becomes attractive and hence the emulsion becomes unstable with respect to flocculation. Note that the change in solvency can be achieved either by adding a nonsolvent to the dispersion medium or by changing the temperature.

C. Bridging Flocculation

Flocculation of droplets by polymer molecules is well known. One of the key mechanisms of flocculation of droplets by polymer molecules is "bridging" of droplets by the polymer molecules. Bridging flocculation can occur in at least two different ways [108]: In one case, the bridging of two or more droplets occurs because segments of the *same* polymer molecule adsorb onto different droplets, thus linking the droplets together (see Fig. 3). In the other case, the bridging of droplets occurs by the interaction of polymer molecules adsorbed onto different droplets, as shown in Fig. 38.

Bridging of droplets by the same polymer molecule (Fig. 3) is favored under the following conditions: (1) The adsorbed polymer molecule extends far enough from the droplet surface so that it can attach to other droplets; (2) there is a sufficient free surface area available on the droplets so that the polymer molecule attached to one droplet can adsorb onto another droplet; and (3) the polymer molecule has two or more adsorbable segments. It should be further noted that there exists an optimum concentration of polymer for this type of bridging flocculation. According to La Mer [41,124], the optimum concentration of polymer corresponds to "half-surface coverage." This condition can be derived as follows: If θ is the fraction of the droplet surface covered by the polymer, then the probability of successful bridging between two droplets is directly proportional to $\theta(1-\theta)$, where $(1-\theta)$ is the fraction of free surface. This probability function, i.e., $\theta(1-\theta)$, has a maximum value at $\theta = 1/2$.

The other type of bridging flocculation shown in Fig. 38 is favored when the following conditions exist: (1) The surface coverage by polymer molecules is very high; (2) the polymer molecules are very long; and (3) the interacting polymer chains have a large affinity for each other.

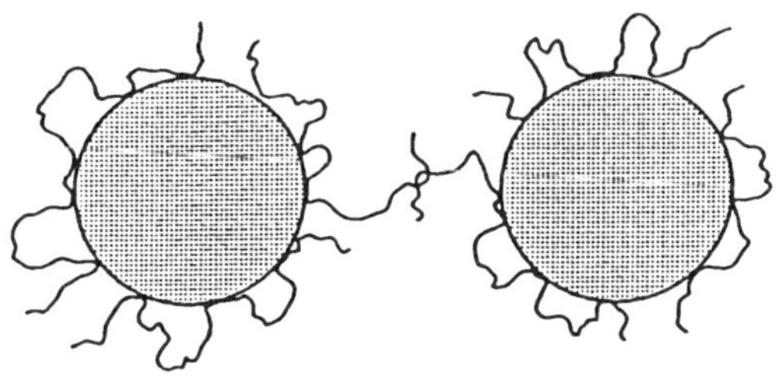

Figure 38. Bridging flocculation of two droplets by two separately adsorbed macromolecules.

The bridging flocculation of droplets can occur in both uncharged and charged droplets. For uncharged droplets, bridging of droplets can be induced by either ionic or nonionic polymers. In the case of charged droplets, the polymer should be either nonionic or ionic, with a charge opposite to that of the droplets (see Fig. 39). If the polymer is nonionic in nature, bridging flocculation of droplets is possible when the adsorbed polymer chain is very long so that it extends far enough from the droplet surface to span the distance over which electrical double layer repulsion operates. The range of electrical double layer repulsion is very much dependent on the ionic strength of the dispersion medium. At low ionic strength, the electrical double layer is quite thick and, therefore, high-molecular-weight polymers are required for bridging flocculation. When the

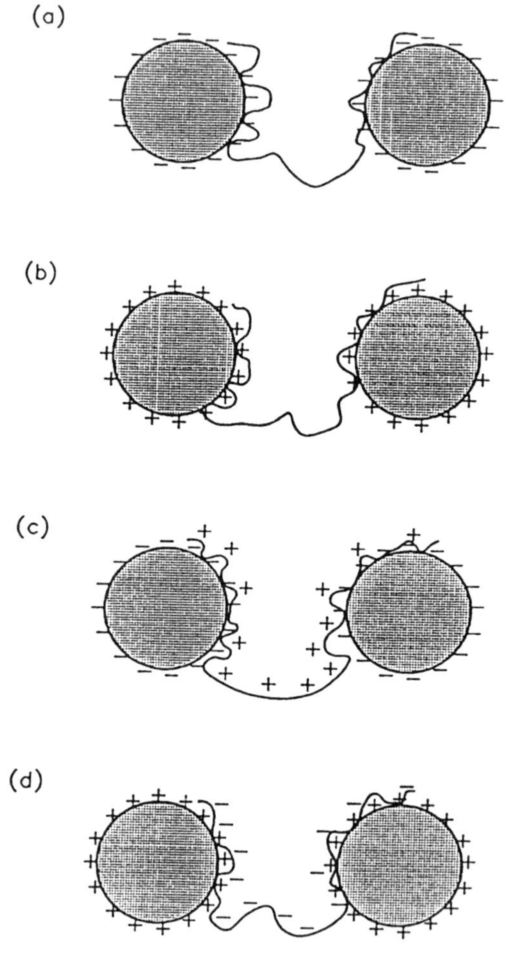

Figure 39. Bridging flocculation. (a) Negatively charged droplets bridged by nonionic polymer; (b) positively charged droplets bridged by nonionic polymer; (c) negatively charged droplets bridged by cationic polyelectrolyte; (d) positively charged droplets bridged by anionic polyelectrolyte.

polymer is ionic in nature, bridging of charged droplets is possible provided that the droplets and the polymer are of opposite charge. The dimensions of the polyelectrolyte chains in the dispersion medium are a function of the ionic strength. At low ionic strength, the repulsion between charged groups along the polymer chain is high because screening of charged groups by couterions is limited and, therefore, the polyelectrolyte assumes a highly extended configuration in the dispersion medium. This favors bridging of droplets.

D. Charge Neutralization

For electrically charged droplets, flocculation of droplets can be induced by the addition of an optimum amount of a polyelectrolyte with a charge opposite to that of the droplets. Under such conditions, there are at least two possible mechanisms that lead to flocculation: the bridging mechanism, as discussed earlier, and "irregular charge distribution" on droplets. This latter effect is also referred to as "electrostatic-patch effect." It is believed that polyelectrolytes having a charge opposite to that of droplets adsorb onto the droplet surface in a flat configuration leading to the formation of patches of uneven charge distribution (refer to Fig. 4). Droplets having uneven charge distribution undergo flocculation owing to attraction between positive and negative patches [41]. However, if the polyelectrolyte is added in large quantities, there is a possibility of restabilization of droplets owing to adsorption of excess polyelectrolyte onto the droplet surface.

E. Depletion Flocculation

The addition of a "free" (nonadsorbing) polymer to an emulsion can lead to weak flocculation of droplets because of the so-called "depletion effect." The depletion effect can be explained as follows: When the surfaces of droplets dispersed in a polymer solution approach to distances closer than the diameter (twice the radius of gyration) of the polymer molecules, the polymer molecules are forced out of the interparticle space (see Fig. 40), resulting in a polymer-free (depleted) zone between the droplets. The exclusion of polymer molecules from the interparticle space occurs mainly because of the loss in configurational entropy of the polymer chains. However, the solution outside the interparticle space retains its bulk concentration of polymer, and, therefore, it exerts an inward force on the droplets because of the osmotic pressure. This results in flocculation of droplets.

It is possible to calculate the free energy due to depletion, G_{dep}, using theoretical analysis. Asakura and Oosawa [125,126] were the first to derive the expression for G_{dep}. Consider the close approach of two droplets (see Fig. 41). When $b \geq 2R_G$, the depletion interaction is negligible (R_G is the radius of gyration of the polymer molecule). When $b < 2R_G$, attraction between droplets occurs because of the depletion effect. The free energy of interaction can be calculated from:

$$G_{dep} = -\Pi \int_V dV \qquad G_{dep} = -\Pi \, \Delta V \tag{147}$$

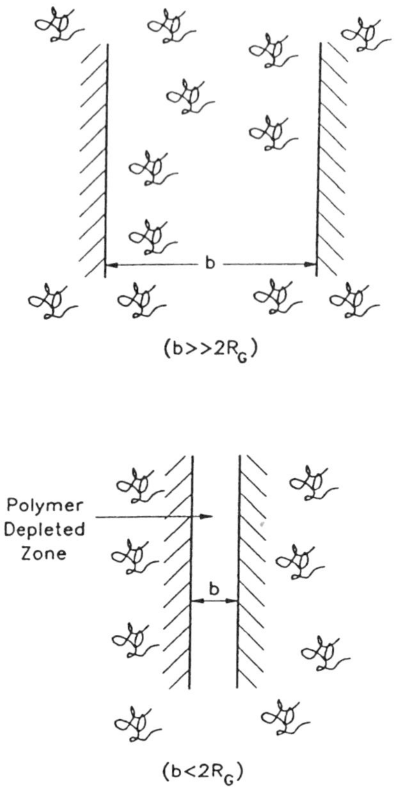

Figure 40. Schematic representation of the close approach of two parallel surfaces in a polymer solution. (From Ref. 34.)

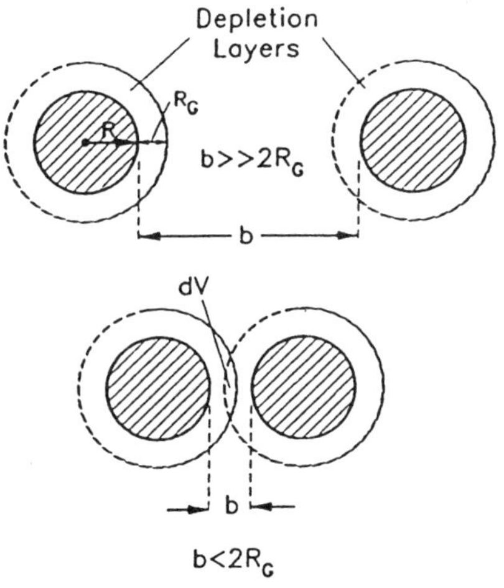

Figure 41. Schematic representation of the close approach of the depletion layers associated with two droplets.

where Π is the osmotic pressure of the medium and ΔV is the overlap volume (lens-shaped) of the depletion layers on the close approach of droplets. The overlap volume (ΔV) is zero when $b \geq 2R_G$, and is given as

$$\Delta V = \frac{\pi}{12}(2R_G - b)^2(6R + 4R_G + b) \tag{148}$$

$$G_{dep} = -\Pi\left(\frac{\pi}{12}\right)(2R_G - b)^2(6R + 4R_G + b) \tag{149}$$

The osmotic pressure Π of the solution is given by van't Hoff's law, provided that the concentration of polymer in the solution C is small and that the molecular weight of the polymer is not too large:

$$\frac{\Pi}{C} = R_0 T B_1 = \frac{R_0 T}{M} \tag{150}$$

where B_1 is the first virial coefficient and M is the molecular weight of the polymer (see Eq. 19).

$$G_{dep} = -\frac{R_0 TC}{M}\left(\frac{\pi}{12}\right)(2R_G - b)^2(6R + 4R_G + b) \tag{151}$$

If $R \gg R_G$, that is, if the size of the polymer molecules is small compared with the size of droplets, Eq. (151) can be approximated as

$$G_{dep} = -\frac{R_0 T}{M} C\left(\frac{\pi}{12}\right)(2R_G - b)^2(6R) \tag{152}$$

We can rewrite Eq. (152) in terms of the volume fraction of polymer ϕ_p in the solution, which is given by

$$\phi_p = \frac{N(4/3)\pi R_G^3}{V} \tag{153}$$

where N is the total number of polymer molecules and V is the total volume of polymer solution. The ratio N/V can be expressed in terms of concentration C (g/cm^3) as

$$\frac{N}{V} = \frac{C}{M} N_A \tag{154}$$

where N_A is Avogadro's number. From Eqs. (153) and (154):

$$\frac{C}{M} N_A = \frac{3\phi_p}{4\pi R_G^3} \tag{155}$$

Substituting Eq. (155) into Eq. (152) gives the following result:

$$G_{dep} = -kT\left(\frac{\pi}{12}\right)\left(\frac{3\phi_p}{4\pi R_G^3}\right)(2R_G - b)^2(6R) \tag{156}$$

Equation (156) could be further rewritten as

$$\frac{G_{dep}}{kT} = -\frac{3}{2}\phi_p \beta x^2 \qquad 0 \leq x \leq 1 \tag{157}$$

where $\beta = R/R_G (\gg 1)$ and $x = (2R_G - b)/2R_G$. When $b = 2R_G$, $x = 0$ and G_{dep} is zero. When $b = 0$, $x = 1$ and the free energy of depletion interaction is equal to $(-3/2)\phi_p \beta kT$. Note that the magnitude of G_{dep} increases with an increase in size of the polymer molecules (R_G). Hence, the molecular weight of polymer and the solvency of the dispersion medium are important factors. The magnitude of G_{dep} also increases with an increase in the free polymer concentration (ϕ_p). Furthermore, a large number of rheological studies [114,127–130] indicate that depletion flocculation becomes significant only when the (free) polymer volume fraction exceeds a certain critical value ϕ_p^+. This critical ϕ_p^+ is a function of the molecular weight of the polymer. The value ϕ_p^+ generally decreases with an increase in the molecular weight of the free polymer.

F. Depletion Stabilization

The addition of a free (nonadsorbing) polymer to an emulsion can lead to either flocculation (depletion flocculation) or stabilization (depletion stabilization), depending on the concentration of the free polymer. At low concentrations of polymer, depletion flocculation is dominant, whereas at high concentrations of polymer, depletion stabilization is the dominant phenomenon. The mechanism of depletion stabilization can be explained as follows: Let us consider the approach of two surfaces (say, parallel plates) immersed in a polymer solution from infinite separation to a distance closer than the diameter of the polymer coils (see Fig. 40). At separation distances greater than twice the radius of gyration of polymer coil, there is negligible interaction between the surfaces. However, when the separation distance b is less than $2R_G$, repulsion occurs between the surfaces, because the closer approach of surfaces is thermodynamically unfavored. When $b < 2R_G$, the polymer coils are either distorted or "squeezed out" from the region between the surfaces. In either case, the total free energy of the system increases. If the polymer coils are distorted and their motion is restricted, there is a loss in configurational entropy of polymer chains; hence, an increase in the total free energy of the system occurs. On the other hand, if the polymer coils are forced out of the region between the surfaces, the solvent—i.e., the dispersion medium between the surfaces—is rendered inaccessible to free polymer. And whenever this happens (that is, a good solvent is rendered inaccessible to free polymer), the total free energy of the system increases. The separation of polymer molecules from the solvent could also be viewed as a "demixing" process involving demixing of the polymer molecules from a good solvent. An increase in the free energy of the system implies that a repulsion between the surfaces occurs. However, note that once the polymer coils are squeezed out of the region between the surfaces (i.e., $b \leq R_G$), an attraction between the surfaces occurs because of the osmotic pressure of bulk polymer solution [34].

Figure 42 shows a schematic diagram of the free energy of depletion interaction between two droplets. At distances larger than $2R_G$, there is negligible interaction between the droplets. In the range $R_G < b < 2R_G$, a maximum in free energy occurs. Upon further decrease in the separation distance, that is, $b < R_G$, one observes a minimum in free energy. Depending on the magnitude of the maximum in free energy, the emulsion may undergo depletion flocculation

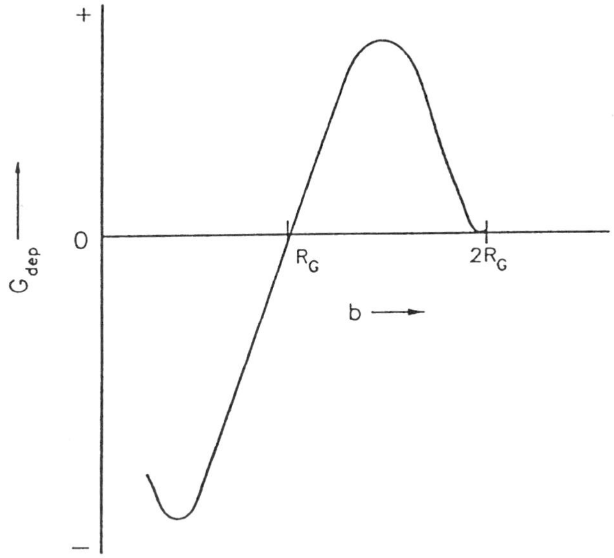

Figure 42. Schematic representation of G_{dep} for two droplets immersed in a polymeric dispersion medium. (From Ref. 34.)

or depletion stabilization. If the polymer concentration is relatively small, the magnitude of the maximum in free energy is not large. And if this maximum free energy is less than the thermal energy of the droplets, the droplets would undergo flocculation. However, at high polymer concentrations, the maximum in free energy may be much larger than the thermal energy of the droplets and, therefore, the droplets would not flocculate (depletion stabilization).

V. Rheology of Emulsions in the Absence of Free Polymer

Some aspects of the rheology of emulsions has been summarized by Sherman [130a] and by Davis [130b] in, respectively, Vols. 1 and 3 of this encyclopedia. In this section, the rheology of emulsions in the *absence* of free polymer is reviewed in detail. Both electrostatically stabilized and sterically stabilized (polymer or surfactant) emulsions will be discussed.

In the absence of free or excess polymer in the dispersion medium, the dispersion medium (oil or water) behaves like a Newtonian fluid. However, the rheological behavior of emulsions can still be quite complex, depending on the morphology or microstructure of the emulsion produced. The emulsions produced could be either stable (unflocculated) or unstable (flocculated), depending on the nature of various components and the interactions between them.

A. Emulsions of Low-Viscosity Newtonian Dispersion Media

When the viscosity of the dispersion medium is low as compared with the viscosity of the dispersed phase, the droplets can be treated as rigid particles; that is, the effects of internal circulation and deformation of droplets can be neglected. Under such conditions, the rheological behavior of emulsions is similar to that of solids-in-liquid suspensions.

1. *Stable (Unflocculated) Emulsions of "Hard Spheres"*

The emulsion droplets can be considered as "hard spheres" when both repulsive and attractive forces between the droplets are negligible and the adsorbed polymer or surfactant layer thickness (δ) is very small relative to the droplet size (that is, δ/R is very small). In such cases, only hydrodynamic and Brownian forces govern the rheological behavior of emulsions. For a dispersion (emulsion/suspension) of hard spheres, the rheological equation of state can be written [131] as:

$$\eta = f(\dot{\gamma}, \eta_c, R, n, \rho_d, \rho_c, t, kT) \tag{158}$$

where $\dot{\gamma}$ is the shear rate, η_c is the continuous phase viscosity, R is the particle radius, n is the number density of particles, ρ_d and ρ_c are the densities of dispersed and continuous phases, respectively, t is time, k is the Boltzmann constant, and T is absolute temperature.

Using dimensional analysis, Eq. (158) can be rewritten [131] as

$$\eta_r = f(\phi, t_r, \rho_r, Re, Pe) \tag{159}$$

where η_r is relative viscosity, ϕ is the volume fraction of the dispersed phase ($= 4\pi nR^3/3$), t_r is reduced time [defined as $t(kT)/\eta_c R^3$], ρ_r is the relative density ρ_d/ρ_c, Re is the Reynolds number (defined as $R^2\dot{\gamma}\rho_c/\eta_c$), and Pe is the Péclet number, which represents the ratio between the time scale on which the structure of the dispersion is destroyed by the shear and the time scale of Brownian motion that restores the structure.

The Péclet number could be written as

$$Pe = \frac{\dot{\gamma}\eta_c R^3}{kT} \tag{160}$$

although Krieger [131] uses the dispersion or emulsion viscosity (instead of η_c) in the definition; that is,

$$Pe = \frac{\dot{\gamma}\eta R^3}{kT} = \frac{\tau R^3}{kT} \tag{161}$$

We will refer to this latter definition as the effective Péclet number.

Under conditions of steady state ($t_r \to \infty$), creeping flow ($Re \to 0$), and neutrally buoyant emulsion ($\rho_r \to 1$), Eq. (159) becomes

$$\eta_r = f(\phi, Pe) \tag{162}$$

Thus, the relative viscosity of different hard-sphere emulsions is the same at the same volume fraction of dispersed phase and Péclet number, even though the emulsions may differ significantly in droplet size, viscosity of the dispersion medium, and temperature at which the measurements are carried out.

Figure 43 shows the relative viscosity versus shear stress data of Krieger [131] for various different-sized monodisperse suspensions of hard spheres, at a fixed volume fraction of 0.50. It is clear from the figure that these hard-sphere suspensions exhibit non-Newtonian shear-thinning behavior. The shear-thin-

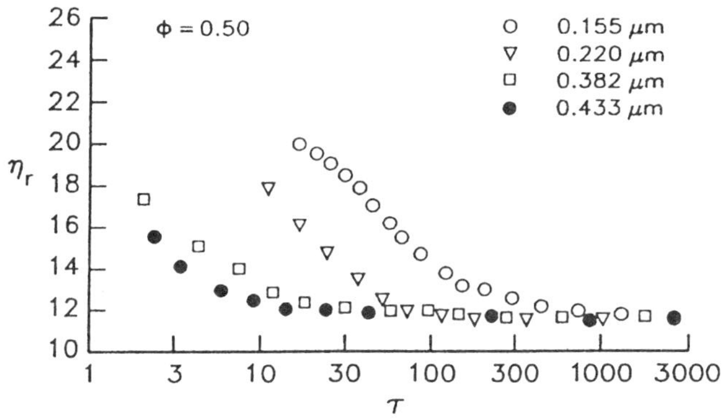

Figure 43. Relative viscosity versus shear stress for 50% monodispersions of cross-linked polystyrene spheres of various sizes in *m*-cresol. (After Ref. 131.)

ning observed here is due to competition between hydrodynamic and Brownian forces. Brownian forces tend to randomize the particles; as a consequence, particles collide with each other and form temporary doublets, triplets, etc. Hydrodynamic forces, on the other hand, result in ordering of the particles into layers and tend to destroy doublets and higher aggregates. At low shear rates, Brownian force dominates; consequently, there occurs a significant amount of rotation of doublets and triplets, etc. A rotating doublet (or higher aggregate) causes more dissipation of energy as compared with the same number of isolated spheres, and, therefore, the viscosity of the emulsion is high at low shear rates. With the increase in shear rate (or shear stress), the number of rotating doublets or higher aggregates decreases because the shear forces are strong enough to break such aggregates; therefore, the viscosity decreases with the increase in shear rate.

Furthermore, the relative viscosity is a function of particle size only in the intermediate stress range; the relative viscosity increases with the decrease in particle size. At extremely low and very high shear stresses, the relative viscosities of hard-sphere suspensions are independent of the particle size.

Figure 44 shows the viscosity versus shear stress data of Rodriguez et al. [132] for monodisperse suspensions of hard spheres at various concentrations of dispersed phase. The flow behavior of these suspensions is similar to that of the suspensions studied by Krieger and co-workers; that is, the flow curves show three different regions: the lower Newtonian region in the low shear stress range, where the viscosity is constant; the shear-thinning region at intermediate stresses, where the viscosity decreases with the increase in shear stress; and the upper Newtonian region in the high stress range, where the viscosity becomes constant again. Also, the difference between the low and high shear limiting viscosities increases with the increase in dispersed-phase concentration; that is, dispersions become more shear-thinning with the increase in ϕ.

The relative viscosity versus effective Péclet number plot for hard-sphere suspensions is shown in Fig. 45 [131]. The dispersed-phase concentration is fixed at 50 vol %. The excellent superposition of results for different dispersions

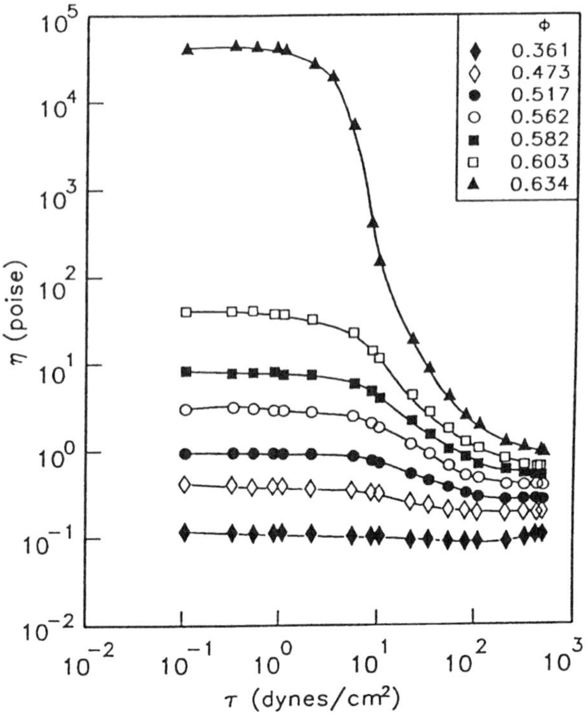

Figure 44. Viscosity versus shear stress data for various differently concentrated monodispersions of latex 84 in bromoform. (After Ref. 132.)

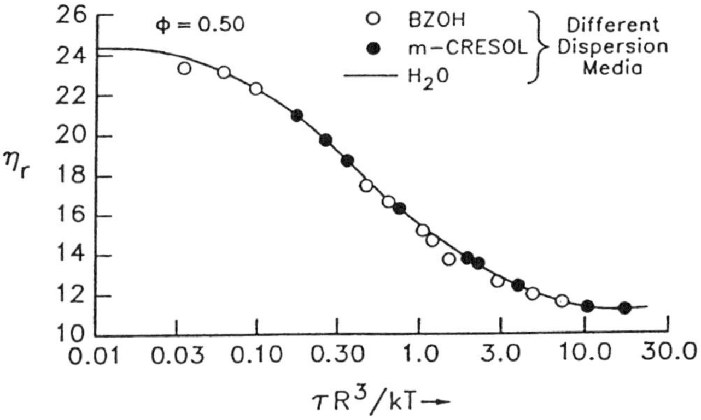

Figure 45. Relative viscosity versus effective Péclet number plot for hard-sphere monodispersions of various sizes in different dispersion media. (From Ref. 131.)

(different particle size, different dispersion media) confirms the fact that for monodisperse hard-sphere dispersions, the relative viscosity is a unique function of the effective Péclet number. It is interesting to note that a similar unique correlation is obtained if the relative viscosity is plotted as a function of the

Péclet number (instead of the effective Péclet number). This is clearly shown in Fig. 46 [133], where the data of Fig. 45 are replotted on the basis of the Péclet number.

The existence of a unique correlation between relative viscosity and Péclet number (or effective Péclet number) for monodisperse hard-sphere dispersions has been confirmed by several other workers [134]. For example, Fig. 47 shows relative viscosity versus Péclet number ($\eta_c R^3 \dot{\gamma}/kT$) data for several hard-sphere

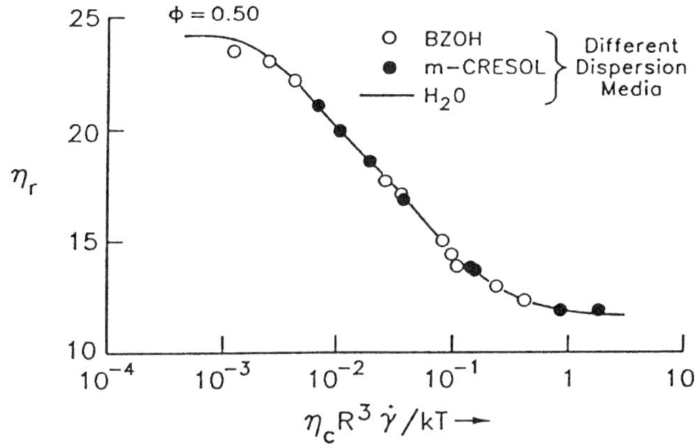

Figure 46. Relative viscosity versus Péclet number plot for hard-sphere monodispersions of various sizes in different dispersion media. (From Ref. 133.)

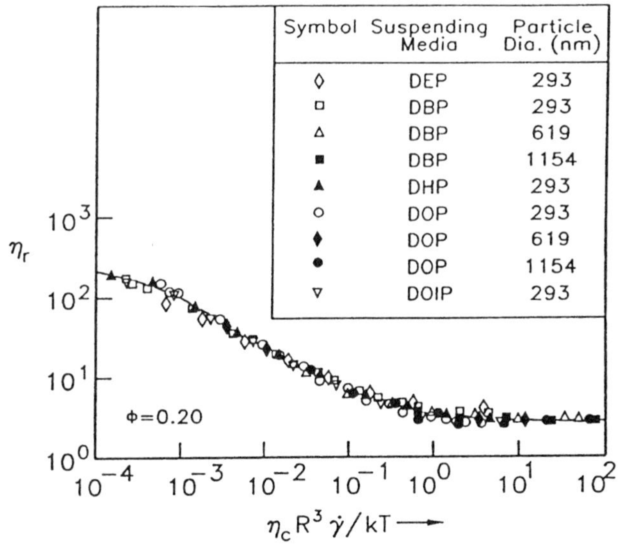

Figure 47. Relative viscosity versus Péclet number plot for nearly hard-sphere monodispersions of polyvinyl chloride spheres of various sizes in several solvents. (From Ref. 134.)

suspensions investigated by Willey and Macosko [134]. The relative viscosity versus Péclet number data for different suspensions fall on the same curve.

It is important to mention here that hard-sphere suspensions and emulsions can exhibit a much more complex rheological behavior (yield stress, elasticity, dilatancy, etc.) near close-packing concentrations, that is, when particles are nearly touching each other. For example, Fig. 48 shows viscosity versus shear stress data of Pal [135] for hard-sphere oil-in-water emulsions. Because the particle size of these emulsions is relatively large (>1 μm), Brownian motion is expected to be of less importance. Therefore, these emulsions are nearly Newtonian up to a dispersed-phase concentration as high as 55.14 vol %. The emulsions with ϕ of 59.61% and 65.15% exhibit only slight pseudoplastic behavior with no yield stress. However, there occurs a dramatic change in the rheological behavior when ϕ reaches 72.21%. The emulsion at $\phi = 72.21\%$ appears to have a significant yield stress. Figure 49 shows similar behavior for hard-sphere suspensions [136]. At high concentrations ($\phi > 0.64$), a dramatic change in the rheological behavior occurs; the suspensions at high concentrations clearly have a yield stress.

Figure 50 shows the storage modulus (G′) as a function of frequency for differently concentrated hard-sphere suspensions [136]. The suspensions behave like inelastic fluids at low concentrations ($\phi < 0.62$); G′ is rather small and its magnitude increases rapidly with frequency. However, at high concentrations, hard-sphere suspensions exhibit elastic solid like behavior; G′ is very high and it is nearly independent of frequency.

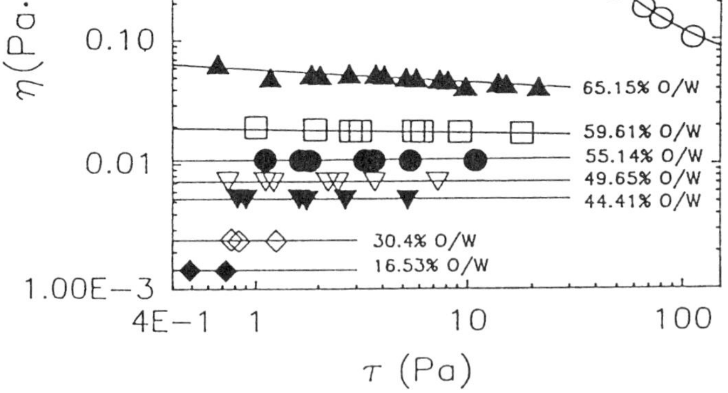

Figure 48. Viscosity versus shear stress data of Pal for nearly hard-sphere oil-in-water emulsions. (After Ref. 135.)

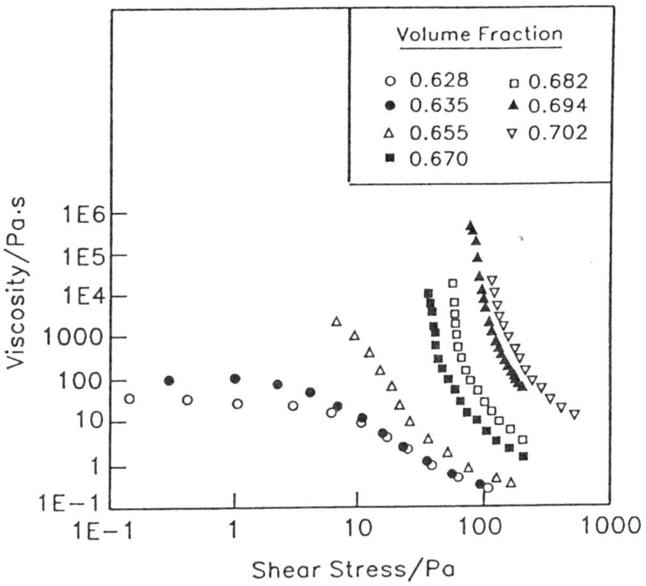

Figure 49. Viscosity versus shear stress data of Jones et al. for hard-sphere suspensions. (After Ref. 136.)

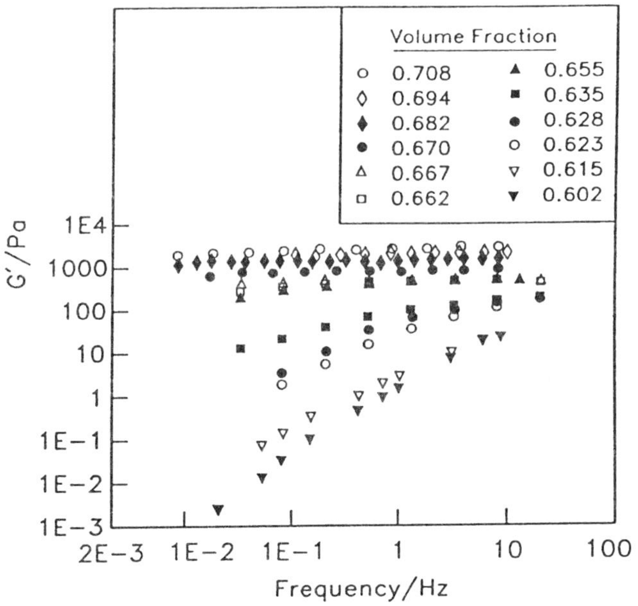

Figure 50. Storage modulus (G′) as a function of frequency for various differently concentrated hard-sphere suspensions. (From Ref. 136.)

According to Hoffman [137], dispersions of monodisperse hard spheres can exhibit shear-thickening (dilatant) behavior at high concentrations (near close-packing concentrations). It is generally believed that shear-thickening in dispersions is caused by transition from a two-dimensional layered arrangement of particles to a random three-dimensional form. At low shear rates, the dispersion particles arrange themselves into layers of two-dimensional hexagonally packed spheres that pass one over another in the direction of flow. As the shear rate is increased to some critical value, the ordered layers begin to disrupt into a random three-dimensional arrangement, and, therefore, viscosity begins to increase with the shear rate. Barnes [138] argues that shear-thickening in dispersions of monodisperse hard spheres can occur even at lower concentrations, although the critical shear rate is much higher—that is, the onset of shear-thickening is shifted to higher shear rates (Péclet number) with a decrease in particle concentration. Also, the smaller the particle size, the higher is the critical shear rate at which shear-thickening begins.

So far we have not discussed the effect of polydispersity on the rheological behavior of hard-sphere dispersions. Several studies indicate that polydispersity can seriously influence the rheological behavior of hard-sphere dispersions, particularly at high concentrations (near close-packing concentrations). Concentrated polydisperse dispersions generally have lower yield stress, lower viscosities, less elastic behavior, and little shear-thickening effect as compared with monodisperse systems. As an example, Fig. 51 shows the data of Chong et al. [139] for bimodal suspensions. The zero shear-rate relative viscosity is plotted as a function of volume fraction of solids for several different particle-size ratios. Each bimodal suspension contained 25 vol % of the total solids as small spheres. The large particles used in this study had a volume-surface average diameter of 236 μm; the small spheres had volume-surface average diameters of 112.5, 73.8 and 33 μm, respectively. Therefore, the diameter ratios of small to large spheres (d/D) were 0.477, 0.313, and 0.138, respectively. It is clear from the figure that: (1) the effect of polydispersity increases with the increase in concentration (that is, the difference between the relative viscosities of monodisperse and bimodal

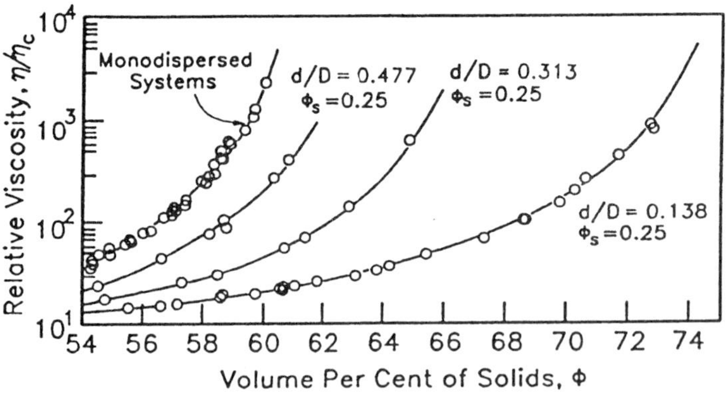

Figure 51. Relative viscosity versus dispersed-phase concentration data for bimodal suspensions. (After Ref. 139.)

dispersions is much larger at higher concentrations); and (2) at a given ϕ in the high concentration range, the relative viscosity decreases dramatically as the particle size ratio decreases.

Figure 52 shows another interesting result obtained by Chong et al. [139]. If the total volume fraction of the dispersed phase in a bimodal dispersion is kept constant and the volume percent of small spheres in the dispersion is varied, the relative viscosity goes through a minimum. Similar results have been obtained by other workers. For example, Fig. 53 shows the plots obtained by Rodriguez et al. [132] for relative viscosity versus volume percent of small spheres in a bimodal dispersion at various different values of total volume fraction. The effect of polydispersity becomes important only at high total concentrations ($\phi \geq 0.562$) and the relative viscosity goes through a minimum.

Viscosity equations for dilute hard-sphere dispersions. Several authors have computed the relative viscosity of dilute hard-sphere dispersions on a theoretical basis. Here we will discuss only the work of Einstein [68,69] and Batchelor [140–144].

First, note that the volume-averaged stress tensor $\langle \boldsymbol{\sigma} \rangle$ and the volume-average rate-of-deformation tensor $\langle \mathbf{d} \rangle$ in a dispersion can be written [140] as

$$\langle \boldsymbol{\sigma} \rangle = \frac{1}{V} \int_V \boldsymbol{\sigma} \, dV \tag{163}$$

$$\langle \mathbf{d} \rangle = \frac{1}{V} \int_V \mathbf{d} \, dV \tag{164}$$

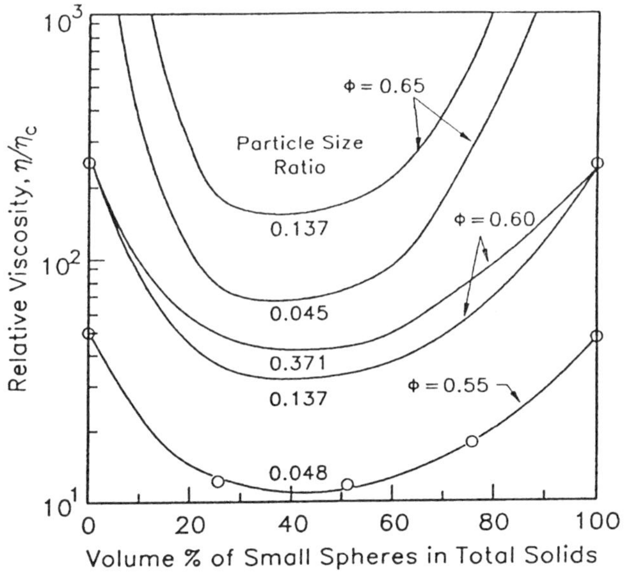

Figure 52. Variation of relative viscosity of bimodal suspensions with the volume percent of small spheres in total solids. (From Ref. 139.)

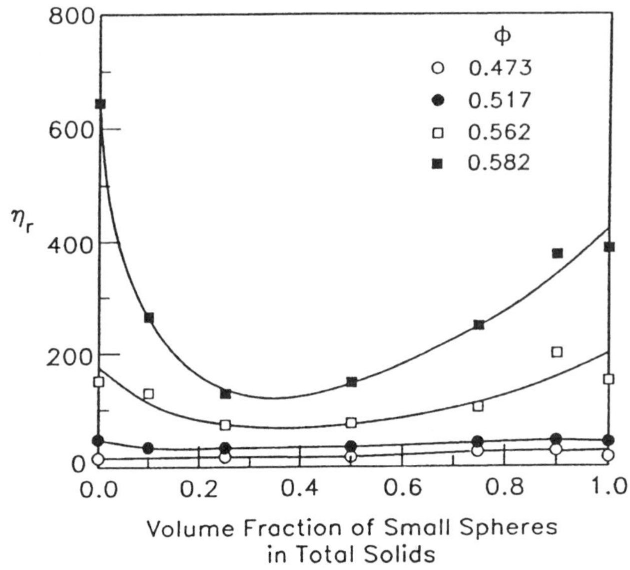

Figure 53. Variation of zero shear-rate relative viscosity of bimodal dispersion with the volume fraction of small spheres in total solids. (From Ref. 132.)

where V is the representative volume of the dispersion large enough to contain many particles, $\boldsymbol{\sigma}$ is the local stress tensor, and $\mathbf{d}$ is the local rate-of-deformation tensor, defined as

$$\mathbf{d} = \frac{1}{2}[\nabla\mathbf{U} + (\nabla\mathbf{U})^{\mathrm{T}}] \tag{165}$$

where $\mathbf{U}$ is the local velocity.

For hard-sphere dispersions in the absence of Brownian motion, the average stress tensor can be written [140–142] as

$$\langle\boldsymbol{\sigma}\rangle = \frac{1}{\mathrm{V}}\int_{\mathrm{V_f}}(-\mathrm{P}\boldsymbol{\delta} + 2\eta_c\mathbf{d})\,\mathrm{dV} + \frac{1}{\mathrm{V}}\sum_{i=1}^{\mathrm{N}}\int_{\mathrm{V_i}}\boldsymbol{\sigma}\,\mathrm{dV} \tag{166}$$

where P is the local pressure, $\boldsymbol{\delta}$ is the unit tensor, $\mathrm{V_f}$ is the volume of the fluid portion of V, η_c is the dispersion medium viscosity, $\mathrm{V_i}$ is the volume of a particle with center at $\mathbf{x}_i$, and N is the number of particles in volume V.

From Eq. (166), it can be further shown [140–142] that

$$\langle\boldsymbol{\sigma}\rangle = 2\eta_c\langle\mathbf{d}\rangle + \frac{1}{\mathrm{V}}\sum_i \mathbf{S}_i + \mathrm{I.T.} \tag{167}$$

where I.T. denotes isotropic term, $\langle\mathbf{d}\rangle$ is the average rate-of-deformation tensor in the dispersion, and $\mathbf{S}_i$ is the stresslet exerted on the fluid by each particle. For rigid particles [140–142]:

$$\mathbf{S}_i = \int_{\mathrm{A_i}}\left[(\mathbf{x}-\mathbf{x}_i)\boldsymbol{\sigma}\cdot\mathbf{n} - \frac{1}{3}\boldsymbol{\delta}(\mathbf{x}-\mathbf{x}_i)\cdot\boldsymbol{\sigma}\cdot\mathbf{n}\right]\mathrm{dA} \tag{168}$$

where $\boldsymbol{\sigma}$ is the contact stress (tensor) in the suspending fluid over the particle surface, A_i is the surface area of a particle with center at $\mathbf{x}_i$, and $\mathbf{n}$ is unit vector normal to the particle surface (directed outward).

If the dispersion is *very dilute* so that there is no *hydrodynamic interaction* between the particles, the integral in Eq. (168) can be evaluated using the results for an isolated sphere obtained by solving Stokes equations. The result is

$$\mathbf{S}_i = \frac{20}{3}\pi\eta_c R^3\langle\mathbf{d}\rangle \tag{169}$$

where R is the particle radius. Thus, the average stress tensor for very dilute dispersions of hard spheres is

$$\langle\boldsymbol{\sigma}\rangle = -\langle P\rangle\boldsymbol{\delta} + \left(2\eta_c + \frac{20}{3}\pi\eta_c R^3 N\right)\langle\mathbf{d}\rangle \tag{170}$$

where N is the number density (number of particles per unit volume of dispersion) given by the following relation:

$$N = \frac{\phi}{(4/3)\pi R^3} \tag{171}$$

in which ϕ is the volume fraction of dispersed phase. Thus,

$$\langle\boldsymbol{\sigma}\rangle = -\langle P\rangle\boldsymbol{\delta} + 2\eta_c\left(1 + \frac{5}{2}\phi\right)\langle\mathbf{d}\rangle \tag{172}$$

To obtain the effective viscosity η of dispersion, one also can write

$$\langle\boldsymbol{\sigma}\rangle = -\langle P\rangle\boldsymbol{\delta} + 2\eta\langle\mathbf{d}\rangle \tag{173}$$

Comparing Eqs. (172) and (173) gives the celebrated Einstein relation for the relative viscosity of a very dilute dispersion:

$$\eta_r = \frac{\eta}{\eta_c} = 1 + \frac{5}{2}\phi \tag{174}$$

This Einstein formula is valid for hard-sphere dispersions only for ϕ less than about 0.02 [141]. In a slightly more concentrated suspension, the particles can no longer be treated as independent because two-particle interactions become important. At still higher concentrations, multiparticle interactions have to be considered.

Batchelor and Green [141] calculated the average stress in dispersions with two-particle hydrodynamic interactions. Batchelor [142] also considered the effect of Brownian motion on the average stress. Brownian motion contributes to average stress in two ways: The direct influence of Brownian motion comes from the stress generated in the fluid by particles moving solely under the influence of the thermodynamic forces. Brownian motion also affects the form of the particle pair-distribution function, that is, the relative positions of particles. This latter effect is referred to as the indirect contribution by Brownian motion. Batchelor's expression for the average stress at low Péclet number ($Pe \to 0$) is given by

$$\langle\boldsymbol{\sigma}\rangle = \langle\boldsymbol{\sigma}\rangle^H + \langle\boldsymbol{\sigma}\rangle^B \tag{175}$$

where

$$\langle \boldsymbol{\sigma} \rangle^{\mathrm{H}} \simeq 2\eta_{\mathrm{c}} \langle \mathbf{d} \rangle \left(1 + \frac{5}{2}\phi + 5.2\phi^2 \right) + \mathrm{I.T.} \tag{176}$$

$$\langle \boldsymbol{\sigma} \rangle^{\mathrm{B}} = 2\eta_{\mathrm{c}} \langle \mathbf{d} \rangle (0.97\phi^2) + \mathrm{I.T.} \tag{177}$$

Here, $\langle \boldsymbol{\sigma} \rangle^{\mathrm{H}}$ refers to the stress contribution due to bulk motion (hydrodynamic contribution) (including the indirect influence of Brownian motion through the particle pair-distribution function) and $\langle \boldsymbol{\sigma} \rangle^{\mathrm{B}}$ is the direct contribution of Brownian motion. Thus, the average stress in hard-sphere dispersions at low Péclet number ($\mathrm{Pe} \rightarrow 0$) is given as

$$\langle \boldsymbol{\sigma} \rangle = 2\eta_{\mathrm{c}} \left(1 + \frac{5}{2}\phi + 6.2\phi^2 \right) \langle \mathbf{d} \rangle + \mathrm{I.T.} \tag{178}$$

This gives the following expression for the relative viscosity at low Péclet numbers:

$$\eta_{\mathrm{r}} = 1 + \frac{5}{2}\phi + 6.2\phi^2 \tag{179}$$

Equation (179) for relative viscosity is generally found to be valid up to a dispersed-phase volume fraction of about 0.1 to 0.15. For example, Fig. 54 shows a comparison of Woutersen and de Kruif data [145] for hard-sphere monodispersions with Batchelor's equation. The equation gives excellent prediction of relative viscosities in the concentration range investigated ($\phi \approx 0.15$).

Viscosity equations for concentrated hard-sphere dispersions. In the preceding section, we dealt with either very dilute dispersions ($\phi < 0.02$), in which all particle interactions were absent, or dilute dispersions ($\phi < 0.15$), in which two-particle (pair) hydrodynamic interactions played a major role and higher-order hydrodynamic interactions (three (or more) particle interactions) were negligible. However, in more concentrated dispersions, higher-order interactions cannot be neglected. The general expression for relative viscosity in the presence of multiparticle interactions can be written as

$$\eta_{\mathrm{r}} = 1 + \mathrm{K}_1\phi + \mathrm{K}_2\phi^2 + \mathrm{K}_3\phi^3 + \mathrm{K}_4\phi^4 \ldots \tag{180}$$

where each coefficient $\mathrm{K_i}$ represents i-body interactions (K_1 is 2.5 for hard-sphere dispersions, and its value does not depend on Brownian motion; K_2 is 6.2 at low Péclet numbers). Three (or more) body interaction coefficients are extremely difficult to calculate from a theoretical point of view. Because of the theoretical difficulties involved in the calculation of $\mathrm{K_i}$ ($\mathrm{i} \geq 3$), several authors have used other "not-so-rigorous" theoretical methods to develop the viscosity equations for concentrated dispersions. Among these methods, the popular ones are the "cell-model" approach and the "effective medium" approach. First, we will discuss the "cell-model" approach.

In the cell model approach, the particles of the dispersion are envisioned to reside in well-defined unit cells in which the flow behavior is the same. The size of the unit cell is dependent on the dispersed-phase concentration and particle size. The flow behavior in a unit cell is determined by solving Stokes equations

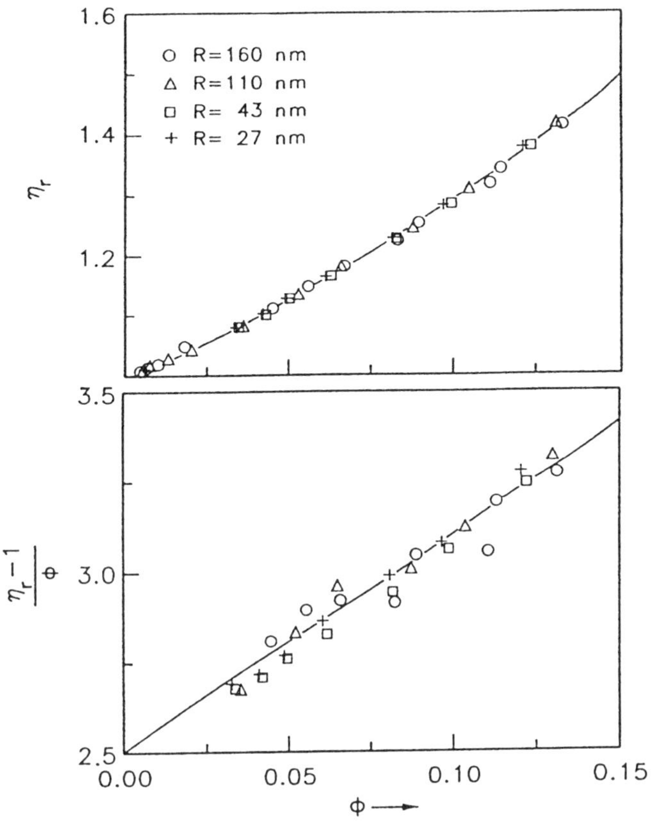

Figure 54. Relative viscosity of unimodal hard-sphere silica dispersions of various sizes. The solid curves represent Batchelor's equation. (From Ref. 145.)

around a single reference particle surrounded by fluid; the interaction effect between the reference particle and the particles in the neighboring cells is taken into account solely through the boundary conditions at the cell boundary. Different authors have used different boundary conditions to solve this problem.

Simha [146] chose the unit cell to be spherical in shape and containing a reference particle at its center. The cell boundary was taken to be a rigid surface. The solution of Stokes equation inside such a rigid concentric enclosure resulted in the following viscosity equation:

$$\eta_r = 1 + \frac{5}{2} I(\lambda)\phi \tag{181}$$

where λ is the ratio of the particle radius (R) to the cell radius (R_C) and $I(\lambda)$ is a complex function of λ given by the expression

$$I(\lambda) = \frac{4(1-\lambda^7)}{4(1+\lambda^{10}) - 25\lambda^3(1+\lambda^4) + 42\lambda^5} \tag{182}$$

$I(\lambda)$ increases slowly until $\lambda \simeq 0.5$; between 0.5 and 1.0, it increases much more rapidly. In order to calculate the relative viscosity of the dispersion from Eq. (182), one needs to know λ, i.e., the relationship between the cell radius and the particle radius. According to Simha,

$$\lambda = \frac{\phi^{1/3}}{f} \qquad \text{when } \phi \text{ is small} \tag{183}$$

$$\lambda = \frac{\phi^{1/3}}{f(1 - \phi^{1/3}/f)} \qquad \text{when } \phi \to \phi_m \tag{184}$$

where f is an empirical parameter ($1.3 < f < 2$) that varies with ϕ. At high values of ϕ, $f = (8\phi_m)^{1/3}$.

Happel [147] also derived a viscosity equation using the cell-model approach. Like Simha, he chose the cell to be spherical in shape. However, he assumed that the cell boundary is a frictionless surface where the normal velocity vanishes. Based on this reasoning, he derived the following equation by solving the Stokes equations in a unit cell:

$$\eta_r = 1 + 5.5\phi I(\lambda) \tag{185}$$

where

$$I(\lambda) = \frac{4\lambda^7 + 10 + (84/11)\lambda^2}{10(1 - \lambda^{10}) - 25\lambda^3(1 - \lambda^4)} \tag{186}$$

Unlike Simha, Happel chose R_C (hence λ) to be such that the volume fraction of particle in the unit cell is equal to the actual volume fraction of particles in the dispersion; i.e.,

$$\lambda = \frac{R}{R_C} = (\phi)^{1/3} \tag{187}$$

The cell model of Happel is often referred to as the "free-surface" model, because the cell boundary is taken to be a frictionless surface.

Frankel and Acrivos [148] utilized the cell model approach to derive a viscosity equation for highly concentrated hard-sphere dispersions. However, their approach is different from that of Simha [146] and Happel [147] in that they use hydrodynamic lubrication theory to analyze the viscous dissipation in a cell rather than obtaining an exact solution to the Stokes equations. The Frankel and Acrivos equation is

$$\eta_r = \frac{9}{8}\left[\frac{(\phi/\phi_m)^{1/3}}{1 - (\phi/\phi_m)^{1/3}}\right] \tag{188}$$

This equation is valid only in the limit of $\phi \to \phi_m$.

One problem with the cell model approach is that the results derived from it are strongly dependent on somewhat arbitrary choices of the size, shape, and boundary conditions of the cell.

Several authors have utilized an alternative approach, the "effective medium" approach to derive the viscosity equations for hard-sphere dispersions. This approach is explained as follows: Let us consider V cm^3 of a dispersion with dispersed-phase concentration ϕ_1. If ΔV_d amount of dispersed phase is added further to this dispersion, the new concentration (ϕ) becomes

$$\phi = \frac{\phi_1 V + \Delta V_d}{V + \Delta V_d} \simeq \phi_1 + \phi_2 \tag{189}$$

where $\phi_2 = \Delta V_d/(V + \Delta V_d)$. If we now assume that the starting dispersion behaves like an *effective medium* of viscosity $\eta(\phi_1)$ toward the newly added dispersed phase (ΔV_d), the final dispersion will have a relative viscosity $\eta_r(\phi_2)$ given as

$$\eta_r(\phi_2) = \frac{\eta(\phi)}{\eta(\phi_1)} = \frac{\eta_r(\phi_1 + \phi_2)}{\eta_r(\phi_1)} \tag{190}$$

Thus,

$$\eta_r(\phi_1 + \phi_2) = \eta_r(\phi_1)\eta_r(\phi_2) \tag{191}$$

The solution of this functional relation is

$$\eta_r = \exp(K\phi) \tag{192}$$

where K is a constant whose value is 2.5 (from Einstein's formula). This equation was originally proposed by Arrhenius in 1887 [149]. In the derivation of the Arrhenius equation, we made the approximation given in Eq. (189). The exact relationship between ϕ, ϕ_1, and ϕ_2 is

$$\phi = \phi_1\left(\frac{V}{V + \Delta V_d}\right) + \phi_2 \tag{193}$$

or

$$\phi = \phi_1(1 - \phi_2) + \phi_2 \tag{194}$$

Therefore, the correct functional relation should be

$$\eta_r(\phi_1 + \phi_2 - \phi_1\phi_2) = \eta_r(\phi_1)\eta_r(\phi_2) \tag{195}$$

The solution of this functional relationship is

$$\eta_r = (1 - \phi)^{-K} \tag{196}$$

where K is 2.5 from Einstein's formula. This viscosity equation was first derived independently by Roscoe [150] and by Brinkman [151]. Although this relation behaves properly at low to moderate volume fractions, it fails to consider the fact that hard-sphere dispersions reach infinite viscosities at close-packing concentration (ϕ_m), which is always less than unity.

Mooney [152] argued that the functional relationship (Eq. 191) should be modified to

$$\eta_r(\phi_1 + \phi_2) = \eta_r\left(\frac{\phi_1}{1 - \phi_2/\phi_m}\right)\eta_r\left(\frac{\phi_2}{1 - \phi_1/\phi_m}\right) \tag{197a}$$

The solution to Eq. (197a) is

$$\eta_r = \exp\left(\frac{K\phi}{1 - \phi/\phi_m}\right) \tag{197b}$$

Again, K should be equal to 2.5 if Einstein's equation is to be satisfied in the limit $\phi \to 0$.

According to Krieger and Dougherty [153], the more appropriate functional relationship governing the viscosity of dispersions is

$$\eta_r(\phi_1 + \phi_2) = \eta_r(\phi_1)\eta_r\left(\frac{\phi_2}{1 - \phi_1/\phi_m}\right) \tag{198}$$

whose solution is

$$\eta_r = \left(1 - \frac{\phi}{\phi_m}\right)^{-K\phi_m} \tag{199}$$

where $K = 2.5$ from Einstein's equation in the limit $\phi \to 0$.

Note that the viscosity equations based on the effective medium approach (or the cell model approach) do not consider the non-Newtonian behavior of dispersions; that is, the shear-thinning effect due to the presence of Brownian motion is ignored. To overcome this difficulty, several authors have assigned the shear-dependence of dispersions to ϕ_m [154–157]. For example, Krieger [131] fitted his data at very low shear rates ($Pe \to 0$) and at very high shear rates ($Pe \to \infty$) using Eq. (199) with the ϕ_m values of 0.57 and 0.68, respectively. Several authors have argued that the effect of polydispersity could also be incorporated through ϕ_m [132,139,154–159]. The maximum packing concentration ϕ_m is expected to vary with the particle size distribution. As an example, Fig. 55 shows how ϕ_m varies with the volume fraction of small spheres in a binary

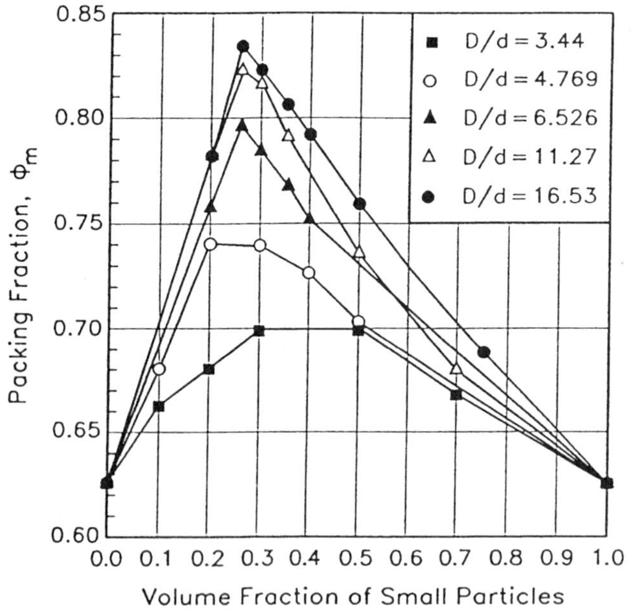

Figure 55. Variation of ϕ_m with the volume fraction of small spheres in a binary mixture of small and large spheres. (From Ref. 158.)

mixture of small and large spheres [158]. The data are shown for different ratios of large- to small-diameter particles (D/d). Clearly, ϕ_m goes through a maximum for a given D/d ratio, and the magnitude of the maximum ϕ_m increases with an increase in D/d. Figure 56 shows the typical relative viscosity plot for the dispersions, plotted on a ϕ/ϕ_m basis [139]. The data shown in this figure include the monodispersed as well as the bidispersed systems of Chong et al. [139]. It is clear from the figure that a universal curve is obtained if data are plotted on a ϕ/ϕ_m basis.

A large number of studies involving hard-sphere suspensions indicate that among the above-mentioned equations, the most successful is one proposed by Krieger and Dougherty [153]. However, this equation as well as all other equations that involve ϕ_m (close-packing concentration where viscosity become infinite) are not readily applicable to emulsions. Pal and Rhodes [154,155] argue that while ϕ_m is a well-defined parameter for solids-in-liquid suspensions, it is not so well defined for emulsions. In the case of suspensions, ϕ_m is determined either directly from experiments [160,161] or by plotting $1/\eta_r$ versus ϕ data and extrapolating the data to a point where the ordinate becomes zero [139,162,163]. But emulsion viscosity rarely becomes infinite (even at very high concentrations), due to the deformable nature of the disperse phase. Therefore, extrapolation of $1/\eta_r$ versus ϕ data to determine ϕ_m gives erroneous results.

According to Pal and Rhodes [155], the viscosity data for hard-sphere (as well as nonhard-sphere) emulsions can be described accurately by the following equation:

$$\eta_r = [1 - K_O K_F(\dot{\gamma})\phi]^{-2.5} \tag{200}$$

where K_O is the factor that takes into account the hydration of droplets (due to adsorbed surfactant/polymer) and K_F $(\dot{\gamma})$ is the factor that takes into account the aggregation of droplets. K_F decreases with the increase in shear rate for non-Newtonian emulsions ($K_F = 1$ for Newtonian emulsions).

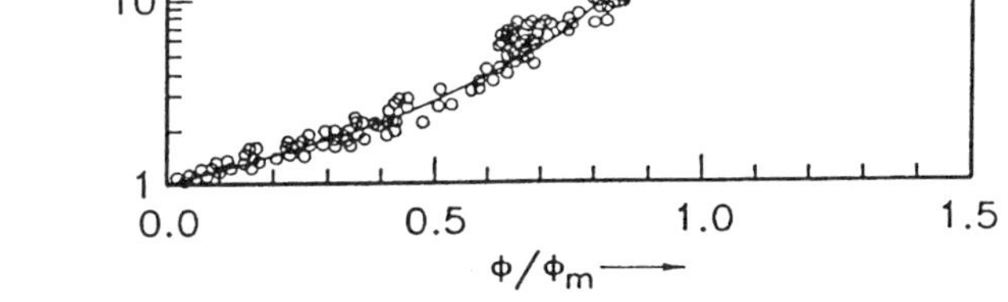

Figure 56. Relative viscosity versus ϕ/ϕ_m plot. (From Ref. 139.)

2. Stable (Unflocculated) Emulsions of "Soft Spheres"

In the preceding section, we dealt with hard-sphere dispersions where only hydrodynamic and Brownian forces played a role. The particles of hard-sphere dispersions undergo no other interaction forces except on contact, where an infinitely strong repulsion occurs to prevent interpenetration. In contrast, "soft-sphere" dispersions are subject to interaction forces (other than hydrodynamic and Brownian) that are felt at finite distances from the particle surface. "Soft spheres" could be thought of as hard spheres with an effective radius $R_{eff} = R + \delta$, where δ is the interaction length or thickness. For example, if the particles are stabilized by electric charge due to adsorbed ionic surfactant or polyelectrolyte, the electrical double-layer thickness can be identified with the interaction length. For sterically stabilized emulsions, the thickness of the adsorbed polymer (surfactant) layer can be considered as the interaction length. However, note that during the approach of two particles, the interaction layer can be deformed and penetrated (hence, the name "soft spheres").

Dispersions of charged particles. Electrostatic forces can have a dramatic influence on the viscosity of aqueous emulsions or suspensions of charged particles. For example, Fig. 57 shows relative viscosity versus effective Péclet number data for electrostatically stabilized suspensions studied by Krieger and Eguiluz [164]. The figure reveals the following important information: (1) Unlike hard-sphere dispersions, η_r versus Péclet number correlation is not unique; that is, η_r depends not only on Pe but also on other factors, such as electrolyte concentration. (2) The relative viscosity increases dramatically with the decrease in electrolyte concentration. This is because the electrical double-layer thickness increases with the decrease in electrolyte concentration. (3) At low shear stresses, the viscosity of suspensions with small electrolyte concentration tends to approach a vertical asymptote, indicating the presence of yield

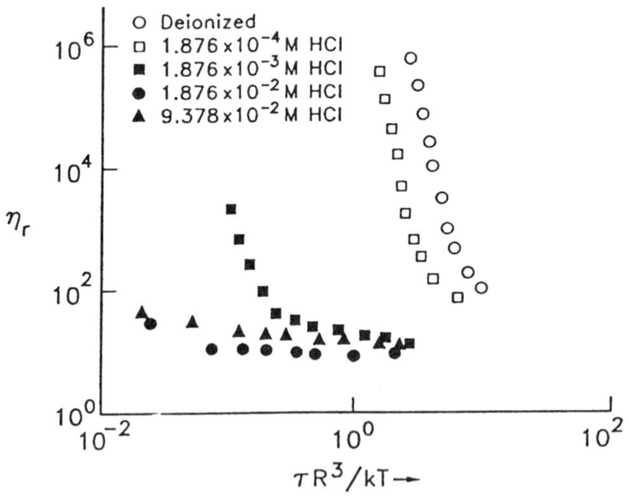

Figure 57. Relative viscosity versus effective Péclet number data for electrostatically stabilized suspensions. (From Ref. 164.)

stress. The magnitude of the yield stress decreases with the increase in electrolyte concentration. At high concentrations of electrolyte, the yield stress disappears.

Note further that electrostatically stabilized dispersions can develop yield stress even at very low concentrations of the dispersed phase, provided that the electrolyte concentration of the continuous phase is negligibly small and that the particles are very small. For instance, Fig. 58 shows the data of Krieger and Eguiluz [164] for deionized latex (electrostatically stabilized) at various volume fractions. The particle radius is about 0.11 μm. Even at a dispersed-phase concentration of 5 vol %, the suspension appears to have a yield stress. This indicates that particle–particle interaction is significant even at low volume fractions of the dispersed phase. This is very possible if the electrical double-layer thickness (δ) relative to the particle radius (R) is quite large. When δ/R is large, the effective volume fraction of the dispersion, ϕ_{eff}, can be several orders of magnitude higher than the actual volume fraction, ϕ. For example, if the particle radius is 100 nm (0.1 μm) and the electrical double-layer thickness is of the same size (100 nm), then the effective volume fraction calculated from Eq. (110) is 8 times the actual volume fraction. Consequently, these dispersions are expected to exhibit strong interparticle interactions and hence high yield stress. Note further that dispersions of charged particles exhibit not only yield stress but also strong viscoelastic behavior when overlapping of double layers occurs [143,165].

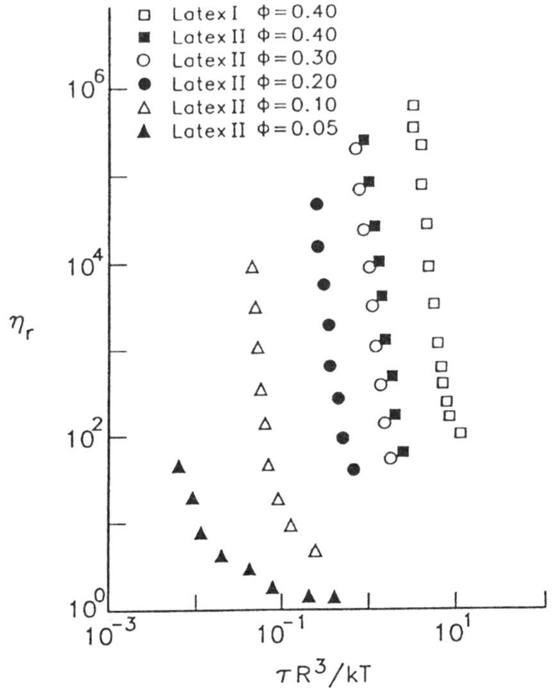

Figure 58. Data for electrostatically stabilized latex at various volume fractions. (After Ref. 164.)

Dispersions of charged particles are also known to exhibit shear-thickening behavior [137,138,165–169]. Figure 59 shows the viscosity data of Boersma et al. [165] for nearly monodispersions of silica particles in a glycerol/water mixture. As can be seen, the viscosity increases with the shear rate at high values of shear rate, indicating shear-thickening (dilatant) behavior. Figure 60 shows the effect of polydispersity on the shear-thickening behavior of electrostatically stabilized dispersions [169]. In the case of polydisperse dispersions the increase in viscosity at the critical shear rate (where the onset of shear-thickening occurs) is less dramatic and more gradual than with monodisperse systems, which exhibit an abrupt increase in viscosity.

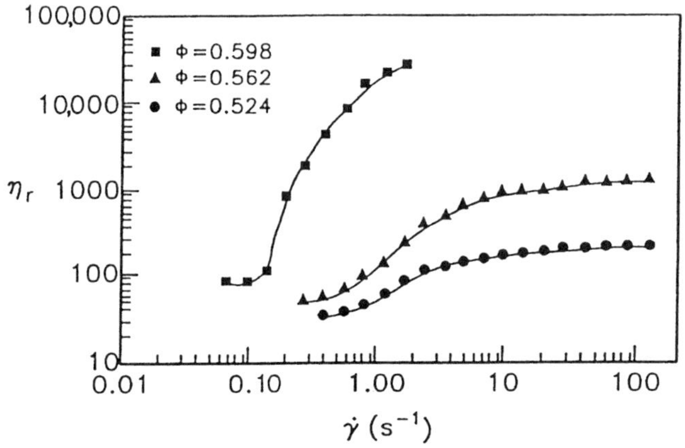

Figure 59. Viscosity data for nearly monodispersions of silica particles in glycerol/water mixture. (After Ref. 165.)

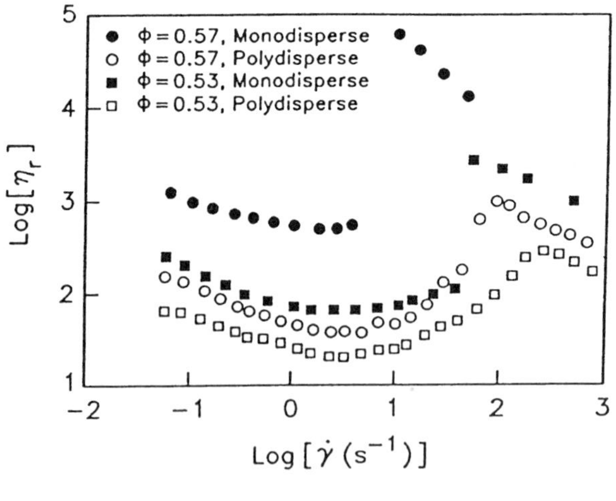

Figure 60. Effect of polydispersity on the shear-thickening behavior of electrostatically stabilized dispersions. (From Ref. 169.)

According to Boersma et al. [169], one can develop a criterion to predict the critical shear rate, at which the onset of shear-thickening occurs in electrostatically stabilized dispersions, by equating the magnitude of the interparticle force with the magnitude of the shear force. Their criterion is

$$\dot{\gamma}_{\text{critical}} = \frac{2\pi\varepsilon_0\varepsilon_r\psi_0^2}{6\pi\eta_c R^2}\left(\frac{\kappa h}{2}\right) \tag{201}$$

where $\dot{\gamma}_{\text{critical}}$ is the shear rate at which the onset of shear-thickening occurs, ε_0 is the permittivity of vacuum, ε_r is the relative dielectric constant of the medium, ψ_0 is the surface potential (which can be approximated with the zeta-potential of the particles in the medium), R is the particle radius, h is the distance between two particles, and κ is the reciprocal Debye-layer thickness (Debye–Hückel parameter) given by

$$\kappa = \sqrt{\frac{2CN_A z^2 e_0^2}{\varepsilon_0\varepsilon_r kT}} \tag{202}$$

where C is the concentration of electrolyte, N_A is Avogadro's number, z is the valency of the ions, e_0 is the elementary unit of charge, k is Boltzmann's constant, and T is the absolute temperature.

Dispersions of polymer-coated (sterically stabilized) spheres. The presence of an adsorbed layer of polymer (and/or surfactant) on the particle surface can have a significant influence on the viscosity of dispersions. As discussed in the previous sections, the effective volume fraction ϕ_{eff} of the dispersions is given by $\phi(1 + \delta/R)^3$, where δ is the adsorbed-layer thickness. Thus, polymerically stabilized dispersions with small particles and a thick adsorbed layer of polymer exhibit much higher viscosities than hard-sphere dispersions because of a higher effective volume fraction at the same value of ϕ. Also, upon gradual increase in ϕ to values where interpenetration and/or compression of adsorbed layers occur, the rheological behavior of dispersions becomes quite complex. Such dispersions exhibit not only yield stress but also viscoelastic behavior. Though hard-sphere dispersions also exhibit yield stress and viscoelasticity, as discussed earlier, this usually occurs near close-packing concentrations. In the case of polymerically stabilized "soft-sphere" dispersions, yield stress and viscoelasticity are observed at significantly lower values of ϕ, depending on the value of the ratio δ/R.

Figure 61 compares the viscosity data for polymerically stabilized dispersions with hard-sphere dispersion [170]. For particles of size 77.5 nm, the average thickness of adsorbed polymer layer is 6 nm ($\delta/R \sim 0.077$). For particles with the radii 305 nm and 503 nm, the average values of δ are 11 nm and 19 nm, respectively (corresponding δ/R values are 0.036 and 0.038). As can be seen from the figure, the viscosities of polymerically stabilized dispersions are much higher than those of the hard-sphere dispersions, when comparison is made at the same values of ϕ. Dispersions with a particle radius of 77.5 nm have the highest viscosities because of the high value of δ/R (hence the high ϕ_{eff}). Figure 62 shows the plots of yield stress as a function of ϕ for the same dispersions. The yield stress in dispersions with small particles is much higher

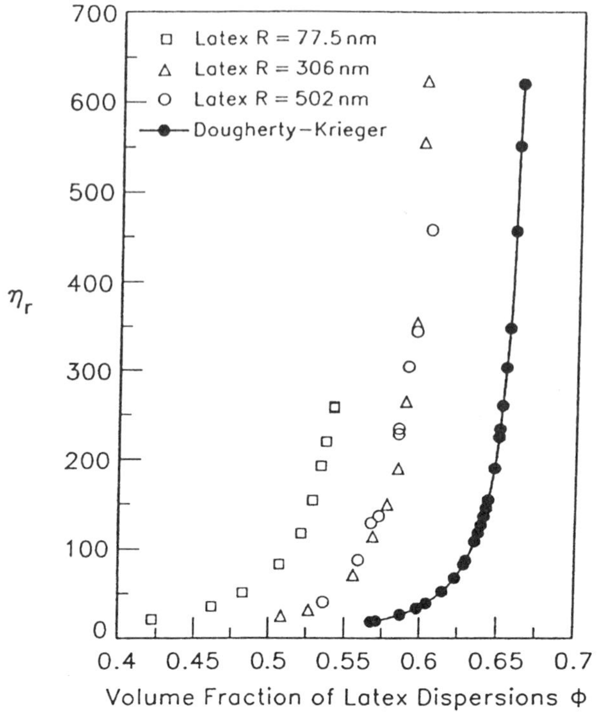

Figure 61. Viscosity data for polymerically stabilized dispersions compared with hard-sphere dispersion data. (From Ref. 170.)

than that in large-particle dispersions (at the same value of ϕ). Also, dispersions with smaller particles develop yield stress at lower values of ϕ. Again, this is owing to the fact that dispersions with small particles have δ/R almost twice the value of the other dispersions and, therefore, the interaction of adsorbed polymeric layers occurs at lower values of ϕ.

Figure 63 shows data for the storage modulus (G') and the loss modulus (G'') for polymerically stabilized dispersions at various values of ϕ [171]. At $\phi = 0.426$, $G'' > G'$, indicating a predominantly viscous response. However, when $\phi \geq 0.483$, G' is higher than G'' over the whole range of frequency, indicating predominantly elastic behavior. As mentioned earlier, hard-sphere dispersions also generally exhibit elastic behavior, but usually at much higher values of ϕ (near close-packing concentration). The main reason for observing the viscoelastic behavior at relatively much lower values of ϕ in the present case is the overlapping of adsorbed polymer layers.

Sterically stabilized dispersions also exhibit shear-thickening behavior at high values of ϕ [137,138,166–168]. Figure 64 shows the viscosity and first normal stress difference (N_1) data for sterically stabilized dispersions studied by Willey and Macosko [134] at a ϕ value of 0.54. The presence of first normal stress difference indicates that these dispersions are viscoelastic, which is to be expected because of the overlapping of adsorbed polymer layers. Also, at a

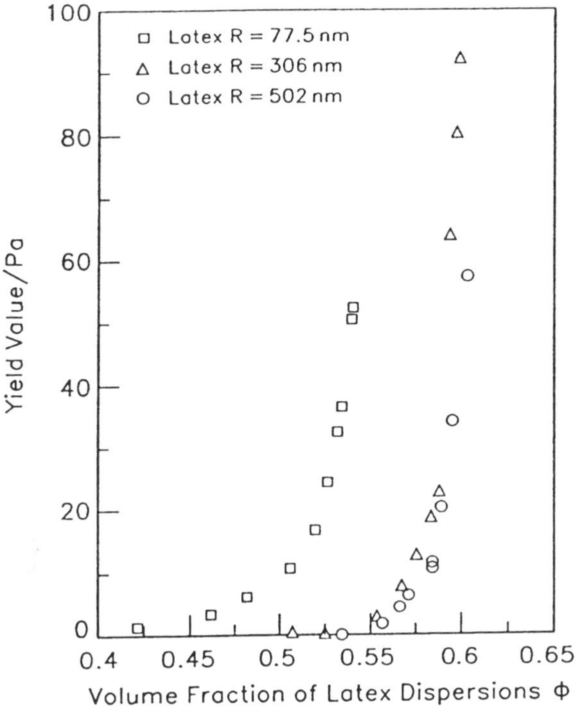

Figure 62. Yield stress as a function of ϕ for polymerically stabilized dispersions. (From Ref. 170.)

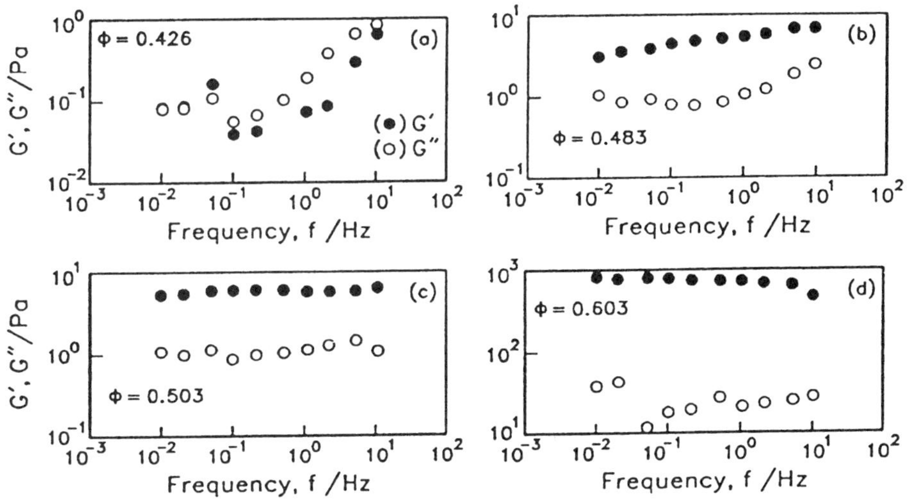

Figure 63. Storage modulus (G′) and loss modulus (G″) data for polymerically stabilized dispersions at various values of ϕ. (From Ref. 171.)

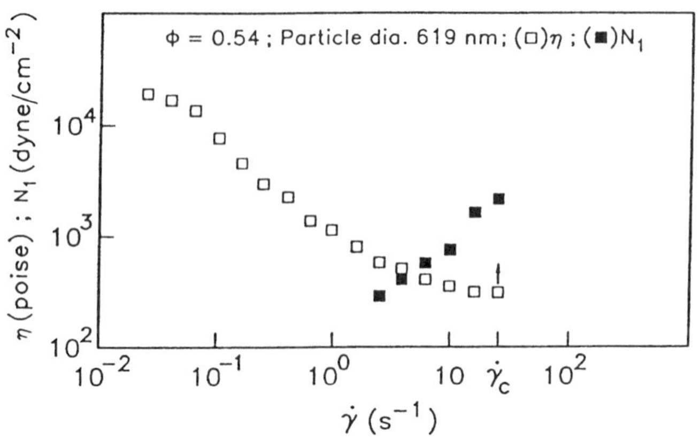

Figure 64. Viscosity and first normal stress difference (N_1) data for sterically stabilized dispersions. (From Ref. 134.)

critical shear rate $\dot{\gamma}_c$ of about 25 sec^{-1}, an abrupt increase in shear stress (viscosity) occurs, indicating shear-thickening behavior.

Finally, note that sterically stabilized dispersions do not exhibit a unique correlation of η_r versus Pe (at any given value of ϕ) as found in the case of hard-sphere dispersions, particularly when δ/R is large. As an example, Fig. 65 shows η_r versus effective Péclet number data for the sterically stabilized dispersions discussed earlier (refer to Figs. 61 and 62) at $\phi = 0.535$ [170]. While the data for dispersions with small values of δ/R superpose, the data for dispersions with large δ/R fall considerably above the other dispersions; dispersions with large δ/R exhibit relatively high viscosities because of the overlapping of adsorbed polymer layers. However, it is possible that η_r versus Pe data for dispersions with different values of δ/R would superpose, provided the comparison is made at the same ϕ_{eff} (instead of ϕ) and that the effective radius (instead of R) is used in the calculation of the Péclet number.

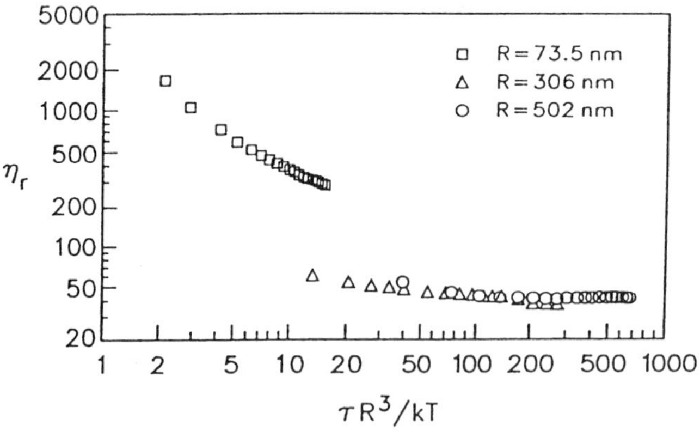

Figure 65. η_r versus effective Péclet number data for sterically stabilized dispersions at $\phi = 0.535$. (From Ref. 170.)

Viscosity equations for dispersions of "soft-spheres." Several authors have proposed theoretical and empirical viscosity equations for "soft-sphere" dispersions. However, most of the theoretical work is restricted to dilute electrostatically stabilized dispersions [133,143,144,172–178].

It is now well-known that in the case of electrostatically stabilized dispersions, the presence of an electric charge on the particles results in an increase in the viscosity owing to three distinguishable electroviscous effects. These are referred to as the primary, secondary, and tertiary electroviscous effects [179].

The primary electroviscous effect is associated with the distortion of the electrical double layer from spherical symmetry by the shear field. The Maxwell stresses generated due to an asymmetric electric field tend to restore the symmetry of the double layer, thereby opposing the flow. This results in extra dissipation of energy and hence an increase in viscosity. The primary electroviscous effect is present even in infinitely dilute dispersion, and, therefore, the Einstein coefficient (2.5 for hard spheres) is altered.

At finite concentrations, however, electrostatic interactions between the particles become important. The increase in viscosity due to Coulombic interactions between the double layers of different particles is referred to as the "secondary electroviscous effect." The presence of electrical charge on the particles contributes to the secondary electroviscous effect in two ways: First, the electrostatic repulsion between the particles affects the microstructure of the dispersion (the spatial distribution of particles). Second, when two particles approach each other in a shear field, the electrostatic repulsion between the particles affects their trajectories. As shown in Fig. 66, the electrostatic repulsion forces the particles across fluid streamlines, thereby dissipating additional energy [174]. This additional dissipation is found to be proportional to the displacement of particles normal to the undisturbed streamlines. In essence, the collision doublets (or multiplets) in electrostatically stabilized dispersion have larger radii than doublets of uncharged particles, resulting in additional dissipation [172].

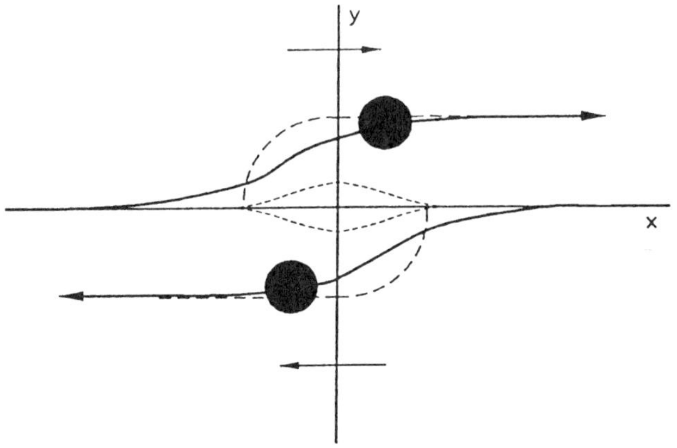

Figure 66. Interactions between two spherical particles in a shear flow. - - - uncharged particles with hydrodynamic interactions; —— charged particles without hydrodynamic interactions; – – according to Ref. 176. (From Ref. 174.)

The shear-thinning behavior observed in electrostatically stabilized dispersions (refer to Fig. 57) could be explained in terms of the secondary electroviscous effect. At rest, the microstructure of dispersion (that is, the spatial distribution of particles) is determined by Brownian motion and electrostatic forces. This microstructure is disturbed when shear force is applied to the dispersion. At low shear rates, the electrostatic repulsion force dominates over the viscous forces; therefore, the radii of the rotating collision doublets is quite large, resulting in high viscosities. With the increase in shear rate, the viscous forces dominate over the electrostatic repulsion force; consequently, the radii of the collision doublets is reduced. This results in lower viscosities [172,174,175].

The tertiary electroviscous effect is associated with the distortion of the particle itself as a result of electrostatic forces [131]. This effect is very important in solutions of polyelectrolytes. The electrostatic repulsion between charged groups on the polyelectrolyte chain determines the dimension and conformation of the polymer molecule. Owing to changes in the chemical environment (such as electrolyte concentration and valency, pH), the electrostatic repulsion between charged groups of the polyelectrolyte changes, resulting in the change of polymer size. This tertiary electroviscous effect is totally absent in rigid particle systems. However, if the particles consist of adsorbed polyelectrolytes, the tertiary effect becomes relevant because the dimension of the adsorbed polyelectrolyte layer changes due to changes in chemical environment (electrolyte concentration, pH, etc.).

Von Smoluchowski [180] was the first to consider the primary electroviscous effect in dispersions. He developed the following equation for dispersion viscosity:

$$\eta_r = 1 + \frac{5}{2}\phi\left[1 + \frac{(\varepsilon_c\zeta/2\pi)^2}{\eta_c K_c R^2}\right] \tag{203}$$

where K_c is the specific conductivity of the continuous phase, ε_c is the dielectric constant of the continuous phase, and ζ is the zeta potential. Thus, the increase in η_r due to the primary electroviscous effect is

$$\frac{5}{2}\frac{\phi}{\eta_c K_c R^2}\left(\frac{\varepsilon_c\zeta}{2\pi}\right)^2$$

Krasny-Ergen [181] modified the preceding equation to

$$\eta_r = 1 + \frac{5}{2}\phi\left[1 + \frac{1.5}{\eta_c K_c R^2}\left(\frac{\varepsilon_c\zeta}{2\pi}\right)^2\right] \tag{204}$$

This equation is identical in form to Eq. (203) except that the electroviscous contribution is multiplied by a factor of 1.5. In deriving Eq. (204), Krasny-Ergen [181] made the assumption that the thickness of the double layer is small as compared with the particle size.

Booth [182] derived a more general equation for the primary electroviscous effect. His equation for a 1:1 electrolyte in which all ions have equal mobilities is given by

$$\eta_r = 1 + \frac{5}{2}\phi\left[1 + \frac{\pi(\kappa R)^2(1+\kappa R)^2 Z}{\eta_c K_c R^2}\left(\frac{\varepsilon_c\zeta}{2\pi}\right)^2\right] \tag{205}$$

where κ is the reciprocal Debye length and Z is a complicated function of κR (κR is the ratio of particle radius to double-layer thickness). In the limiting cases of small κR and large κR, the expressions for Z are as follows: For small κR, i.e. a large double-layer thickness,

$$Z = \frac{1}{200\pi(\kappa R)} + \frac{11(\kappa R)}{3200\pi} \tag{206}$$

For large κR, i.e. a small double-layer thickness,

$$Z = \frac{3}{2\pi(\kappa R)^4} \tag{207}$$

Substitution of Eq. (207) for Z into Booth's equation (Eq. 205) gives the Krasny-Ergen expression (Eq. 204).

All of Eqs. (203)–(207) are valid only for very dilute systems, where only the primary electroviscous effect is relevant. As ζ approaches zero, these equations reduce to the Einstein equation.

Recently, Russel [133,143,144,174,175] developed a theory for the zero-shear viscosity of electrostatically stabilized dispersions at finite concentrations. However, the dispersions considered were not very concentrated, so only pair interactions were important. Russel's equation is

$$\eta_r = 1 + \frac{5}{2}(\beta\phi) + \frac{5}{2}(\beta\phi)^2 + \frac{B\phi^2}{(\kappa R)^5} \tag{208}$$

where β is the factor that takes into account the primary electroviscous effect and is given by Booth's equation (Eq. 205) and B is given as

$$B = \frac{3}{40}\left(\ln \frac{\alpha}{\ln \alpha}\right)\left[\ln \frac{\alpha}{\ln(\alpha/\ln \alpha)}\right]^4 \tag{209}$$

In this expression, α is a measure of the ratio of electrostatic force to Brownian diffusion and is given by

$$\alpha = \frac{4\pi\varepsilon_c\psi_0^2(\kappa R)\exp(2\kappa R)}{kT} \tag{210}$$

In deriving his viscosity equation (Eq. 208), Russel assumed that $\alpha \gg 1$ so that electrostatic repulsion between the particles kept the dispersion stable.

The coefficient of the ϕ^2 term in the Russel equation (Eq. 208) is compared with the experimental data in Fig. 67. It is clear that Russel's theory predicts the data very well. Note that the coefficient of the ϕ^2 term could be as high as 1000.

Little theoretical work has been done to include multiparticle interactions in electrostatically stabilized dispersions; the same can be said about sterically stabilized (i.e., polymer-coated) dispersions. However, several authors [172] have applied the hard-sphere equations to "soft-sphere" dispersions (both electrostatically and sterically stabilized) with the modification that ϕ is replaced by

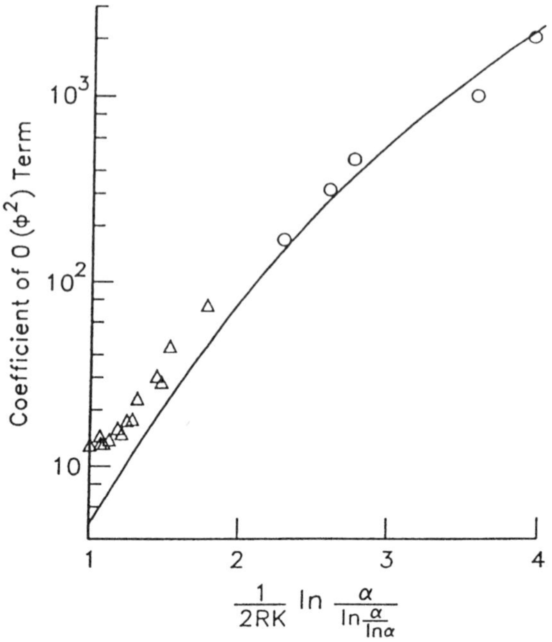

Figure 67. Comparison of experimental data for the coefficient of the ϕ^2 term with the theoretical prediction. (From Ref. 133.)

ϕ_{eff}. For example, the Mooney equation (Eq. 197) when applied to "soft-sphere" dispersions becomes

$$\eta_r = \exp\left(\frac{K\phi_{\mathrm{eff}}}{1 - \phi_{\mathrm{eff}}/\phi_m}\right) \tag{211}$$

or

$$\frac{\phi_{\mathrm{eff}}}{\ln \eta_r} = \frac{1}{K} - \frac{\phi_{\mathrm{eff}}}{K\phi_m} \tag{212}$$

where K = 2.5. If $\phi_{\mathrm{eff}} = K_0\phi$, Eq. (212) could be written as

$$\frac{\phi}{\ln \eta_r} - \frac{1}{K_0K} - \frac{\phi}{K\phi_m} \tag{213}$$

According to Eq. (213), one should obtain a straight line when $\phi/(\ln \eta_r)$ is plotted as a function of ϕ. The slope of the line is $-(1/K\phi_m)$ and its intercept is $(1/K_0K)$. Since K = 2.5, one can calculate ϕ_m and K_0 from the slope and intercept.

The equation proposed by Pal and Rhodes (Eq. 200) could also be applied to "soft-sphere" dispersions. According to this equation:

$$\left(\frac{1}{\eta_r}\right)^{0.4} = 1 - K_0K_F\phi \tag{214}$$

Thus, the intercept of $(1/\eta_r)^{0.4}$ versus ϕ data is expected to be unity, and the slope is $(-K_0K_F)$. For Newtonian dispersions, $K_F = 1$ and therefore the slope becomes ($-K_0$). As an example, Fig. 68 shows van der Waarden's data [183] for electrostatically stabilized Newtonian emulsions. These data when analyzed in terms of Eq. (214) give straight lines with intercepts of unity (refer to Fig. 69). Figure 70 shows the same data plotted as η_r versus $\phi_{eff}(= K_0\phi)$. From the figure, it is clear that Eq. (214) fits the data very well.

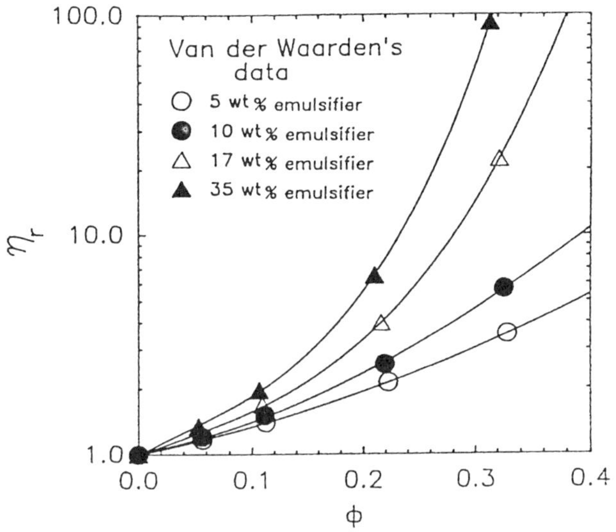

Figure 68. Van der Waarden's data for electrostatically stabilized Newtonian emulsions. (After Ref. 183.)

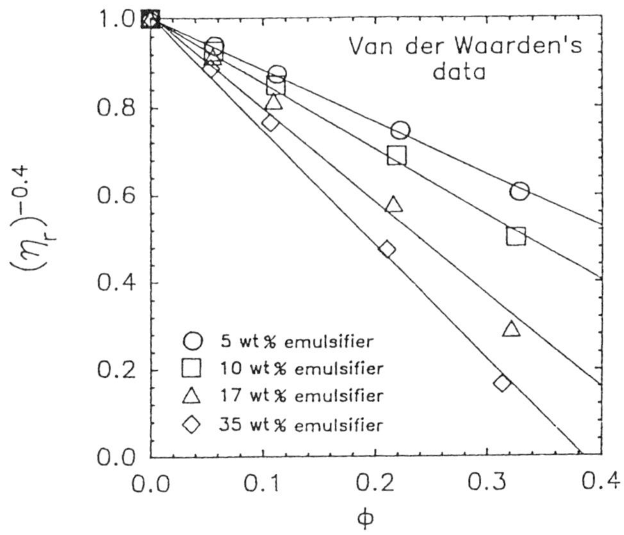

Figure 69. Van der Waarden's data analyzed in terms of Eq. (214). (After Ref. 183.)

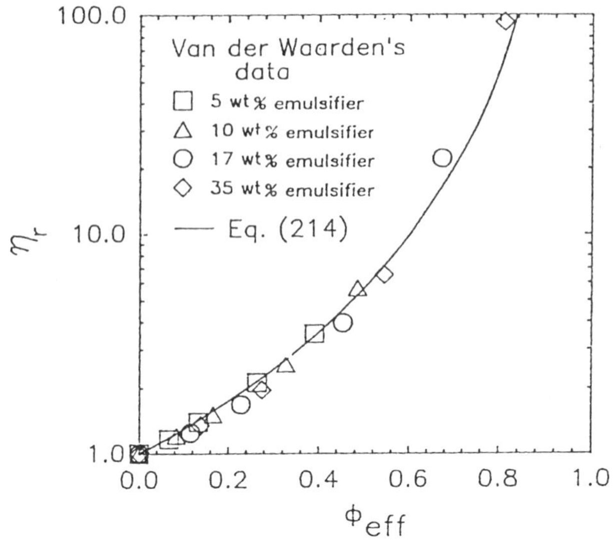

Figure 70. Van der Waarden's data plotted as η_r versus $\phi_{\text{effective}}$ (note that $\phi_{\text{effective}} = K_0\phi$). (After Ref. 183.)

3. Flocculated Emulsions

Flocculation of emulsions could occur under certain conditions. For example, if the electrical double-layer thickness or adsorbed-surfactant- (polymer-) layer thickness is significantly smaller than the distance over which van der Waals's attraction is strong, flocculation of droplets would occur in the secondary minima. Of course, rapid flocculation of droplets in the case of electrostatically stabilized dispersions can be induced by the addition of electrolytes, which lead to compression of the electrical double layer, or by the addition of surfactant or polymer of opposite charge. In all these cases, flocculation of droplets usually occurs in the primary minima, and, therefore, the flocculation is strong. In the case of sterically stabilized emulsions, flocculation of droplets can be induced by changing the solvency of the dispersion medium, which leads to compression of the adsorbed-polymer-layer thickness. As an example, Fig. 71 shows the effect of solvent quality on the relative viscosity versus shear rate plots of sterically stabilized dispersions of PVC particles, at a volume fraction of 0.20 [133,134]. When the dispersion medium is a poor solvent for the adsorbed layer, flocculation of particles takes place and flocculated dispersions exhibit much higher viscosities than stable unflocculated dispersions.

Flocculated dispersions generally exhibit pseudoplastic behavior in that the apparent viscosity decreases with an increase in shear rate. Figure 72 shows a typical flow curve for flocculated dispersions [184]. To describe such a flow curve, we need three parameters, viz., Bingham yield stress (τ_0), plastic viscosity (η_{pl}), and critical shear rate ($\dot{\gamma}_0$). The Bingham yield stress is the intercept of the linear part of the flow curve on the stress axis, and the plastic viscosity is the

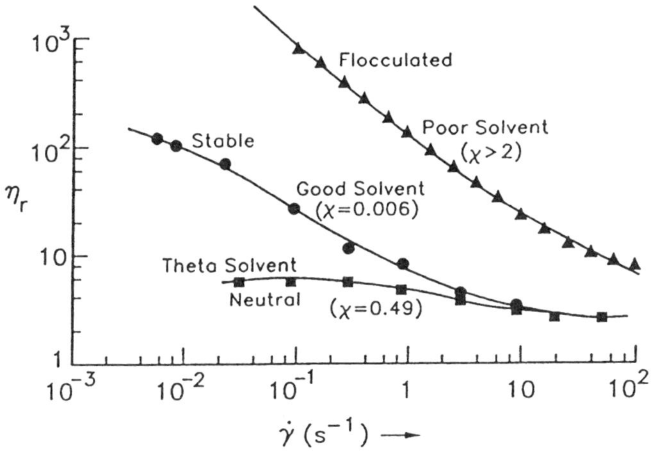

Figure 71. The effect of solvent quality on the relative viscosity versus shear rate plots. (From Refs. 133 and 134.)

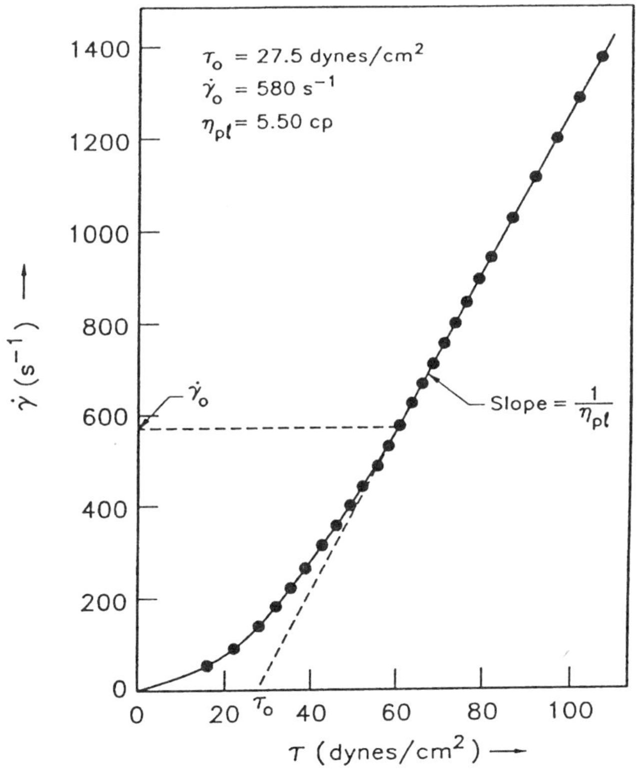

Figure 72. A typical flow curve for flocculated dispersions. (From Ref. 184.)

slope of the linear part of τ vs. $\dot{\gamma}$ plot. The linear part of the flow curve can be represented by

$$\tau = \tau_0 + \eta_{pl}\dot{\gamma} \qquad (215)$$

Note that the Bingham model (Eq. 215) is not the only one available to fit the data above $\dot{\gamma} > \gamma_0$. Several other models have been proposed in the literature, such as the Herschel–Bulkley model [87] and the Casson model [89].

Following is a brief discussion about several theories developed to explain the flow behavior of flocculated systems.

Goodeve's impulse theory of viscosity of flocculated dispersions. Goodeve [185] postulated that the presence of net attractive forces between the dispersed particles results in the formation of a "scaffolding structure"; that is, the flocculated system consists of a network of particles aggregated in the form of chains and cross chains. When the dispersion is sheared, the links (bonds) between the particles are stretched and eventually they rupture. Assuming that the particle–particle links obey Hooke's law when stretching occurs, Goodeve calculated the momentum flux or shear stress transmitted from a fast-moving layer of particles to a slower, adjacent layer. Hooke's Law is given by

$$F = Gx \qquad (216)$$

where F is the force, G is the modulus of elasticity, and x is the extension of the link. If ℓ is the distance between the adjacent layers of particles and the velocity gradient ($\dot{\gamma}$) in the dispersion is uniform, then the extension x at any given time t (after the beginning of the extension) is given by

$$x = \ell\dot{\gamma}t \qquad (217)$$

Therefore, the force is equal to

$$F = G\ell\dot{\gamma}t \qquad (218)$$

If the mean life of the link is t_0, then the magnitude of impulse (I) transmitted from a fast-moving layer to a slower, adjacent layer is

$$I = \int_0^{t_0} F\,dt = \frac{G\ell\dot{\gamma}}{2}\,t_0^2 \qquad (219)$$

Thus, the momentum flux is

$$\tau = I \times \text{rate of impulses per unit area}$$
$$\tau = I\left(\frac{n\ell}{t_0}\right) = \left(\frac{G\ell^2\dot{\gamma}}{2}\right)nt_0 \qquad (220)$$

where n is the number of links per unit volume.

If the links between the particles are weak such that the link breakage requires a small amount of energy as compared with kT (where k is Boltzmann's constant and T is absolute temperature), the links are broken mainly by thermal energy. The average life of such links that break due to thermal energy is

$$t_0 = \frac{1}{\nu}e^{E/kT} \qquad (221)$$

where E is the energy required to break the link and ν is the frequency with which redistribution of energy takes place between the link and its surroundings.

However, if the energy required to break the link is high compared with kT, the link is broken mainly by shear. The average life of links that are broken by shear is

$$t_0 = \frac{F_c}{G\ell\dot{\gamma}} \tag{222}$$

where F_c is the critical value of force at which a link breaks. Note that this equation follows from Eq. (218).

The momentum flux due to the transmission of impulses from a fast-moving layer to a slower, adjacent layer can now be calculated. The momentum flux due to the transmission of impulses by low-energy links is

$$\tau_m = \left(\frac{G\ell^2\dot{\gamma}}{2}\right)\left(\frac{n_m}{\nu}\right)e^{E/kT} \tag{223}$$

where n_m is the number of low-energy links per unit volume. Note that the low-energy links are mainly those between the molecules of the dispersion medium. Equation (223) indicates that τ_m is directly proportional to the shear rate ($\dot{\gamma}$), implying Newtonian behavior. The momentum flux due to particle–particle interactions (particle–particle links are assumed to be high-energy links) is

$$\tau_p = \left(\frac{G\ell^2\dot{\gamma}}{2}\right)\frac{n_p F_c}{G\ell\dot{\gamma}} = \frac{\ell n_p F_c}{2} \tag{224}$$

where n_p is the number of links between particles per unit volume. This momentum flux is independent of shear rate. Now, the total momentum flux is

$$\tau = \tau_p + \tau_m = \left(\frac{\ell n_p F_c}{2}\right) + \left(\frac{G\ell^2 n_m}{2\nu}\, e^{E/kT}\right)\dot{\gamma} \tag{225}$$

This equation describes an ideal Bingham plastic behavior:

$$\tau = \tau_0 + \eta_{pl}\dot{\gamma} \tag{226}$$

where τ_0 is the yield stress, given by $(\ell n_p F_c/2)$, and η_{pl} is the plastic viscosity, given by $(G\ell^2 n_m e^{E/kT}/2\nu)$. The yield stress is dependent on particle–particle interactions, whereas the plastic viscosity is dependent on the nature of the dispersion medium (the links between the dispersion-medium molecules).

Gillespie's extension of Goodeve's impulse theory. According to Goodeve's impulse theory, the relationship between the shear stress and the shear rate for flocculated dispersed systems is given by Eq. (225). As just stated, Eq. (225) could be rewritten in the form of Eq. (226), which describes Bingham plastic behavior. In deriving Eq. (225), it was assumed that particle–particle links are high-energy links that are broken by shear only. The effect of thermal energy on the number of particle–particle links and on the link lifetime was ignored. Gillespie [186,187] extended Goodeve's impulse theory by incorporating thermal effects on particle–particle links.

The momentum flux due to particle–particle links is

$$\tau_p = \left(\frac{G\ell^2\dot{\gamma}}{2}\right) n_p t_0 \tag{227}$$

If ν is the average number of links per particle, then

$$\frac{d\nu}{dt} = \alpha N_0 - \beta\nu \tag{228}$$

where α is the rate constant for link formation, N_0 is the number of basic particles per unit volume, and β is the rate constant for link rupture. At equilibrium

$$\frac{d\nu}{dt} = 0 \tag{229}$$

$$\rightarrow \quad \nu = \frac{\alpha N_0}{\beta} \tag{230}$$

Since two particles share each link, this implies that

$$n_p = \frac{\alpha N_0^2}{\beta} \tag{231}$$

Combining Eqs. (227) and (231) and taking t_0 equal to $1/\beta$ leads to

$$\tau_p = \frac{G\ell^2\dot{\gamma}\alpha N_0^2}{2\beta^2} \tag{232}$$

Note that in the derivation given by Gillespie [186], the numerical factor in the denominator of the right-hand side of Eq. (232) is 4 rather than 2; that is,

$$\tau_p = \frac{G\ell^2\dot{\gamma}\alpha N_0^2}{4\beta^2} \tag{233}$$

The reason for this discrepancy is not clear. If we now assume that the rate constant for link formation is given by

$$\alpha = (\alpha_0 + \alpha_1\dot{\gamma})n_L \tag{234}$$

where α_0 is the rate constant for flocculation due to Brownian motion, $\alpha_1\dot{\gamma}$ is the rate constant for flocculation due to shear, and n_L is the number of links formed between two contacting particles, then

$$\tau_p = \frac{G\ell^2\dot{\gamma}}{4\beta^2}(\alpha_0 + \alpha_1\dot{\gamma})n_L N_0^2 \tag{235}$$

The rate constant for link rupture can be written in a similar manner:

$$\beta = \beta_0 + \beta_1\dot{\gamma} \tag{236}$$

where β_0 is the rate constant for link rupture due to thermal effects and $\beta_1\dot{\gamma}$ is the rate constant for link rupture due to shear. Thus,

$$\tau_p = \frac{G\ell^2\dot{\gamma}(\alpha_0 + \alpha_1\dot{\gamma})}{4\beta_1^2(\beta_0/\beta_1 + \dot{\gamma})^2} n_L N_0^2 \tag{237}$$

The rate constant for link rupture due to shear ($\beta_1\dot{\gamma}$) is

$$\beta_1\dot{\gamma} = \frac{1}{t_0} = \frac{\ell\dot{\gamma}}{x_c} \tag{238}$$

$$\rightarrow \quad \beta_1 = \frac{\ell}{x_c} \tag{239}$$

where x_c is the critical separation for rupture due to shear. Therefore,

$$\tau_p = \frac{E_A N_0^2(\alpha_0 + \alpha_1\dot{\gamma})\dot{\gamma}}{2(\beta_0/\beta_1 + \dot{\gamma})^2} \tag{240}$$

where

$$E_A = \frac{1}{2}Gx_c^2 \cdot n_L = \varepsilon_A n_L \tag{241}$$

In these equations, E_A is the total binding energy between two particles and ε_A is the binding energy of a single link. Combining Eqs. (226) and (240), gives

$$\tau = \eta_{pl}\dot{\gamma} + \frac{E_A N_0^2(\alpha_0 + \alpha_1\dot{\gamma})\dot{\gamma}}{2(\beta_0/\beta_1 + \dot{\gamma})^2} \tag{242}$$

This equation is further simplified to the Bingham plastic relation at high shear rate:

$$\tau = \eta_{pl}\dot{\gamma} + \tau_0 \tag{226}$$

where τ_0, the yield stress, is given by

$$\tau_0 = \frac{1}{2}E_A N_0^2 \alpha_1 \tag{243}$$

According to Gillespie, α_1 is related to the particle radius (R) as follows:

$$\alpha_1 = \frac{32R^3}{3} \tag{244}$$

Since the volume fraction of dispersed particles, ϕ, is given by

$$\phi = N_0\left(\frac{4}{3}\pi R^3\right) \tag{245}$$

then

$$\rightarrow \quad \tau_0 = \frac{3\phi^2}{\pi^2 R^3}E_A \tag{246}$$

Thus, the yield stress is expected to increase with the volume fraction of the dispersed phase and the binding energy between particles. With the increase in particle size, the yield stress is expected to decrease.

Gillespie and Wiley [188] later showed that using rheological measurements (yield stress) and surface tension data, one can easily calculate the Hamaker constant and the minimum distance of separation between droplets (particles) in a flocculated system.

Gillespie and Wiley [188] used the following relationship between the Hamaker constant (A) and various surface energies:

$$A = 96\pi b^2(\gamma_c + \gamma_d - 2\boldsymbol{\sigma}) \tag{247}$$

where b is the minimum distance of separation between particles, γ_c is the surface tension of the continuous phase, γ_d is the surface tension of the dispersed phase, and $\boldsymbol{\sigma}$ is the interfacial tension. This relationship follows readily from the work of Bradley [189] and Hamaker [190].

For identical spheres with the separation b, the attractive energy is given by

$$E_A = \frac{-AR}{12b} \tag{248}$$

From Eqs. (247) and (248):

$$b = \frac{-E_A}{8\pi R\delta} \tag{249}$$

$$A = \frac{3}{2\pi\delta}\left(\frac{E_A}{R}\right)^2 \tag{250}$$

where $\delta = \gamma_c + \gamma_d - 2\boldsymbol{\sigma}$. Equations (249) and (250) can be rewritten in terms of the yield stress; i.e.,

$$b = \frac{\pi R^2 \tau_0}{24\phi^2\delta} \tag{251}$$

$$A = \frac{\pi^3 R^4 \tau_0^2}{6\delta\phi^4} \tag{252}$$

Casson's model. Casson [89] assumed that in a flocculated system of particles, the flocs exist in the form of long chains. These chains break down into smaller fragments if the tension experienced by the chains due to the shear force exceeds the cohesive force of the chain due to mutual attraction between the particles. Smaller chains, on the other hand, tend to unite to form longer chains upon collision. Thus, at any given shear rate, a dynamic equilibrium exists between chain buildup and breakdown processes. With the increase in shear rate, the equilibrium is shifted in the direction of better dispersion; i.e., the average chain size decreases with the increase in shear rate. The average axial ratio J (defined as ℓ/r, where ℓ is half the length of the chain and r is the radius of cross section) is given by

$$J = \alpha + \frac{\beta}{\sqrt{\eta_c \dot{\gamma}}} \tag{253}$$

Here α and β are constants for a given dispersed system; they depend on the magnitude of the interparticle forces and have positive values.

The rate of energy dissipation by the chains, which are assumed to be inflexible cylindrical rods, in unit volume of dispersion is found to be

$$\mathring{E}_{chains} = \frac{2\eta_c\dot{\gamma}^2\ell^3 N\pi a^3 J}{(aJ+1)^2} \tag{254}$$

where N is the number of rods per unit volume of dispersion and a is a constant. Since

$$N = \frac{J^2\phi}{2\pi\ell^3} \tag{255}$$

the expression for $\mathring{E}_{chains}$ becomes

$$\mathring{E}_{chains} = \frac{\eta_c\dot{\gamma}^2\phi a^3 J^3}{(aJ+1)^2} \tag{256}$$

Casson assumed that the axial ratio $J > 1$; therefore, Eq. (256) can be rewritten as

$$\mathring{E} = \eta_c\dot{\gamma}^2\phi aJ \tag{257}$$

Now, the total rate of energy dissipation ($\mathring{E}_T$) in a unit volume of dispersion includes both $\mathring{E}_{chains}$ and $\mathring{E}_{liquid}$; i.e.,

$$\mathring{E}_T = \mathring{E}_{chains} + \mathring{E}_{liquid} = (\eta_c\dot{\gamma}^2\phi aJ) + \eta_c\dot{\gamma}^2(1-\phi) \tag{258}$$

Also, $\mathring{E}_T$ can be replaced by $\eta\dot{\gamma}^2$, where η is dispersion viscosity; that is,

$$\eta\dot{\gamma}^2 = (\eta_c\dot{\gamma}^2\phi aJ) + \eta_c\dot{\gamma}^2(1-\phi) \tag{259}$$

$$\rightarrow \quad \eta = \eta_c[(\phi aJ) + (1-\phi)] \tag{260}$$

Upon substituting the expression for J, Eq. (260) becomes

$$\eta = \eta_c[1 + (a\alpha - 1)\phi] + \frac{(\sqrt{\eta_c})a\beta\phi}{\sqrt{\dot{\gamma}}} \tag{261}$$

This viscosity equation is valid only for very dilute dispersions of cylindrical chains. It does not include chain–chain interactions. Using the effective medium approach of Brinkman [151] and Roscoe [150], Casson extended Eq. (261) to concentrated dispersions. The new equation is given as follows:

$$(\eta)^{1/2} = \left(\frac{\eta_c}{(1-\phi)^A}\right)^{1/2} + \frac{B}{A}\left[\left(\frac{1}{1-\phi}\right)^{A/2} - 1\right] \tag{262}$$

where $A = a\alpha - 1$ and $B = a\beta/\sqrt{\dot{\gamma}}$.

Multiplying Eq. (262) by $\sqrt{\dot{\gamma}}$, gives

$$\tau^{1/2} = \tau_0^{1/2} + \eta_\infty^{1/2}\dot{\gamma}^{1/2} \tag{263}$$

where τ_0 is the yield stress, given by

$$\tau_0 = \left(\frac{a\beta}{a\alpha - 1}\right)^2\left[\left(\frac{1}{1-\phi}\right)^{(a\alpha-1)/2} - 1\right]^2 \tag{264a}$$

and η_∞ is the high-shear limiting viscosity, given by

$$\eta_\infty = \frac{\eta_c}{(1-\phi)^{a\alpha-1}} \tag{264b}$$

From Eq. (263), the plot of $\sqrt{\tau}$ versus $\sqrt{\dot{\gamma}}$ should be a straight line with slope of $\sqrt{\eta_\infty}$ and intercept of $\sqrt{\tau_0}$.

Cross's model. Cross [80] derived a viscosity equation by assuming that particles aggregate into random chains. The average size of the chains decreases with the increase in shear rate, so at very high shear rates, the system is completely deflocculated. If L is the number of links that an average chain has at any shear rate and N_0 is the total number of single particles (linked or separate) per unit volume, then

$$\frac{dL}{dt} = \alpha N_0 - (\beta_0 + \beta_1 \dot{\gamma}^n)L \tag{265}$$

where α is the rate constant for link formation, β_0 is the rate constant for link rupture due to Brownian motion, and $\beta_1 \dot{\gamma}^n$ is the rate constant for link rupture due to shear (β_1 and n are constants). At equilibrium, $dL/dt = 0$. Therefore,

$$L = \frac{\alpha N_0}{\beta_0 + \beta_1 \dot{\gamma}^n} \tag{266}$$

When $\dot{\gamma} = 0$, the number of links in an average chain is L_0, and hence

$$\frac{L}{L_0} = \frac{1}{1 + (\beta_1/\beta_0)\dot{\gamma}^n} \tag{267}$$

Cross then assumed that the viscosity of the dispersion is directly proportional to the average chain length (i.e., L):

$$\eta = \eta_\infty + CL \tag{268}$$

where C is a constant and η_∞ is the limiting high-shear viscosity (note that $L = 0$ when $\dot{\gamma} \to \infty$). From Eq. (268), $\eta_0 - \eta_\infty = CL_0$ and therefore,

$$\frac{L}{L_0} = \frac{\eta - \eta_\infty}{\eta_0 - \eta_\infty} \tag{269}$$

From Eqs. (267) and (269):

$$\frac{\eta - \eta_\infty}{\eta_0 - \eta_\infty} = \frac{1}{1 + (\beta_1/\beta_0)\dot{\gamma}^n} \tag{270}$$

or

$$\frac{\eta - \eta_\infty}{\eta_0 - \eta_\infty} = \frac{1}{1 + K\dot{\gamma}^n} \tag{271}$$

where $K = \beta_1/\beta_0$. This equation is well known as the Cross equation.

Elastic Floc model of Firth and Hunter. As mentioned earlier, the flow curve for flocculated dispersions often appears like the curve shown in Fig. 72; that is, three parameters are required to describe the flow behavior: τ_0, η_{pl}, and $\dot{\gamma}_0$. The model developed by Firth and Hunter [191–193] is capable of predicting these three parameters $(\tau_0, \eta_{pl}, \dot{\gamma}_0)$ from knowledge of the volume fraction of particles (ϕ), the particle radius (R), and the interparticle forces.

The Firth and Hunter model assumes that the basic flow unit at $\dot{\gamma} > \dot{\gamma}_0$ is an "elastic" floc. At a given shear rate above $\dot{\gamma}_0$, there exists a dynamic equilibrium between the growth and breakup of flocs. The floc consists of a number of smaller "domains," which are referred to as "flocculi." The average number of flocculi in a floc (and hence the floc size) decreases with an increase in the shear rate, but the quantity C_{FP}, defined as

$$C_{FP} = \frac{\text{volume fraction of flocs } (\phi_F)}{\text{volume fraction of particles } (\phi)} \tag{272}$$

remains constant for all $\dot{\gamma} > \dot{\gamma}_0$. C_{FP} is a measure of the openness of floc structure; the higher the C_{FP}, the more open is the structure of a floc; i.e., more continuous phase liquid is entrapped in the floc. For very compact flocs, C_{FP} is approximately unity; otherwise, $C_{FP} > 1$. The value of C_{FP} is generally determined by the highest shear rate, $\dot{\gamma}_{max}$, to which the dispersion is ever subjected; i.e., C_{FP} is dependent on the shear history of the dispersion. However, C_{FP} remains constant in the shear rate range of $\dot{\gamma}_0 < \dot{\gamma} < \dot{\gamma}_{max}$. Below $\dot{\gamma}_0$, large, open aggregates of flocs are formed and, therefore, C_{FP} increases with the decrease in shear rate.

According to Firth and Hunter, the rate of energy dissipation per unit volume in a flowing dispersion (assuming $\dot{\gamma} > \dot{\gamma}_0$) is given by

$$\mathring{E} = \eta\dot{\gamma}^2 = \mathring{E}_v + \mathring{E}_R + \mathring{E}_{FR} + \mathring{E}_{DR} \tag{273}$$

where $\mathring{E}_v$ is the rate of energy dissipation due to the dispersion medium flowing around the floc, $\mathring{E}_R$ is the rate of energy dissipation due to rotation of the floc, $\mathring{E}_{FR}$ is the rate of energy dissipation due to floc rupture, and $\mathring{E}_{DR}$ is the rate of energy dissipation due to rupture of the floc–floc doublets. The last two energy terms ($\mathring{E}_{FR}$ and $\mathring{E}_{DR}$) are related to the effects of collisions between flocs. Upon collision, the flocs may either coalesce to form a larger floc or form a floc–floc doublet. When coalescence or doublet formation occurs, some energy is released as heat into the surrounding liquid. However, rupture of a large floc to smaller flocs or rupture of a floc–floc doublet requires energy that must be supplied by the shear field. These energies are $\mathring{E}_{FR}$ and $\mathring{E}_{DR}$, respectively.

The rate of energy dissipation due to rupture of a floc–floc doublet ($\mathring{E}_{DR}$) consists of two parts: $\mathring{E}_{DL}$ is the rate of energy dissipation due to rupture of the links between two flocs forming a doublet, and $\mathring{E}_{DE}$ is the rate of energy dissipation due to stretching of the links between particles *within* the flocs as the tension from the shear field is transmitted to the contact area of the floc–floc doublet.

Firth and Hunter further assumed that the flocs are nearly spherical in shape and that the internal structure of a floc consists of an open network of "girders" constructed from chains of particles. The schematic diagram of a floc before and after stretching is given in Fig. 73. The number of particle chains crossing a unit area of floc in a given direction is given by an empirical relationship:

$$n_c = \frac{0.25}{C_{FP}R^2} \tag{274}$$

where R is the particle radius.

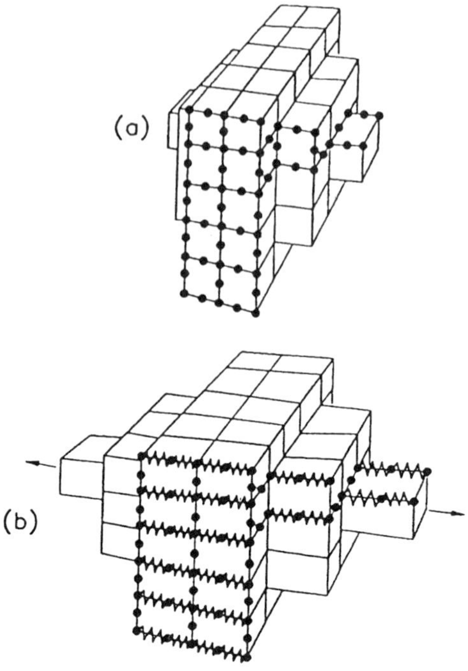

Figure 73. Schematic representation of a floc before and after stretching. (From Ref. 193.)

In order to derive the expressions for $\dot{\gamma}_0$, C_{FP}, η_{pl}, and τ_0 in terms of the colloidal properties of the dispersion (ϕ, R, zeta-potential), it is necessary to calculate the attractive force (or potential energy) between the particles. Firth and Hunter assumed that the interaction between the particles can be described by DLVO theory of colloid stability. This theory considers van der Waals attraction as well as electrostatic repulsion between the particles, although in the case of flocculated systems van der Waals attraction dominates over electrical double-layer repulsion at all separations. The total potential energy curve (G_T) assumed by Firth and Hunter is shown in Fig. 74. This curve can be described as follows:

When $b \geq b_1$

$$G_T(b) = -\left(\frac{AR}{12b}\right) + Rf(\varepsilon, \kappa b)\xi^2 \tag{275}$$

When $b_0 \leq b \leq b_1$

$$G_T(b) = -|G_0| + \frac{L}{2}(b - b_0)^2 \tag{276}$$

where b_0 is the distance of closest approach between two particles, b_1 is the separation distance at which the attractive force between the particles is maximum, ξ is the surface potential, $|G_0|$ is the magnitude of minimum potential energy, L is the elastic constant of the bond between neighboring particles, and

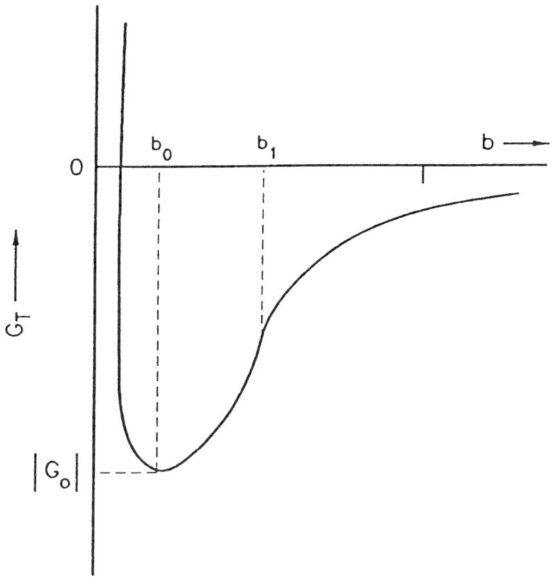

Figure 74. Total potential energy curve assumed by Firth and Hunter. (From Ref. 193.)

$f(\varepsilon, \kappa b)$ is a function of the dielectric constant (ε) and the reciprocal Debye length κ, which depends strongly on electrolyte concentration. The expression for $f(\varepsilon, \kappa b)$ at low potentials is given by

$$f(\varepsilon, \kappa b) = 2\pi\varepsilon \ln [1 + \exp(-\kappa b)] \tag{277}$$

provided that the particles approach each other under conditions of thermodynamic equilibrium; i.e., the surface potential remains constant during interaction.

Thus, the maximum force required to separate the particles, F_m, is

$$F_m = \left(\frac{dG_T}{db}\right)_{b=b_1} = R\left[\frac{A}{12b_1^2} + B(b_1)\xi^2\right] \tag{278}$$

where

$$B(b_1) = \left(\frac{\partial f}{\partial b}\right)_{b=b_1} = \frac{-2\pi\varepsilon\kappa}{1 + \exp(\kappa b_1)} \tag{279}$$

Based on Eq. (278), one can now easily derive various important relationships. Since flocs are assumed to be elastic, they undergo stretching and compression during rotation in the shear field. The maximum tensile stress in the floc at a given $\dot{\gamma}$ is

$$\boldsymbol{\sigma}_m = 5\eta_c\dot{\gamma} \tag{280}$$

The critical shear rate $\dot{\gamma}_0$, by definition, is the shear rate at which the shear field strength is just enough to break every floc–floc doublet formed by collisions between the flocs. (At $\dot{\gamma} < \dot{\gamma}_0$, the shear field is not capable of disrupting the floc–floc doublets.) Thus, a stress balance gives

$$\boldsymbol{\sigma}_{m,0} = 5\eta_c\dot{\gamma}_0 = n_F \cdot F_m \tag{281}$$

where n_F is the number of particle links (per unit area) between the colliding flocs. By substituting Eq. (278) for F_m, we obtain

$$\dot{\gamma}_0 = \frac{n_F R}{5\eta_c}\left[\frac{A}{12b_1^2} + B(b_1)\xi^2\right] \tag{282}$$

As discussed earlier, the value of C_{FP} is determined by the maximum shear rate $\dot{\gamma}_{max}$ to which the dispersion is subjected. According to Firth and Hunter, the flocs are just capable of sustaining the maximum tensile stress at $\dot{\gamma}_{max}$. Therefore, from stress balance:

$$\boldsymbol{\sigma}_{m,max} = 5\eta_c\dot{\gamma}_{max} = n_c \cdot F_m \tag{283}$$

where n_c is as it was defined in Eq. (274). Upon substituting Eqs (274) and (278) for n_c and F_m, the following relation for C_{FP} is obtained:

$$C_{FP} = \frac{1}{20R\eta_c\dot{\gamma}_{max}}\left[\frac{A}{12b_1^2} + B(b_1)\xi^2\right] \tag{284}$$

In order to derive the expression for Bingham yield stress, it is necessary to calculate $\mathring{E}_{DR}$, i.e., the rate of energy dissipation per unit volume of dispersion due to rupture of the floc–floc doublets. The stress at which rupture of a doublet takes place is given by

$$\boldsymbol{\sigma} = n_F \cdot F_m \tag{285}$$

This same stress is transmitted through the flocs as well. Thus, the force to which particle–particle bonds within the flocs are subjected can be readily determined by stress balance:

$$F = \frac{n_F}{n_c} F_m \tag{286}$$

From the total potential energy relations discussed earlier (Eqs. 275 and 276), the force between the particles is given by

$$\begin{aligned} F &= \frac{dG_T}{db} \\ &= R\left[\frac{A}{12b^2} + B(b)\xi^2\right] \qquad \text{when } b \geq b_1 \\ &= L(b - b_0) \qquad \text{when } b_0 \leq b \leq b_1 \end{aligned} \tag{287}$$

where B(b) is defined in Eq. (279) and L is given by

$$L = \frac{R}{b_1 - b_0}\left[\frac{A}{12b_1^2} + B(b_1)\xi^2\right] \tag{288}$$

Substituting the expressions for F (Eq. 287) and F_m (Eq. 278) into Eq. (286), one can calculate the extension of particle–particle bonds within the floc when the bonds are subjected to force F:

$$b = b_0 + \frac{n_F}{n_c}(b_1 - b_0) \tag{289}$$

For this extension, the energy stored in each bond is given by

$$\begin{aligned} G_T(b) + |G_0| &= \frac{L}{2}\left(\frac{n_F}{n_c}\right)^2 (b_1 - b_0)^2 \\ &= \frac{R}{2}\left(\frac{n_F}{n_c}\right)^2 (b_1 - b_0)\left[\frac{A}{12b_1^2} + B(b_1)\xi^2\right] \end{aligned} \tag{290}$$

In order to obtain the energy dissipated in stretching each bond to a separation distance b, one has to add E_L to Eq. (290). E_L is the energy dissipated owing to the movement of liquid into the narrow gap of separating spheres. Since there are $n_C/2R$ bonds per unit volume of a floc, the total energy dissipated due to stretching of particle–particle bonds in two flocs (forming a doublet) is

$$\left\{E_L + \frac{R}{2}\left(\frac{n_F}{n_c}\right)^2 (b_1 - b_0)\left[\frac{A}{12b_1^2} + B(b_1)\xi^2\right]\right\}\left(\frac{n_c}{2R}\right)\left(\frac{8}{3}\pi a^3\right)$$

where a is the floc radius. If we now multiply this expression by the frequency of collisions between the flocs per unit volume, we can calculate $\mathring{E}_{DE}$, i.e., the rate of energy dissipation per unit volume of dispersion due to stretching of links between particles within the flocs:

$$\mathring{E}_{DE} = \left(\frac{3\phi_F^2\dot{\gamma}}{\pi^2 a^3}\right)\left\{E_L + \frac{R}{2}\left(\frac{n_F}{n_c}\right)^2 (b_1 - b_0)\left[\frac{A}{12b_1^2} + B(b_1)\xi^2\right]\right\}\left(\frac{n_c}{2R}\right)\left(\frac{8}{3}\pi a^3\right) \tag{291}$$

where $3\phi_F^2\dot{\gamma}/\pi^2 a^3$ is the frequency of collisions between the flocs per unit volume. Equation (291) can be rewritten as

$$\mathring{E}_{DE} = \frac{C_{FP}\phi^2\dot{\gamma}}{\pi R^3}\left[E_L + 8R^5 n_F^2 C_{FP}^2 (b_1 - b_0)Q\right] \tag{292}$$

where

$$Q = \frac{A}{12b_1^2} - \frac{2\pi\varepsilon\kappa\xi^2}{1 + \exp(\kappa b_1)} \tag{293}$$

The rate of energy dissipation per unit volume of dispersion due to rupture of the floc–floc doublets, $\mathring{E}_{DR}$, is given by:

$$\mathring{E}_{DR} = \mathring{E}_{DL} + \mathring{E}_{DE} \tag{294}$$

Since the number of links between particles within the flocs is much larger than the number of links between flocs, $\mathring{E}_{DL} \ll \mathring{E}_{DE}$ and, therefore,

$$\mathring{E}_{DR} \simeq \mathring{E}_{DE} \tag{295}$$

Thus, the total rate of energy dissipation per unit volume of dispersion for $\dot{\gamma} > \dot{\gamma}_0$ can be written as (refer to Eq. 273):

$$\mathring{E} = \mathring{E}_v + \mathring{E}_R + \mathring{E}_{FR} + \mathring{E}_{DE} \tag{296}$$

According to Firth and Hunter, the rate of energy dissipation due to rotation of flocs ($\mathring{E}_R$), as well as the rate of energy dissipation due to rupture of flocs ($\mathring{E}_{FR}$), is negligible as compared to the other terms in Eq. (296) and, therefore,

$$\mathring{E} \simeq \mathring{E}_v + \mathring{E}_{DE} = \tau\dot{\gamma} \tag{297}$$

Since at high shear rates ($\dot{\gamma} > \dot{\gamma}_0$) $\tau = \tau_0 + \eta_{pl}\dot{\gamma}$, one can write

$$\mathring{E}_v + \mathring{E}_{DE} = \tau_0\dot{\gamma} + \eta_{pl}\dot{\gamma}^2 \tag{298}$$

Firth and Hunter further assumed that the rate of energy dissipation due to the dispersion medium flowing around the flocs (i.e., $\mathring{E}_v$) is equal to $\eta_{pl}\dot{\gamma}^2$. As a consequence,

$$\tau_0 \simeq \frac{\mathring{E}_{DE}}{\dot{\gamma}} = \frac{C_{FP}\phi^2}{\pi R^3}[E_L + 8R^5 n_F^2 C_{FP}^2(b_1 - b_0)Q] \tag{299}$$

When E_L is significantly larger than the energy required to stretch a particle–particle bond, we can neglect the second term within the brackets of Eq. (299); i.e.,

$$\tau_0 = \frac{C_{FP}\phi^2 E_L}{\pi R^3} = \frac{\phi^2 E_L Q}{20\pi R^4 \eta_c \dot{\gamma}_{max}} \tag{300}$$

When E_L is insignificant as compared to the energy required to stretch a particle–particle bond:

$$\tau_0 = \frac{8C_{FP}^3\phi^2 R^2 n_F^2(b_1 - b_0)Q}{\pi}$$
$$\tau_0 = \left[\frac{n_F^2(b_1 - b_0)}{1000\pi\eta_c^3\dot{\gamma}_{max}^3}\right]\frac{\phi^2}{R}Q^4 \tag{301}$$

Finally, the plastic viscosity η_{pl} can be calculated from either the Einstein relation or other semiempirical relations, provided that the floc volume fraction ϕ_F is used in these relations. For example, the Einstein relation [68,69] gives

$$\eta_{pl} = \eta_c(1 + 2.5\phi_F) = \eta_c(1 + 2.5C_{FP}\phi) \tag{302}$$

Obviously, this relation is restricted to low concentrations of dispersed phase. At higher volume fractions, one could use any of the following equations:

Mooney equation [152]:

$$\eta_{pl} = \eta_c \exp\left(\frac{2.5\phi_F}{1 - K\phi_F}\right) = \eta_c \exp\left(\frac{2.5C_{FP}\phi}{1 - KC_{FP}\phi}\right) \tag{303}$$

where K is about 1.4.

Dougherty–Krieger equation [153]:

$$\eta_{pl} = \eta_c\left(1 - \frac{\phi_F}{\phi_m}\right)^{-2.5\phi_m} = \eta_c\left(1 - \frac{C_{FP}\phi}{\phi_m}\right)^{-2.5\phi_m} \tag{304}$$

where ϕ_m is $\simeq 0.7$.

Thomas equation [163]:

$$\eta_{pl} = \eta_c[1 + 2.5\phi_F + 10.05\phi_F^2 + 2.73 \times 10^{-2} \exp(16.6\phi_F)] \tag{305}$$

or

$$\eta_{pl} = \eta_c[1 + 2.5C_{FP}\phi + 10.05(C_{FP}\phi)^2 + 2.73 \times 10^{-2} \exp(16.6C_{FP}\phi)] \tag{306}$$

Improved elastic floc model. Hunter and co-workers [184,194–197] later improved the elastic floc model of Firth and Hunter. According to these authors, the more appropriate relation between n_c and C_{FP} is as follows:

$$n_c = \frac{C_{FP} + 0.7}{gC_{FP}(C_{FP} - 1)R^2} \tag{307}$$

where g is a constant with a value of the order of 10. From Eqs. (307) and (283):

$$\frac{C_{FP}(C_{FP} - 1)}{C_{FP} + 0.7} = \frac{1}{5\eta_c g\dot{\gamma}_{max}R^2} F_m \tag{308}$$

In the C_{FP} range of 1.5–7 or so, this expression can be approximated as

$$C_{FP} \simeq 1.5 + \frac{1}{5\eta_c g\dot{\gamma}_{max}R}\left[\frac{A}{12b_1^2} + B(b_1)\xi^2\right] \tag{309}$$

Van de Ven and Hunter [194] envisioned that the energy term E_L consists of two parts: $E_L = E_s + E_f$, where E_s is the energy dissipation in overcoming the viscous drag experienced by each particle inside the floc when the floc is stretched and E_f is the energy required in the internal movement of liquid within the floc during stretching of a floc. Note that the floc undergoes a change in shape and volume during stretching.

Using Stokes law, E_s per sphere was found to be

$$E_s = \frac{0.9\lambda\eta_c\dot{\gamma}a^2\delta^2}{R} \tag{310}$$

where δ is the extension of particle–particle bonds and λ is a constant that takes into account the shielding effect of particles.

The energy dissipated per sphere due to the internal movement of liquid within the floc during the stretching of a floc was found to be

$$E_f = 0.9\pi^2\lambda\eta_c\dot{\gamma}a^2\delta \tag{311}$$

Therefore, the energy dissipation term E_L per sphere in the floc becomes

$$E_L = 0.9\lambda\eta_c\dot{\gamma}\delta a^2\left(\frac{\delta}{R} + \pi^2\right) \tag{312}$$

The number of spheres in a floc can be calculated from

$$n_s = \frac{a^3}{C_{FP}R^3} \tag{313}$$

Consequently, the energy dissipation term E_L per floc becomes

$$E_L = \frac{0.9\lambda\eta_c\dot{\gamma}\delta a^5}{C_{FP}R^3}\left(\frac{\delta}{R} + \pi^2\right) \tag{314}$$

Van de Ven and Hunter [194] also argued that in the absence of floc–floc interactions, the rate of energy dissipation per unit volume of the dispersion is given by

$$\mathring{E}_v = \eta_{pl}\dot{\gamma}^2 + \frac{\lambda^*\lambda\eta_c\dot{\gamma}^2 a^2\delta\phi}{R^3} \tag{315}$$

where the first term on the right-hand side takes into account the energy dissipation due to the dispersion medium flowing around the flocs (assumed to be impervious) and the second term takes into account the energy dissipation due to the internal movement of liquid within the flocs. Note that an isolated floc undergoes deformation and changes in shape and volume as it rotates in the shear field. In Eq. (315), λ^* is another constant.

Following Firth and Hunter's arguments, the total rate of energy dissipation per unit volume of dispersion is

$$\mathring{E} \simeq \mathring{E}_v + \mathring{E}_{DR} \simeq \mathring{E}_v + \mathring{E}_{DE} = (\tau_0 + \eta_{pl}\dot{\gamma})\dot{\gamma} \tag{316}$$

Combining Eqs. (315) and (316) gives

$$\tau_0 = \frac{1}{\dot{\gamma}}\left(\mathring{E}_{DE} + \frac{\lambda^*\lambda\eta_c\dot{\gamma}^2 a^2\delta\phi}{R^3}\right) \tag{317}$$

where $\mathring{E}_{DE}$ is given by

$$\mathring{E}_{DE} = \left(\frac{3\alpha_0\phi_F^2\dot{\gamma}}{\pi^2 a^3}\right)\left[2E_L + \frac{R}{2}\left(\frac{n_F}{n_c}\right)^2(b_1 - b_0)Q(2n_b)\right] \tag{318}$$

In this expression, α_0 is the orthokinetic capture efficiency and n_b is the number of bonds per floc, given by

$$n_b = \frac{a^3}{R^3}\left[\frac{C_{FP} + 0.7}{C_{FP}(C_{FP} - 1)}\right] \tag{319}$$

Upon substituting Eq. (314) for E_L and Eq. (319) for n_b into Eq. (318) for $\mathring{E}_{DE}$, one obtains

$$\mathring{E}_{DE} = \frac{6\alpha_0 C_{FP}\phi^2\dot{\gamma}}{\pi^2 R^3}\left[0.9\lambda\eta_c\dot{\gamma}\delta a^2\left(\frac{\delta}{R} + \pi^2\right) + \left(\frac{C_{FP} + 0.7}{C_{FP} - 1}\right)\frac{R}{2}\left(\frac{n_F}{n_c}\right)^2(b_1 - b_0)Q\right] \tag{320}$$

According to van de Ven and Hunter [194], in most practical situations, the energy dissipated in the internal movement of liquid within the flocs is significantly larger than the energy required to stretch the particle–particle bonds in a floc; therefore,

$$\mathring{E}_{DE} \simeq \frac{6\alpha_0 C_{FP}\phi^2\dot{\gamma}}{\pi^2 R^3}\left[0.9\lambda\eta_c\dot{\gamma}\delta a^2\left(\frac{\delta}{R} + \pi^2\right)\right] \tag{321}$$

Since $\delta/R \ll \pi^2$, Eq. (321) can be further approximated as

$$\mathring{E}_{DE} \simeq \frac{5.4\alpha_0 C_{FP}\phi^2\dot{\gamma}^2\lambda\eta_c\delta a^2}{R^3} \tag{322}$$

Thus, the yield stress is given by

$$\begin{aligned}\tau_0 &= \frac{\lambda\eta_c\dot{\gamma}a^2\delta\phi}{R^3}(\lambda^* + 5.4\alpha_0 C_{FP}\phi) \simeq 5.4\alpha_0\lambda\eta_c\dot{\gamma}\,\frac{a^2\delta}{R^3}\phi^2 C_{FP} \\ \tau_0 &= z\eta_c\delta\phi^2 C_{FP}\end{aligned} \tag{323}$$

where $z = 5.4\alpha_0\lambda\dot{\gamma}a^2/R^3$. According to this equation, it appears that τ_0 is a function of shear rate. But it should be noted that the parameter z is found to be independent of shear rate [196]. Experimental results generally tend to agree with Eq. (323). Also, using Eq. (323) one can estimate the floc size, provided that a reasonable value of δ is assumed [194].

Models for viscoelastic properties of flocculated dispersions. Not only do flocculated dispersions exhibit yield stress and shear-thinning behavior, but they also show viscoelastic behavior at small deformations [198–207]. Van den Tempel [198] derived an expression for the instantaneous elastic modulus of flocculated dispersions at low shear stress.

Van den Tempel assumed that the particles aggregate into random chains and that these chains are interlinked so as to form a three-dimensional network of particles. If there are N_0 particles per unit volume of the dispersion, then

$$\left(\frac{4}{3}\right)\pi R^3 N_0 = \phi \tag{324}$$

or

$$\pi D^3 N_0 = 6\phi \tag{325}$$

Assuming that the distance between neighboring particles in a chain is very small as compared to the particle diameter D, the total length (L_t) of all the chains per unit volume of the dispersion is

$$L_t = N_0 D = \frac{6\phi}{\pi D^2} \tag{326}$$

Let us now consider a cube-shaped element of unit volume in a flocculated dispersion. The total chain length in the cube is $6\phi/\pi D^2$. Assuming that this length is equally divided in three mutually perpendicular directions, the total chain length that supports stresses in any one direction is equal to $2\phi/\pi D^2$. This implies that any one face of the cube-shaped element of unit volume cuts through $2\phi/\pi D^2$ chains.

Van den Tempel further assumed that the van der Waals attractive force between consecutive particles in a chain is

$$F = -\frac{AD^{1.5}}{8b^{2.5}} \tag{327}$$

where b is the separation distance between neighboring particles. If the cube-shaped sample of dispersion is stretched, the separation distance between the particles increases. Let b_0 be the distance between the surfaces of neighboring particles when the element is unstrained and let b be the distance when the element is stretched. Thus, the relative extension of the cubical sample is

$$\varepsilon = \frac{b - b_0}{D} \tag{328}$$

and the shear strain corresponding to extension ε is

$$\gamma = 2\varepsilon = \frac{2(b - b_0)}{D} \tag{329}$$

The Young modulus (E) is related to uniaxial stress ($\boldsymbol{\sigma}$) and longitudinal strain (ε) as follows:

$$E = \frac{d\boldsymbol{\sigma}}{d\varepsilon} \tag{330}$$

The shear modulus (G) is given by

$$G = \frac{1}{3}E = \frac{1}{3}\left(\frac{d\boldsymbol{\sigma}}{d\varepsilon}\right) = \frac{2}{3}\left(\frac{d\boldsymbol{\sigma}}{d\gamma}\right) \tag{331}$$

Since the cohesive force of a single chain is given by Eq. (327) and there are $2\phi/\pi D^2$ chains that support the stress in any given direction, this implies that

$$\boldsymbol{\sigma} = -\left(\frac{AD^{1.5}}{8b^{2.5}}\right)\left(\frac{2\phi}{\pi D^2}\right) = -\frac{A\phi}{4\pi D^{0.5} b^{2.5}} \tag{332}$$

Therefore,

$$G = \frac{2}{3}\left(\frac{d\boldsymbol{\sigma}}{db}\right)\left(\frac{db}{d\gamma}\right) = \frac{2}{3}\left(\frac{2.5A\phi}{4\pi D^{0.5} b^{3.5}}\right)\left(\frac{D}{2}\right) \tag{333}$$

$$G = \frac{5A\phi D^{0.5}}{24\pi b^{3.5}} \quad \text{and} \quad G_0 = \frac{5A\phi D^{0.5}}{24\pi b_0^{3.5}} \tag{334}$$

where G_0 is the elastic modulus of the unstrained sample.

Thus, at constant ϕ, G_0 is expected to increase with particle diameter. However, experimental results do not support this conclusion. Sherman [199] derived a different expression for the shear modulus by assuming that the particles aggregate into spherical flocs (rather than random chains):

$$G_0 = \frac{\phi(1 + 1.828\alpha)A}{36\pi D^3 b_0^3} \tag{335}$$

where α is the volume fraction of continuous phase in the floc. According to this expression, G_0 is expected to decrease with an increase in particle size ($G_0 \propto D^{-3}$). This observation is in agreement with experimental results.

Strenge and Sonntag [200] also developed an expression for G_0, assuming that particles aggregate into chains:

$$G_0 = \frac{A\phi(D + b_0)}{8\pi D^2 b_0^2} \tag{336}$$

For $D \gg b_0$, Eq. (336) reduces to

$$G_0 = A \frac{\phi}{8\pi D b_0^2} \tag{337}$$

Like the other two models, this model predicts a linear relationship between G_0 and ϕ, but the diameter dependence is different ($G_0 \propto 1/D$). However, the experimental data show a much stronger dependence on particle size, as discussed earlier.

Van de Ven and Hunter [202] developed an expression for the storage modulus (G') of flocculated dispersions using the elastic floc model. Because flocs are assumed to be elastic in nature, they are capable of storing energy upon deformation. Consequently, the dispersion is expected to have a nonzero storage modulus (G').

The complex viscosity (η^*) of a viscoelastic material can be written as

$$\eta^* = \eta' - j\eta'' = \eta' - \frac{jG'}{\omega} \tag{338}$$

where the dynamic viscosity (η') is related to viscous energy dissipation in the material and the storage modulus (G') is related to the storage and recovery of elastic energy.

If a viscoelastic material is subjected to sinusoidal strain with amplitude γ_0 and frequency ω:

$$\gamma = \gamma_0 \sin \omega t \tag{339}$$

the resulting stress will be out of phase with the strain (stress is in advance of strain):

$$\begin{aligned} \tau &= \tau_0 \sin(\omega t + \delta) \\ &= (\tau_0 \cos \delta) \sin \omega t + (\tau_0 \sin \delta) \cos \omega t \\ \tau &= \tau' + \tau'' \end{aligned} \tag{340}$$

where τ_0 is the amplitude of the stress response, δ is the phase angle, $\tau'(= \tau_0 \cos \delta \sin \omega t)$ is the stress related to the storage and recovery of elastic energy, and $\tau''(= \tau_0 \sin \delta \cos \omega t)$ is the stress related to viscous energy dissipation. The storage modulus (G') is given by

$$G' = \frac{\text{in-phase stress amplitude}}{\text{strain amplitude}} = \frac{\tau_0 \cos \delta}{\gamma_0} \tag{341}$$

Therefore,

$$\tau' = \gamma_0 G' \sin \omega t \tag{342}$$

The rate at which energy is stored and recovered per unit volume of the material is

$$\mathring{E} = \tau' \dot{\gamma} = G' \gamma_0 \dot{\gamma} \sin \omega t \tag{343}$$

$$\rightarrow \quad \langle \mathring{E} \rangle = \frac{2G' \gamma_0 \dot{\gamma}}{\pi} \tag{344}$$

where $\langle \mathring{E} \rangle$ is the average rate at which energy is stored and recovered.

As shown earlier (Eq. 290), the energy stored during the stretching of a single particle–particle bond in a floc (E_e) is given by

$$E_e = \frac{R}{2}\left(\frac{n_F}{n_c}\right)^2 (b_1 - b_0)\left[\frac{A}{12b_1^2} + B(b_1)\xi^2\right] \tag{345}$$

The angular speed of rotation (Ω rad/sec) of a floc is given by

$$\Omega = \frac{\dot{\gamma}}{2} \tag{346}$$

and the time required for one full rotation of a floc is $4\pi/\dot{\gamma}$ sec. Since the particle–particle bonds are stretched twice during the rotation of a floc, the average rate at which energy is stored per bond is equal to $E_e\dot{\gamma}/2\pi$. Upon multiplying this energy storage rate with the number of bonds per unit volume of the dispersion (n_v), we obtain the average rate at which energy is stored and recovered per unit volume of the dispersion; i.e.,

$$\langle \mathring{E} \rangle = \frac{n_v E_e \dot{\gamma}}{2\pi} = \left(\frac{n_v\dot{\gamma}}{2\pi}\right)\frac{R}{2}\left(\frac{n_F}{n_c}\right)^2 (b_1 - b_0)\left[\frac{A}{12b_1^2} + B(b_1)\xi^2\right] \tag{347}$$

Upon comparing this equation with Eq. (344), we get

$$G' = \left(\frac{n_v}{4\gamma_0}\right)\left(\frac{R}{2}\right)\left(\frac{n_F}{n_c}\right)^2 (b_1 - b_0)\left[\frac{A}{12b_1^2} + B(b_1)\xi^2\right] \tag{348}$$

The number of bonds per unit volume is given by $n_v = N_f n_b$, where N_f is the number of flocs per unit volume of dispersion and n_b is the number of particle–particle bonds per floc:

$$N_f = \frac{3\phi_F}{4\pi a^3} = \frac{3C_{FP}\phi}{4\pi a^3} \tag{349}$$

$$n_b = \frac{a^3}{R^3}\left[\frac{C_{FP} + 0.7}{C_{FP}(C_{FP} - 1)}\right] \simeq \frac{a^3}{R^3 C_{FP}} \tag{350}$$

$$\rightarrow \quad n_v = \frac{3\phi}{4\pi R^3} \tag{351}$$

In these relations, a is the floc radius. Thus,

$$G' = \left(\frac{3\phi}{32\pi R^2 \gamma_0}\right)\left(\frac{n_F}{n_c}\right)^2 (b_1 - b_0)\left[\frac{A}{12b_1^2} + B(b_1)\xi^2\right] \tag{352}$$

This equation generally predicts the same trends as observed experimentally for flocculated dispersed systems; that is, G' is independent of frequency, inversely proportional to the maximum applied strain, and proportional to the volume fraction of the dispersed particles.

B. Emulsions of High-Viscosity Newtonian Dispersion Media

In high-viscosity dispersion mediums, the rheological behavior of emulsions is governed mainly by internal circulation and deformation of droplets.

Taylor [208] extended Einstein's analysis to infinitely dilute emulsions of two immiscible Newtonian liquids. The dispersed droplets were assumed to remain spherical by interfacial tension. According to Taylor's analysis, dilute emulsions behave as *purely viscous* Newtonian fluids with a steady-shear viscosity of

$$\eta = \eta_c\left[1 + \left(\frac{5k+2}{2k+2}\right)\phi\right] \qquad (353)$$

In the limiting case of infinite viscosity ratio (k), Eq. (353) reduces to Einstein's equation, discussed earlier. In the limit of $k \to 0$, Eq. (353) reduces to

$$\eta = \eta_c(1+\phi) \qquad (354)$$

Thus, the viscosity of dilute emulsions (where $0 < k < \infty$) is lower than that of dilute suspensions (where $k \to \infty$), provided there are no additional complications present in the system (such as an electroviscous effect and/or the inhibition of internal circulation due to the presence of surfactant).

Taylor's analysis does not consider the deformation of droplets. The deformation of droplets introduces additional complications in the rheological behavior of emulsions. Emulsions of deformable droplets are expected to exhibit viscoelastic behavior even when the emulsion is quite dilute and the dispersed and continuous phases are purely viscous (inelastic) Newtonian liquids. This happens because the dispersed droplets possess elasticity of shape owing to the restoring force of interfacial tension when they are deformed. In oscillatory flow, the viscoelastic effects are manifested as finite nonzero "storage modulus." In steady shearing flow, the presence of elastic effects is manifested as finite nonzero "normal stresses." Further, note that emulsions of deformable droplets are expected to exhibit shear-thinning behavior even when the dispersed-phase concentration is extremely low. The physical basis for this behavior can be explained as follows.

If we consider a droplet of radius R immersed in a fluid with interfacial tension $\boldsymbol{\sigma}$, then under a steady macroscopic shear flow the droplet is subjected to two opposing effects:

1. A viscous stress of magnitude $\eta_c\dot{\gamma}$ that tends to elongate the droplet ($\dot{\gamma}$ is the shear rate)
2. A stress of magnitude $\boldsymbol{\sigma}/R$ that tends to minimize the surface energy (and hence tends to maintain the droplet in a spherical shape)

Thus, the equilibrium shape of the droplet is governed by a dimensionless group $(\eta_c\dot{\gamma}R/\boldsymbol{\sigma})$. Because the droplet shape (hence the flow pattern in and around the droplet) varies with the shear rate, one should observe shear-thinning viscosity in emulsions.

Oldroyd [209,210] was the first to develop a complete constitutive equation for dilute emulsions of two immiscible Newtonian liquids, taking into consideration the deformation of droplets. For steady shearing flow, Oldroyd's analysis gives the following expressions for shear viscosity and first normal stress difference $(\tau_{11} - \tau_{22})$:

$$\eta = \eta_0 \left(\frac{1 + \lambda_1 \lambda_2 \dot{\gamma}^2}{1 + \lambda_1^2 \dot{\gamma}^2} \right) \tag{355}$$

$$\tau_{11} - \tau_{22} = \frac{2\eta_0 \dot{\gamma}^2 (\lambda_1 - \lambda_2)}{1 + \lambda_1^2 \dot{\gamma}^2} \tag{356}$$

where

$$\lambda_1 = \lambda_0 \left[1 + \phi \frac{19k + 16}{5(k+1)(2k+3)} \right] \tag{357}$$

$$\lambda_2 = \lambda_0 \left[1 - \phi \frac{3(19k + 16)}{10(k+1)(2k+3)} \right] \tag{358}$$

$$\lambda_0 = \frac{(19k + 16)(2k + 3)}{40(k+1)} \left(\frac{\eta_c R}{\sigma} \right) \tag{359}$$

$$\eta_0 = \eta_c \left[1 + \frac{5k + 2}{2(k+1)} \phi + \frac{(5k+2)^2}{10(k+1)^2} \phi^2 \right] \tag{360}$$

Thus, Oldroyd's model predicts dilute emulsions to be shear-thinning (see Eq. 355) and viscoelastic (Eq. 356) even though the two immiscible liquids involved are purely viscous (inelastic) and Newtonian in nature. Note further that the zero-shear viscosity (η_0) predicted from the Oldroyd model (Eq. 360) is the same as that predicted by Taylor's model (Eq. 353) in the limit $\phi \to 0$.

For an oscillatory shear flow of frequency ω, Oldroyd's analysis leads to the following expression for the complex modulus of an emulsion (G^*) in the linear viscoelastic region:

$$G^* = G_c^* \left(\frac{1 + 3\phi H}{1 - 2\phi H} \right) \tag{361}$$

where

$$H = \frac{(4\sigma/R)(2G_c^* + 5G_d^*) + (G_d^* - G_c^*)(16G_c^* + 10G_d^*)}{(40\sigma/R)(G_c^* + G_d^*) + (2G_d^* + 3G_c^*)(16G_c^* + 19G_d^*)} \tag{362}$$

$$G_c^* = j\omega\eta_c \tag{363}$$

$$G_d^* = j\omega\eta_d \tag{364}$$

In these equations, G_c^* and G_d^* are complex moduli of continuous and dispersed phases, respectively. It can be further shown that Eq. (361) gives the following expressions for the storage modulus (G') and the loss modulus (G''):

$$G' = \frac{\eta_0 \omega^2 (\lambda_1 - \lambda_2)}{1 + \omega^2 \lambda_1^2} \tag{365}$$

$$G'' = \frac{\eta_0 \omega (1 + \omega^2 \lambda_1 \lambda_2)}{1 + \omega^2 \lambda_1^2} \tag{366}$$

In summary, Oldroyd's model predicts nonzero first normal stress difference, nonzero storage modulus, and shear-thinning viscosity for dilute emulsions prepared from purely viscous (inelastic) Newtonian liquids. It is interesting to note that the model also predicts that

$$\eta'(\omega)\Big|_{\omega\to\dot{\gamma}} = \frac{G''(\omega)}{\omega}\Big|_{\omega\to\dot{\gamma}} = \eta(\dot{\gamma}) \tag{367}$$

$$\frac{G'(\omega)}{\omega^2}\Big|_{\omega\to\dot{\gamma}} = \left(\frac{\tau_{11} - \tau_{22}}{2\dot{\gamma}^2}\right) = \frac{\eta_0(\lambda_1 - \lambda_2)}{1 + \lambda_1^2 \dot{\gamma}^2} \tag{368}$$

where η' is the dynamic viscosity, defined as G''/ω. Thus, one can potentially predict the steady-shear properties of emulsions from dynamic measurements.

Oldroyd's model is valid only for dilute emulsions. Schowalter et al. [211] and Frankel and Acrivos [212] also developed constitutive equations for *dilute* emulsions of Newtonian liquids. Their equations give predictions similar to Oldroyd's constitutive equation. Choi and Schowalter [213] developed a constitutive equation for moderately concentrated emulsions of Newtonian liquids by taking into consideration hydrodynamic interactions between neighboring droplets. A cell model approach of Simha [146] was used to account for hydrodynamic interactions between neighboring droplets. The Choi–Schowalter constitutive equation gives the following expression for shear viscosity and first normal stress difference $(\tau_{11} - \tau_{22})$ under steady shearing flow:

$$\eta = \eta_0 \left(\frac{1 + \lambda_1 \lambda_2 \dot{\gamma}^2}{1 + \lambda_1^2 \dot{\gamma}^2}\right) \tag{369}$$

$$\tau_{11} - \tau_{22} = \frac{2\eta_0 \dot{\gamma}^2 (\lambda_1 - \lambda_2)}{1 + \lambda_1^2 \dot{\gamma}^2} \tag{370}$$

where h_1, h_2, and η_0 are:

$$h_1 = \frac{(19k + 16)(2k + 3)}{40(k + 1)} \left(\frac{\eta_c R}{\sigma}\right) \left[1 + \frac{5(19k + 16)}{4(k + 1)(2k + 3)} \phi\right] \tag{371}$$

$$h_2 = \frac{(19k + 16)(2k + 3)}{40(k + 1)} \left(\frac{\eta_c R}{\sigma}\right) \left[1 + \frac{3(19k + 16)}{4(k + 1)(2k + 3)} \phi\right] \tag{372}$$

$$\eta_0 = \eta_c \left[1 + \frac{5k + 2}{2(k + 1)} \phi + \frac{5(5k + 2)^2}{8(k + 1)^2} \phi^2\right] \tag{373}$$

For an oscillatory shear flow, the Choi–Schowalter model leads to the following expressions for the storage modulus and the loss modulus in the linear viscoelastic region:

$$G' = \frac{\eta_0 \omega^2 (h_1 - h_2)}{1 + \omega^2 h_1^2} \tag{374}$$

$$G'' = \frac{\eta_0 \omega (1 + \omega^2 h_1 h_2)}{1 + \omega^2 h_1^2} \tag{375}$$

Thus, the Choi–Schowalter model, like Oldroyd's model, predicts emulsions of purely viscous Newtonian liquids to be viscoelastic and shear-thinning. Also, the Choi–Schowalter equations are very similar to Oldroyd equations. However, the parameters h_1, h_2, and η_0 in the Choi–Schowalter equation differ from the corresponding parameters $(\lambda_1, \lambda_2, \eta_0)$ in the Oldroyd equations. For concentrated emulsions, where droplet–droplet hydrodynamic interaction is important, the parameters in the Choi–Schowalter equations are generally more realistic.

According to both the Oldroyd and the Choi–Schowalter models, the storage modulus is proportional to ω^2 and the loss modulus is proportional to ω in the low-frequency range. In the high-frequency range, these models predict that the storage modulus is constant (independent of frequency) and the loss modulus is proportional to the frequency:

$$\begin{aligned} \lim_{\omega \to 0} \left(\frac{G'(\omega)}{\omega^2} \right) &= \eta_0 (h_1 - h_2) && \text{Choi–Schowalter} \\ &= \eta_0 (\lambda - \lambda_2) && \text{Oldroyd} \end{aligned} \tag{376}$$

$$\lim_{\omega \to 0} \left(\frac{G''(\omega)}{\omega} \right) = \eta_0 \tag{377}$$

$$\begin{aligned} \lim_{\omega \to \infty} G'(\omega) &= \frac{\eta_0 (h_1 - h_2)}{h_1^2} && \text{Choi–Schowalter} \\ &= \frac{\eta_0 (\lambda_1 - \lambda_2)}{\lambda_1^2} && \text{Oldroyd} \end{aligned} \tag{378}$$

$$\begin{aligned} \lim_{\omega \to \infty} \left(\frac{G''(\omega)}{\omega} \right) &= \eta_0 \left(\frac{h_2}{h_1} \right) && \text{Choi–Schowalter} \\ &= \eta_0 \left(\frac{\lambda_2}{\lambda_1} \right) && \text{Oldroyd} \end{aligned} \tag{379}$$

VI. Rheology of Emulsions in the Presence of Free Polymer

The presence of free polymer in the dispersion medium not only increases the viscosity of the dispersion medium (i.e., produces a thickening effect) but also imparts non-Newtonian behavior to the medium. In general, the rheological behavior of the dispersion is strongly reflective of the dispersion medium rheology, provided that the presence of polymer does not cause flocculation of dispersed particles. Note that flocculation of particles is possible due to "depletion" or "bridging" effects.

A. Stable (Unflocculated) Dispersions of Rigid Particles

The flow curves of stable (unflocculated) dispersions of *rigid* particles in polymeric dispersion media are generally similar in shape to that of the dispersion medium itself. As an example, Fig. 75 shows the flow curves (viscosity versus shear-rate plots) of stable dispersions of glass spheres in a thermoplastic polymer melt at 150°C [160]. The flow curve for the polymeric dispersion medium (without solids) exhibits two regions: a *lower Newtonian region* at low shear rates, where the viscosity of the medium is constant, and a *shear-thinning region* (power-law region) at moderate shear rates, where the viscosity decreases with an increase in the shear rate. With the addition of particles to the dispersion medium, the shape of the flow curves remains the same but the viscosity increases, i.e., the flow curves shift upwards toward higher viscosities. The flow curves in the figure follow the Carreau model (see Table 1, Eq. 57), as indicated by the solid lines.

Figure 76 shows another set of flow curves, for dispersions of glass beads in aqueous solution of polyethylene oxide [214]. Again, the flow curves of dispersions are similar in shape to that of the polymeric dispersion medium.

The first normal stress difference ($\tau_{11} - \tau_{22}$) as a function of shear stress is shown in Fig. 77 for dispersions of carbon black in polystyrene melt at 180°C [215]. A reduction in normal stresses occurs when particles are added to the non-Newtonian dispersion medium. However, the shape of the normal stress versus shear stress curves is the same (linear) for dispersions and dispersion medium.

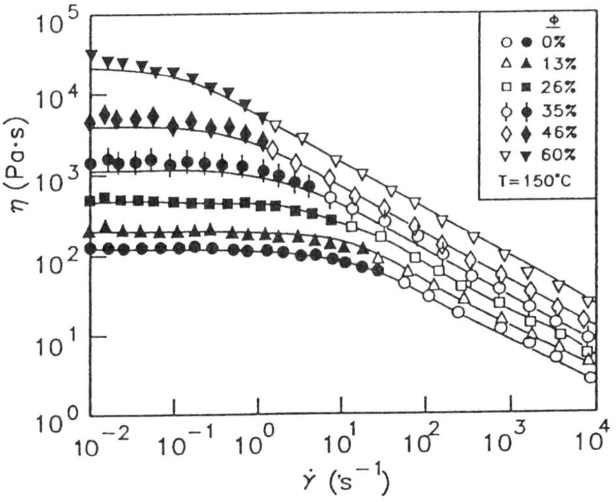

Figure 75. Flow curves for stable dispersions of glass spheres in a thermoplastic polymer melt. (From Ref. 160.)

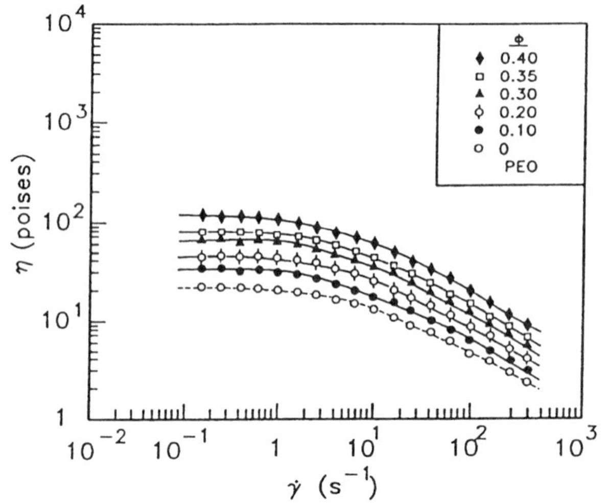

Figure 76. Flow curves for dispersions of glass beads in aqueous solution of polyethylene oxide. (From Ref. 214.)

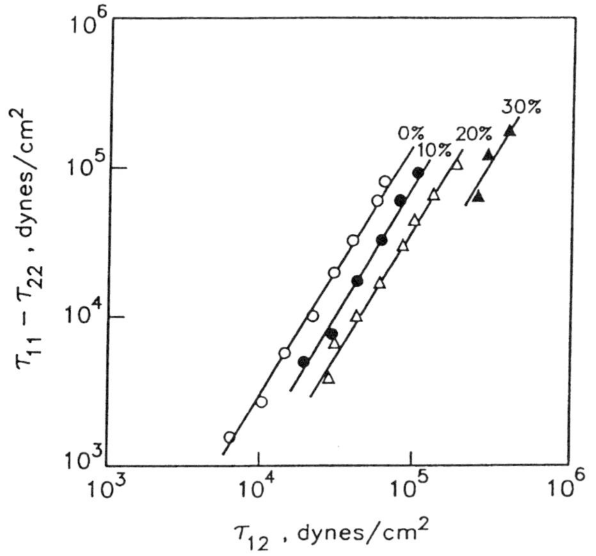

Figure 77. First normal stress difference as a function of shear stress for dispersions of carbon black in polystyrene melt. (From Ref. 215.)

In Fig. 78 are shown the results of oscillatory experiments carried out with dispersions of spherical glass beads in an NBS fluid 40 (polymeric liquid) [216]. Figure 78a shows the real part of the complex viscosity (η') and Fig. 78b the storage modulus (G'), both as a function of frequency. Again, the shape of the plots for dispersions is similar to that of the dispersion medium, although both η' and G' increase with an increase in the volume fraction of particles.

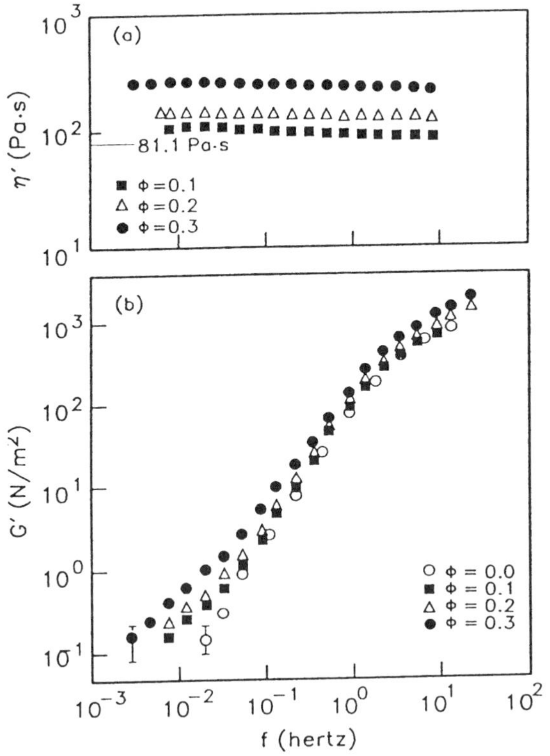

Figure 78. Results of oscillatory experiments carried out with dispersions of spherical glass beads in an NBS fluid 40. (From Ref. 216.)

From the preceding discussion it is clear that the flow behavior of stable (unflocculated) dispersions of rigid particles in non-Newtonian polymeric media is similar to that of the dispersion medium itself, although most of the rheological parameters (except first normal stress difference) increase with an increase in the volume fraction of particles. However, the flow behavior of dispersions is expected to deviate significantly from that of the dispersion medium at very high concentrations of dispersed particles. At high values of ϕ, particle–particle interactions become important.

1. Definition of Relative Viscosity

The relative viscosity of a dispersion is generally defined as the ratio of the dispersion viscosity to the suspending-medium viscosity. In the case of a Newtonian suspending medium, this definition poses no difficulty at all, because the suspending-medium viscosity is constant. However, in the case of non-Newtonian suspending media (polymeric liquids) this definition poses a difficulty in that the viscosity of the suspending medium is not unique and it varies with the shear rate or shear stress [75,76,217–219]. In the published literature on dispersions of solid particles in a non-Newtonian medium, the following three different definitions of relative viscosity have been proposed:

Constant-shear-rate relative viscosity ($\eta_{r,\dot{\gamma}}$). The constant-shear-rate relative viscosity $\eta_{r,\dot{\gamma}}$ is defined as the ratio of dispersion viscosity at a given shear rate $\eta|_{\dot{\gamma}}$ to the suspending-medium viscosity at the same shear rate $\eta_c|_{\dot{\gamma}}$:

$$\eta_{r,\dot{\gamma}} = \frac{\eta|_{\dot{\gamma}}}{\eta_c|_{\dot{\gamma}}} \tag{380}$$

Constant-shear-stress relative viscosity ($\eta_{r,\tau}$). The constant-shear-stress relative viscosity $\eta_{r,\tau}$ is defined as the ratio of dispersion viscosity at a given stress $\eta|_\tau$ to the suspending-medium viscosity at the same shear stress $\eta_c|_\tau$:

$$\eta_{r,\tau} = \frac{\eta|_\tau}{\eta_c|_\tau} \tag{381}$$

True relative viscosity ($\eta_{r,t}$). The true relative viscosity $\eta_{r,t}$ at any given shear rate $\dot{\gamma}$ is defined as the ratio of dispersion viscosity at a given shear rate $\eta|_{\dot{\gamma}}$ to the suspending-medium viscosity at the actual average shear rate $\overline{\dot{\gamma}_L}$ experienced by the suspending medium when the dispersion shear rate is $\dot{\gamma}$:

$$\eta_{r,t}|_{\dot{\gamma}} = \frac{\eta|_{\dot{\gamma}}}{\eta_c|_{\overline{\dot{\gamma}_L}}} \tag{382}$$

Calculation of relative viscosity. The calculation of constant-shear-rate relative viscosity $\eta_{r,\dot{\gamma}}$ is straightforward from the viscosity versus shear rate data for dispersion and dispersion medium. Likewise, the calculation of constant-shear-stress relative viscosity $\eta_{r,\tau}$ is straightforward from the viscosity versus shear stress data for dispersion and dispersion medium. However, calculation of true relative viscosity $\eta_{r,t}$ is not as easy. The main difficulty is that there is no simple way to calculate the average shear rate experienced by the non-Newtonian dispersion medium when the dispersion is subjected to any given macroscopic shear rate. In the case of power-law dispersions, Laven and Stein [219] recently proposed a method to calculate the average shear rate experienced by the dispersion medium. Their method is described in the following paragraphs.

The average rate of energy dissipation in the dispersion ($\mathring{E}$) can be written as:

$$\begin{aligned} \mathring{E} &= \frac{1}{V}\int_{(1-\phi)V} \eta_c \dot{\gamma}_L^2 \, dV = (1-\phi)\overline{\eta_c}\overline{\dot{\gamma}_L^2} \\ \mathring{E} &= (1-\phi)\overline{\eta_c}\overline{\dot{\gamma}_L^2} \end{aligned} \tag{383}$$

where $\dot{\gamma}_L$ is the local shear rate in the dispersion medium, $\overline{\dot{\gamma}_L^2}$ is the average of the square of the local shear rate experienced by the dispersion medium, and $\overline{\eta_c}$ is the dispersion-medium viscosity evaluated at $(\overline{\dot{\gamma}_L^2})^{1/2}$. For dispersions it is also possible to write

$$\mathring{E} = \eta|_{\dot{\gamma}}\dot{\gamma}^2 \tag{384}$$

where $\dot{\gamma}$ is the macroscopic shear rate applied to the dispersion. From Eqs. (383) and (384):

$$\eta|_{\dot{\gamma}}\dot{\gamma}^2 = (1-\phi)\overline{\eta_c \dot{\gamma}_L^2} \tag{385}$$

$$\rightarrow \quad \frac{\eta|_{\dot{\gamma}}}{\eta_c} = \eta_{r,t}|_{\dot{\gamma}} = (1-\phi)\left(\frac{\overline{\dot{\gamma}_L^2}}{\dot{\gamma}^2}\right) \tag{386}$$

We assume that the dispersion follows power-law behavior, i.e.,

$$\begin{aligned} \eta|_{\dot{\gamma}} &= K\dot{\gamma}^{n-1} \\ &= K(\dot{\gamma}^2)^{(n-1)/2} \end{aligned} \tag{387}$$

where K and n are power-law constants. From this relation, we can also write

$$\frac{\eta|_{\dot{\gamma}}}{\eta|_{\sqrt{\overline{\dot{\gamma}_L^2}}}} = \left(\frac{\dot{\gamma}^2}{\overline{\dot{\gamma}_L^2}}\right)^{(n-1)/2} \tag{388}$$

$$\rightarrow \quad \frac{\dot{\gamma}^2}{\overline{\dot{\gamma}_L^2}} = \left(\frac{\eta|_{\dot{\gamma}}}{\eta|_{\sqrt{\overline{\dot{\gamma}_L^2}}}}\right)^{2/(n-1)} \tag{389}$$

From Eqs. (386) and (389):

$$\frac{\dot{\gamma}^2}{\overline{\dot{\gamma}_L^2}} = \left(\frac{(1-\phi)\eta_c}{\eta|_{\sqrt{\overline{\dot{\gamma}_L^2}}}}\right)^{2/(n-1)}\left(\frac{\overline{\dot{\gamma}_L^2}}{\dot{\gamma}^2}\right)^{2/(n-1)} \tag{390}$$

$$\rightarrow \quad \frac{\dot{\gamma}^2}{\overline{\dot{\gamma}_L^2}} = \left(\frac{1-\phi}{\eta_{r,\dot{\gamma}}|_{\sqrt{\overline{\dot{\gamma}_L^2}}}}\right)^{2/(n+1)} \tag{391}$$

This last expression enables us to calculate the average shear rate in the dispersion medium. The following procedure can be used:

1. Assume any value of average shear rate in the dispersion medium $(\overline{\dot{\gamma}_L^2})^{1/2}$.
2. Calculate the constant-shear-rate viscosity for the dispersion at a shear rate of $(\overline{\dot{\gamma}_L^2})^{1/2}$.
3. Calculate the macroscopic shear rate $\dot{\gamma}$ from Eq. (391).

Thus, via these steps we can generate $\dot{\gamma}$ versus $(\overline{\dot{\gamma}_L^2})^{1/2}$ data. The true relative viscosity $\eta_{r,t}|_{\dot{\gamma}}$ can then be calculated at any shear rate $\dot{\gamma}$.

2. *Relative Viscosity of Dilute Dispersions*

Laven and Stein [219] analyzed a large amount of experimental data for dispersions of solid particles in non-Newtonian polymeric media. The dispersions and their suspending media exhibited power-law behavior. Figure 79 shows their

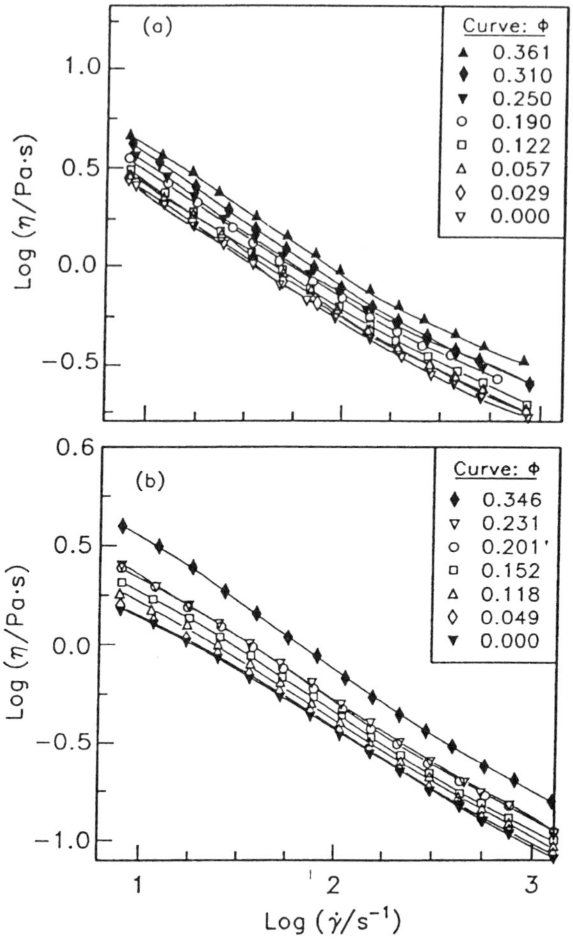

Figure 79. Viscosity data for dispersions of glass spheres in two different dispersion media. (a) Dispersion medium is 2% solution of polyethylene oxide in water; (b) dispersion medium is 2% solution of polyacrylamide in water. (From Ref. 219.)

data for dispersions of glass spheres in two different polymeric media. The dispersions and their suspending media exhibit a linear relationship between viscosity and shear rate on a log–log plot, indicating power-law behavior. The dispersions have almost the same value of the power-law index n as that of their suspending media (the slopes are the same), although the consistency index increases with the addition of particles.

From the experimental viscosity data for dispersions and their suspending media, Laven and Stein [219] calculated the relative viscosities ($\eta_{r,\dot{\gamma}}$, $\eta_{r,\tau}$, and $\eta_{r,t}$) according to the procedures discussed earlier. The Einstein coefficients were determined from the slopes of relative viscosity versus ϕ plots at $\phi = 0$. Figure 80 shows the Einstein coefficient (K_E^t) determined from the true relative viscosity ($\eta_{r,t}$) data. As can be seen, K_E^t is constant independent of the nature of the

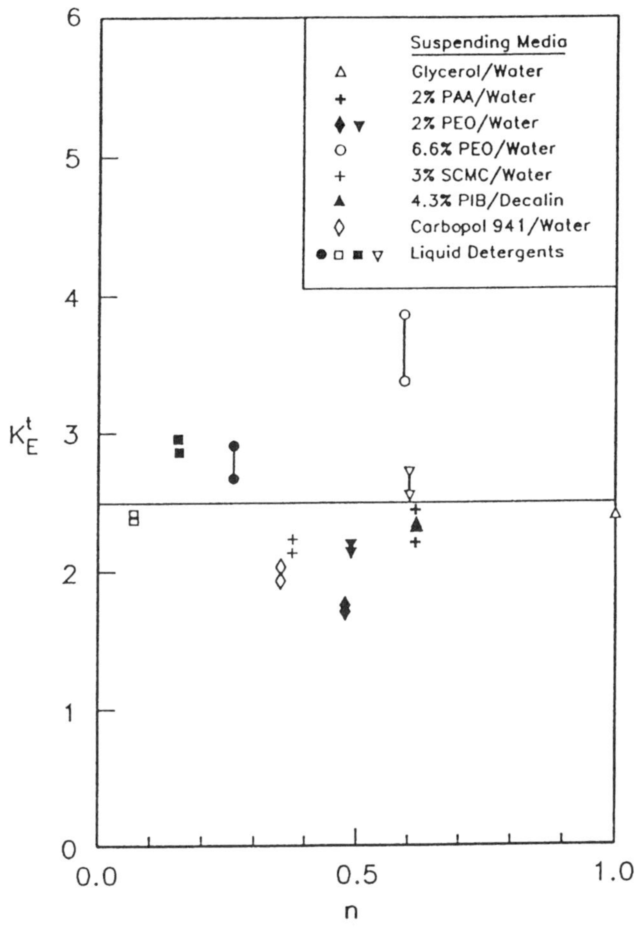

Figure 80. Einstein coefficient (K_E^t) as a function of n. (From Ref. 219.)

dispersion medium. The value of K_E^t at $n = 1$ (i.e., a Newtonian dispersion medium) and at $n = 0.1$ (an extremely shear-thinning dispersion medium) is the same, approximately 2.5, as expected from Einstein's theory.

Figure 81 shows the Einstein coefficient ($K_E^{\dot{\gamma}}$) determined from the constant-shear-rate relative viscosity ($\eta_{r,\dot{\gamma}}$) data. Unlike K_E^t, $K_E^{\dot{\gamma}}$ is a function of the nature of the dispersion medium. $K_E^{\dot{\gamma}}$ decreases linearly with a decrease in the power-law index (n) of the dispersion medium. As expected for Newtonian systems, $K_E^{\dot{\gamma}} \simeq 2.5$ when $n = 1$; at $n = 0$ (infinite pseudoplasticity), $K_E^{\dot{\gamma}} \simeq 0.75$. Using theoretical analysis, it can be easily shown that $K_E^{\dot{\gamma}}$ is a linear function of n, as observed here.

For dilute dispersions, we have already seen that

$$\eta_{r,t} = 1 + K_E^t \phi \tag{392}$$

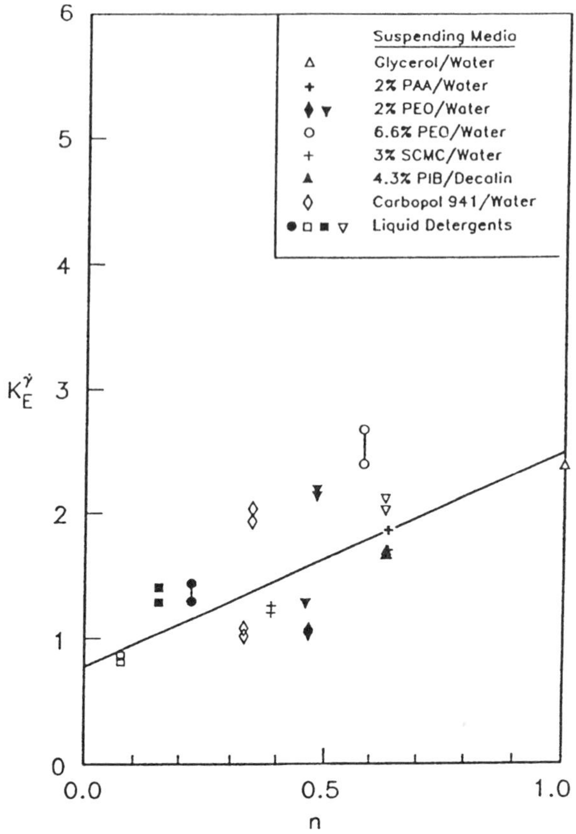

Figure 81. Einstein coefficient ($K_E^{\dot{\gamma}}$) as a function of n. (From Ref. 219.)

where $K_E^t \simeq 2.5$. From Eqs. (386) and (392):

$$\frac{\overline{\dot{\gamma}_L^2}}{\overline{\dot{\gamma}^2}} = \frac{1 + K_E^t \phi}{1 - \phi} \tag{393}$$

Upon substitution of Eq. (393) into Eq. (391):

$$\begin{aligned} \eta_{r,\dot{\gamma}} &= (1 + K_E^t \phi)^{(n+1)/2} (1 - \phi)^{(1-n)/2} \\ &= \left[1 + \left(\frac{n+1}{2}\right) K_E^t \phi + \ldots\right] \left[1 - \left(\frac{1-n}{2}\right) \phi + \ldots\right] \\ &= 1 + \left(\frac{n+1}{2}\right) K_E^t \phi + \left(\frac{n-1}{2}\right) \phi + \ldots \end{aligned} \tag{394}$$

Thus, for dilute dispersions,

$$\eta_{r,\dot{\gamma}} = 1 + \left[\left(\frac{n+1}{2}\right) K_E^t + \left(\frac{n-1}{2}\right)\right] \phi \tag{395}$$

Upon comparison of Eq. (395) with

$$\eta_{r,\dot{\gamma}} = 1 + K_E^{\dot{\gamma}}\phi \tag{396}$$

one obtains

$$K_E^{\dot{\gamma}} = \left[\left(\frac{n-1}{2}\right) + \left(\frac{n+1}{2}\right)K_E^t\right] \tag{397}$$

Since $K_E^t \approx 2.5$, $K_E^{\dot{\gamma}}$ is a linear function of the power-law index (n). When $n = 1$, Eq. (397) gives $K_E^{\dot{\gamma}} = 2.5$, as expected; when $n = 0$, $K_E^{\dot{\gamma}}$ from Eq. (397) is 0.75, as observed experimentally.

Figure 82 shows the plots of $\eta_{r,\dot{\gamma}}$ versus ϕ for different values of n, calculated from Eq. (395) with $K_E^t = 2.5$. The plots indicate that $\eta_{r,\dot{\gamma}}$ becomes less sensitive to ϕ with a decrease in n. Thus, the concentration sensitivity of $\eta_{r,\dot{\gamma}}$ is much more important in the case of Newtonian dispersion medium ($n = 1$).

In Fig. 83, the Einstein coefficients (K_E^{τ}) determined from the constant-shear-stress relative viscosity ($\eta_{r,\tau}$) data are plotted as a function of the power-law index (n). Like $K_E^{\dot{\gamma}}$, K_E^{τ} is dependent on the rheological nature of the dispersion medium. K_E^{τ} increases exponentially to infinity as n decreases from 1 to 0. At $n = 1$, K_E^{τ} is about 2.5, as expected for a Newtonian dispersion medium. Using theoretical analysis, it can be easily shown that K_E^{τ} is a nonlinear function of n, and as $n \to 0$, $K_E^{\tau} \to \infty$.

By the definition of viscosity, the shear stress in the dispersion at any given shear rate ($\dot{\gamma}_1$) is

$$\tau|_{\dot{\gamma}_1} = \eta|_{\dot{\gamma}_1}\dot{\gamma}_1 \tag{398}$$

where $\eta|_{\dot{\gamma}_1}$ is the dispersion viscosity at $\dot{\gamma}_1$. One can rewrite Eq. (398) as

$$\tau|_{\dot{\gamma}_1} = \eta_{r,\dot{\gamma}}|_{\dot{\gamma}_1}\tau_c|_{\dot{\gamma}_1} \tag{399}$$

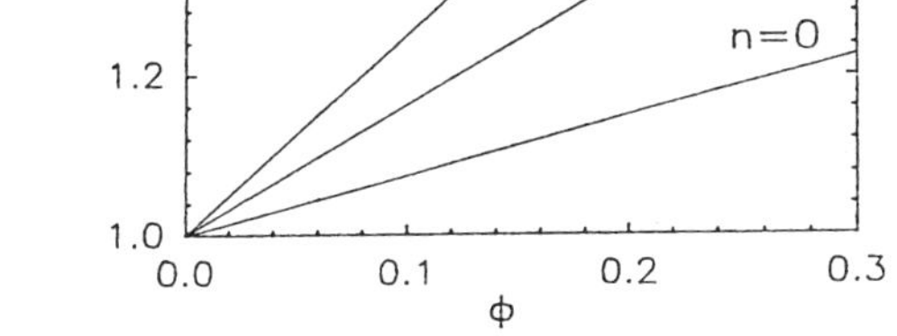

Figure 82. Plots of $\eta_{r,\dot{\gamma}}$ versus ϕ for different values of power-law index n.

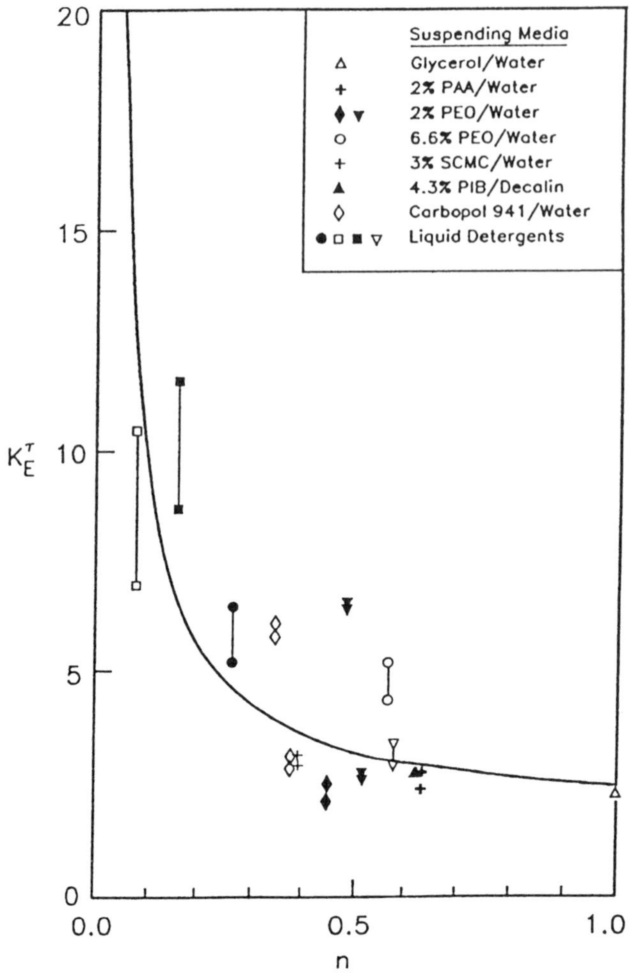

Figure 83. Einstein coefficient (K_E^τ) as a function of n. (From Ref. 219.)

where $\tau_c|_{\dot\gamma_1}$ is the shear stress developed in the pure dispersion medium when it is subjected to $\dot\gamma_1$. For dilute dispersions, Eq. (399) gives

$$\tau|_{\dot\gamma_1} = (1 + K_E^{\dot\gamma}\phi)\tau_c|_{\dot\gamma_1} \tag{400}$$

Thus, when the *same shear rate* $\dot\gamma_1$ is applied to the dispersion and to the pure dispersion medium (without particles), the stresses developed in the two systems are different in magnitude and related to each other by Eq. (400). Now, if we apply the *same shear stress* to the dispersion and to the pure dispersion medium (without particles), the shear rates developed will be different. Let the shear rate developed in the dispersion medium be $\dot\gamma_1$ and the shear rate in the dispersion be $\dot\gamma_2$. Therefore,

$$\tau|_{\dot\gamma_2} = \tau_c|_{\dot\gamma_1} \tag{401}$$

For dilute dispersions, Eqs. (400) and (401) can be combined:

$$(1 + K_E^{\dot{\gamma}}\phi)\tau_c|_{\dot{\gamma}_2} = \tau_c|_{\dot{\gamma}_1} \tag{402}$$

Since the dispersion medium is assumed to follow a power law, Eq. (402) gives

$$(1 + K_E^{\dot{\gamma}}\phi)\dot{\gamma}_2^n = \dot{\gamma}_1^n \tag{403}$$

$$\rightarrow \qquad \frac{\dot{\gamma}_2^2}{\dot{\gamma}_1^2} = (1 + K_E^{\dot{\gamma}}\phi)^{-2/n} \tag{404}$$

This equation gives the relationship between the shear rate developed in the dispersion ($\dot{\gamma}_2$) and the shear rate developed in the pure dispersion medium without solids ($\dot{\gamma}_1$), when they are subjected to the same shear stress. Furthermore, we can write

$$\frac{\eta|_{\dot{\gamma}_2}}{\eta|_{\dot{\gamma}_1}} = \left(\frac{\dot{\gamma}_2^2}{\dot{\gamma}_1^2}\right)^{(n-1)/2} \tag{405}$$

assuming that the dispersion also follows a power law (with the same n as the dispersion medium). Combining Eqs. (404) and (405) gives

$$\begin{aligned} \eta|_{\dot{\gamma}_2} &= \eta|_{\dot{\gamma}_1}(1 + K_E^{\dot{\gamma}}\phi)^{(1-n)/n} \\ &= (\eta_{r,\dot{\gamma}})\eta_c|_{\dot{\gamma}_1}(1 + K_E^{\dot{\gamma}}\phi)^{(1-n)/n} \\ &= \eta_c|_{\dot{\gamma}_1}(1 + K_E^{\dot{\gamma}}\phi)^{1/n} \end{aligned} \tag{406}$$

$$\begin{aligned} \rightarrow \qquad \eta_{r,\tau} &= \frac{\eta|_{\dot{\gamma}_2}}{\eta_c|_{\dot{\gamma}_1}} = (1 + K_E^{\dot{\gamma}}\phi)^{1/n} \\ &\simeq 1 + \frac{K_E^{\dot{\gamma}}}{n}\phi + \ldots \end{aligned} \tag{407}$$

From Eqs. (397) and (407):

$$K_E^{\tau} = \frac{K_E^{\dot{\gamma}}}{n} = \left[\left(\frac{n-1}{2n}\right) + \left(\frac{n+1}{2n}\right)K_E^t\right] \tag{408}$$

This relation predicts a nonlinear relationship between K_E^{τ} and n. Also, when $n \rightarrow 0$, $K_E^{\tau} \rightarrow \infty$; when $n = 1$, $K_E^{\tau} = K_E^t = 2.5$, as expected. The solid curve shown in Fig. 83 is a plot of the relationship between K_E^{τ} and n. Clearly, Eq. (408) describes the experimental data very well.

Figure 84 shows the plots of $\eta_{r,\tau}$ versus ϕ for different values of n, calculated from Eq. (407) with $K_E^t = 2.5$. As can be seen, $\eta_{r,\tau}$ becomes more sensitive to ϕ with a decrease in the power-law index. This trend is opposite to what was observed in the case of constant-shear-rate relative viscosity ($\eta_{r,\dot{\gamma}}$).

3. *Relative Viscosity of Concentrated Dispersions*

Several authors have developed theoretical and empirical equations for the viscosity of concentrated dispersions of particles in non-Newtonian media.

Jarzebski [220] extended the cell model approach of Frankel and Acrivos [148] to dispersions of particles in a non-Newtonian fluid. The dispersion

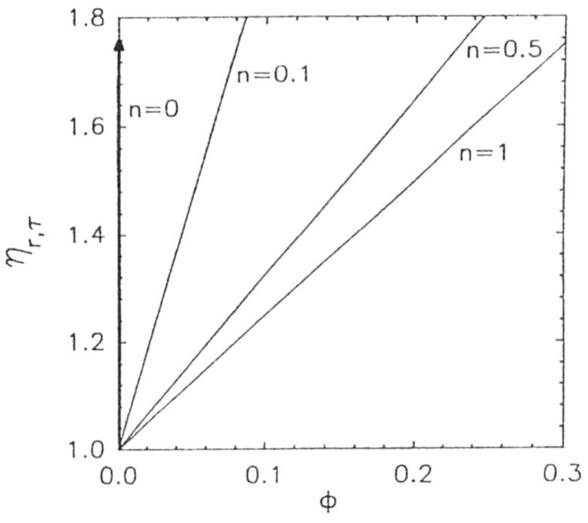

Figure 84. Plots of $\eta_{r,\tau}$ versus ϕ for different values of the power-law index n.

medium and the dispersion were assumed to exhibit power-law behavior. It was further assumed that the power-law index (n) for the dispersion medium and that for the dispersion are the same, although the consistency index (K) increases with the addition of particles. The particles of the dispersion were envisioned to reside in cells of cubical shape, as shown in Fig. 85. For a cubical arrangement of particles, the dimensions of the particle (R) and of the cell (δ) are related to the dispersed-phase (particle) concentration as follows:

$$\frac{R}{\delta} = \frac{1}{2}\left(\frac{(\phi/\phi_m)^{1/3}}{1 - (\phi/\phi_m)^{1/3}}\right) \tag{409}$$

where ϕ_m is the maximum attainable concentration of particles.

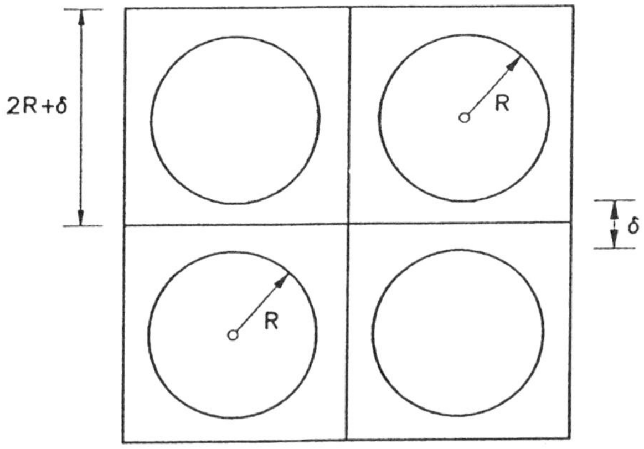

Figure 85. Particles surrounded by cubical cells. (From Ref. 220.)

The rate of energy dissipation in the dispersion was estimated by considering the hydrodynamic interaction between two neighboring particles. It was assumed that most of the energy dissipation in the dispersion is due to the squeezing motion of the fluid between the particles (see Fig. 86). Using lubrication approximation, the equations of motion were solved for a power-law fluid. The *rth* component of velocity was found to be

$$\frac{V_r}{U} = -\frac{r}{2h}\left(\frac{2n+1}{n+1}\right)\left[1-\left(\frac{2z}{h}\right)^{(n+1)/n}\right] \tag{410}$$

From knowledge of the velocity distribution, the rate of energy dissipation over the half-gap volume between the particles was found to be

$$\overset{\circ}{E} = \frac{4\pi K}{(n+1)}\left(\frac{U}{2}\right)^{n+1}\left(\frac{2n+1}{n}\right)^{n}\left(\frac{1}{2}\right)^{n+1} R^{2-n} I \tag{411}$$

where

$$I = \int_0^{\pi/2} \frac{\sin^{n+2}\theta \cos\theta \, d\theta}{[(\delta/2R) + 1 - \cos\theta]^{2n+1}} = \int_0^{\pi/2} \frac{\sin^{n+2}\theta \cos\theta \, d\theta}{[(\phi/\phi_m)^{-1/3} - \cos\theta]^{2n+1}} \tag{412}$$

This integral can be solved numerically. Its value varies with ϕ/ϕ_m and n, as shown in Fig. 87. Now, the average relative velocity between the neighboring particles is

$$U = \left[\frac{1}{12}(2R+\delta)^2 II\right]^{1/2} \tag{413}$$

where II is the second invariant of the rate of deformation tensor in the surrounding fluid. Substituting Eq. (413) for U into Eq. (411), one obtains

$$\overset{\circ}{E} = \frac{4\pi K}{n+1}\left(\frac{1}{2}\right)^{2(n+1)}\left(\frac{2n+1}{n}\right)^{n} R^{2-n}\left[\frac{1}{12}(2R+\delta)^2 II\right]^{(n+1)/2} I \tag{414}$$

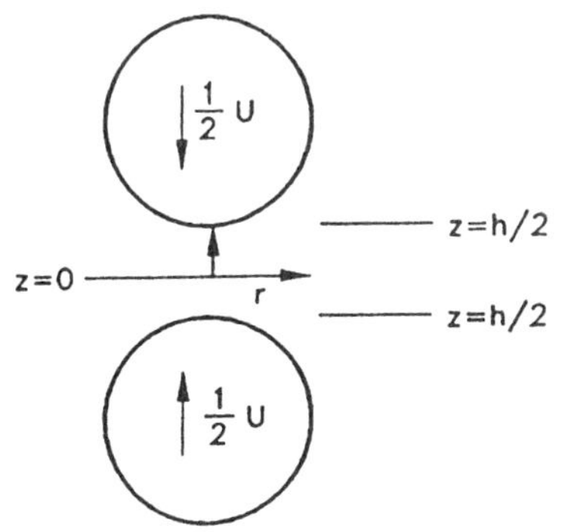

Figure 86. Squeezing motion of the fluid between two neighboring particles. (From Ref. 220.)

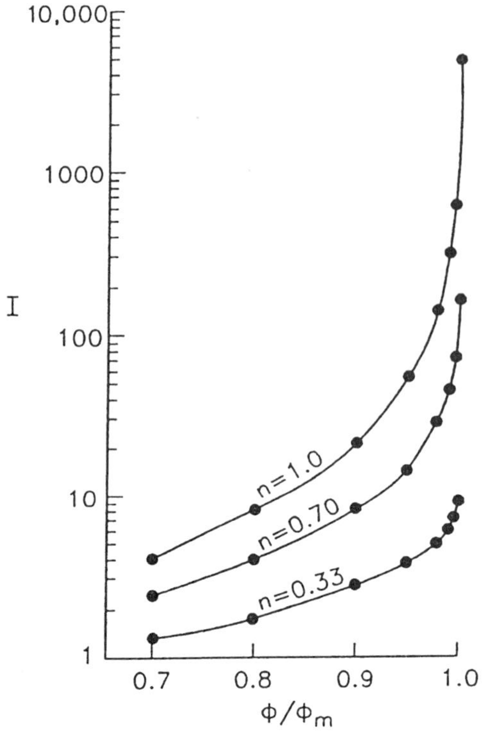

Figure 87. I as a function of ϕ/ϕ_m for different values of n. (From Ref. 220.)

This is the rate of energy dissipation over the half-gap volume between two neighboring particles. Since there are six particles surrounding a given particle in a cubical packing, Eq. (414) should be multiplied by a factor of 6 to obtain the energy dissipation within a single cubical cell:

$$\overset{\circ}{E}_{cell} = \frac{24\pi K}{n+1}\left(\frac{1}{2}\right)^{2(n+1)}\left(\frac{2n+1}{n}\right)^{n} R^{2-n}\left[\frac{1}{12}(2R+\delta)^2 II\right]^{(n+1)/2} I \tag{415}$$

If we now treat the dispersion as a homogeneous fluid (macroscopically), the rate of energy dissipation per unit volume of the dispersion is given by

$$\overset{\circ}{E} = \frac{1}{V}\int_V -(\overline{\overline{\tau}} : \nabla\vec{V})\, dV \tag{416}$$

where $\overline{\overline{\tau}}$ is the deviatoric stress tensor and $\nabla\vec{V}$ is the gradient of velocity field. Assuming dispersion to be a power-law fluid with K* as the consistency index and n as the power-law index (the same as that of the dispersion medium), Eq. (416) leads [220] to

$$\overset{\circ}{E} = \left(\frac{K^*}{n+1}\right)\left(\frac{1}{2}\right)^{(n-1)/2}(II)^{(n+1)/2} V \tag{417}$$

$$\rightarrow \quad \overset{\circ}{E}_{cell} = \left(\frac{K^*}{n+1}\right)\left(\frac{1}{2}\right)^{(n-1)/2} (II)^{(n+1)/2} V_{cell} \tag{418}$$

For a cubical cell, $V_{cell} = 8(R + \delta/2)^3$. Consequently,

$$\overset{\circ}{E}_{cell} = \left(\frac{8K^*}{n+1}\right)\left(\frac{1}{2}\right)^{(n-1)/2} (II)^{(n+1)/2}\left(R + \frac{\delta}{2}\right)^3 \tag{419}$$

Equating Eq. (419) with Eq. (415), and assuming that $\delta \ll R$, gives

$$\frac{K^*}{K} = \pi\left(\frac{3}{32}\right)^{1/2}\left(\frac{2n+1}{n}\right)^n\left(\frac{1}{24}\right)^{n/2} I \tag{420}$$

This equation gives the ratio of consistency index of suspension to consistency index of dispersion medium. Since

$$\begin{aligned} \eta_{r,\dot{\gamma}} &= \frac{\eta|_{\dot{\gamma}}}{\eta_c|_{\dot{\gamma}}} = \frac{K^*\dot{\gamma}^{n-1}}{K\dot{\gamma}^{n-1}} \\ &= \frac{K^*}{K} \end{aligned} \tag{421}$$

the constant-shear-rate relative viscosity is given by

$$\eta_{r,\dot{\gamma}} = \pi\left(\frac{3}{32}\right)^{1/2}\left(\frac{2n+1}{n}\right)^n\left(\frac{1}{24}\right)^{n/2} I \tag{422}$$

According to this equation, the influence of dispersed-phase concentration on $\eta_{r,\dot{\gamma}}$ increases with the increase in the power-law index n. This observation is consistent with the results for dilute dispersions discussed earlier.

Finally, note that if the shape of the cell were assumed to be spherical rather than cubical, Eq. (422) would have to be divided by a factor of $\pi/6$; i.e.,

$$\eta_{r,\dot{\gamma}} = \left(\frac{27}{8}\right)^{1/2}\left(\frac{2n+1}{n}\right)^n\left(\frac{1}{24}\right)^{n/2} I \tag{423}$$

Tanaka and White (221) also developed an expression for constant-shear-rate relative viscosity ($\eta_{r,\dot{\gamma}}$) of concentrated dispersions, using the cell model approach of Frankel and Acrivos [148]. The dispersion medium was assumed to exhibit power-law behavior. The rate of energy dissipation in a dispersion is given by:

$$\overset{\circ}{E} = \eta|_{\dot{\gamma}}\dot{\gamma}^2 \tag{424}$$

When the same shear rate is applied to the dispersion medium, the rate of energy dissipation is given by:

$$\overset{\circ}{E}_c = \eta_c|_{\dot{\gamma}}\dot{\gamma}^2 \tag{425}$$

From the above two equations:

$$\eta_{r,\dot{\gamma}} = \frac{\eta|_{\dot{\gamma}}}{\eta_c|_{\dot{\gamma}}} = \frac{\overset{\circ}{E}}{\overset{\circ}{E}_c} \tag{426}$$

By considering the hydrodynamic interaction between two neighboring particles, Tanaka and White [221] derived the following expression for $\mathring{E}/\mathring{E}_c$:

$$\frac{\mathring{E}}{\mathring{E}_c} = C[(\phi_m/\phi)^{1/3} - 1]^{1-2n} \tag{427}$$

where C is a constant. Thus,

$$\eta_{r,\dot{\gamma}} = C[(\phi_m/\phi)^{1/3} - 1]^{1-2n} \tag{428}$$

Since the right-hand side of this equation is independent of shear rate, the presence of particles does not affect the shear-rate dependence of viscosity. In other words, the addition of particles to a non-Newtonian dispersion medium simply shifts the viscosity versus shear rate curve vertically upward without affecting the shape of the curve (i.e., n remains the same). Also, note that the influence of ϕ on $\eta_{r,\dot{\gamma}}$ increases with the increase in n. This is consistent with the analyses of Laven and Stein [219] and Jarzebski [220]. However, note that at $n = 0.5$, Eq. (428) (from Tanaka and White) predicts that the viscosity of the dispersion is constant, independent of ϕ. This is clearly not in agreement with the experimental observations.

Kawase and Ulbrecht [222] used Happel's free-surface cell model [147] to develop an expression for the viscosity of concentrated dispersions of particles in power-law liquids. In Happel's model, the equations of motion are solved around a single reference particle surrounded by fluid. The hydrodynamic interaction between the reference particle and the neighboring particles is taken into consideration by assuming that each particle is surrounded by a hypothetical fluid cell (spherical in shape) whose outside surface is frictionless. The radius of this hypothetical cell is related to the dispersed-phase concentration and particle size. For the hypothetical cell considered by Kawase and Ulbrecht [222], the ratio of cell radius (R_c) to particle radius (R) is as follows:

$$\ell = \frac{R_c}{R} = \frac{0.605}{\phi} \tag{429}$$

Based on Happel's model, Kawase and Ulbrecht first derived the expression for the drag force on a particle in an assembly of particles dispersed in a power-law fluid. Their expression for the drag force (F_D) is as follows:

$$F_D = 2\pi K \frac{U^n}{R^{n-2}} g_1(n, \ell) \tag{430}$$

where K and n are power-law constants, U is the terminal velocity of a particle, and $g_1(n, \ell)$ is a complicated function of n and ℓ., Note that ℓ is a function of ϕ only as indicated by Eq. (429). Equation (430), for drag force on a particle, is valid for concentrated dispersions. If we take the limit $\phi \to 0$ (i.e., $\ell \to \infty$), Eq. (430) should give the drag force on a single particle (without particle–particle hydrodynamic interactions):

$$F_D|_{\phi \to 0} = 2\pi K \frac{U^n}{R^{n-2}} g_2(n) \tag{431}$$

where $g_2(n)$ is the following function of n:

$$g_2(n) = \left(\frac{27}{4}\right)^{(n-1)/2} \left[\frac{3(-22n^2 + 29n + 2)}{n(n+2)(2n+1)}\right] \tag{432}$$

Assuming that the dispersion follows a power law with the same n as that of the dispersion medium, Kawase and Ulbrecht then determined the relative consistency index (K_r), which is defined as the ratio of the consistency index of the dispersion to that of the dispersion medium. In deriving the expression for K_r, it was further assumed that K_r is equal to the ratio of F_D to $F_D|_{\phi\to 0}$; that is,

$$K_r = \frac{F_D}{F_D|_{\phi\to 0}} = \frac{g_1(n,\ell)}{g_2(n)} \tag{433}$$

Note that K_r is the same as the constant-shear-rate relative viscosity ($\eta_{r,\dot{\gamma}}$) because the power-law index of the dispersion and that of the dispersion medium are assumed to be the same. Therefore,

$$\eta_{r,\dot{\gamma}} = \frac{g_1(n,\ell)}{g_2(n)} \tag{434}$$

Like the previous models, this model predicts that the influence of the volume fraction of particles (ϕ) on the relative viscosity increases with an increase in the power-law index (n). Also, when $n = 1$ Eq. (434) reduces to

$$\eta_r = \frac{1 + 8.203\phi^5}{1 - 2.478\phi + 18.456\phi^5 - 20.326\phi^6} \tag{435}$$

and in the limit $\phi \to 0$, this equation becomes

$$\eta_r = 1 + 2.48\phi \tag{436}$$

which is almost identical with Einstein's relationship.

So far we have discussed only theoretical models. Several authors have proposed empirical viscosity equations for concentrated dispersions of particles in non-Newtonian liquids. Jarzebski [220] proposed the following empirical expressions for the relative consistency index of dispersions (K_r):

$$K_r = \frac{9}{8}\left[\frac{(\phi/\phi_m)^{1/3}}{1 - (\phi/\phi_m)^{1/3}}\right]^n \tag{437}$$

$$\ln K_r = \left(\frac{2.5\phi}{1 - \phi/\phi_m}\right)^n \tag{438}$$

Since the power-law index n for the dispersion and that for its continuous phase were assumed by Jarzebski to be the same, the relative consistency index of Eqs. (437) and (438) is the same as the constant-shear-rate relative viscosity ($\eta_{r,\dot{\gamma}}$). According to Eqs. (437) and (438), which are based on the Frankel and Acrivos [148] equation (Eq. 188) and the Mooney [152] equation (Eq. 197), the influence of particle concentration (ϕ) on K_r, and hence $\eta_{r,\dot{\gamma}}$, increases with an increase in n, as expected. However, Eqs. (437) and (438) have not been tested experimentally.

Kitano et al. [223] suggested that the constant shear-stress relative viscosity $\eta_{r,\tau}$ of dispersions of particles in non-Newtonian media can be described adequately by the following equation:

$$\eta_{r,\tau} = \left(1 - \frac{\phi}{\phi_m}\right)^{-2} \tag{439}$$

This equation was originally proposed by Maron and Pierce [224]. Several authors [225] have found Equation (439) fits their data very well. For example, Fig. 88 shows the relative viscosity data of Poslinski et al. [160], Chong et al. [139], and Kitano et al. [223], plotted as $\eta_{r,\tau}$ versus ϕ/ϕ_m. The figure also shows the plot of Eq. (439). Clearly, this equation fits the data rather well. However, according to Laven and Stein [219], $\eta_{r,\tau}$ is a function not only of volume fraction of particles but also of the rheology of the dispersion medium (refer to Fig. 84).

4. *Viscoelastic Properties of Dispersions*

As discussed in the preceding sections, dispersions of particles in polymeric media exhibit viscoelastic behavior, particularly if the dispersion medium itself is viscoelastic.

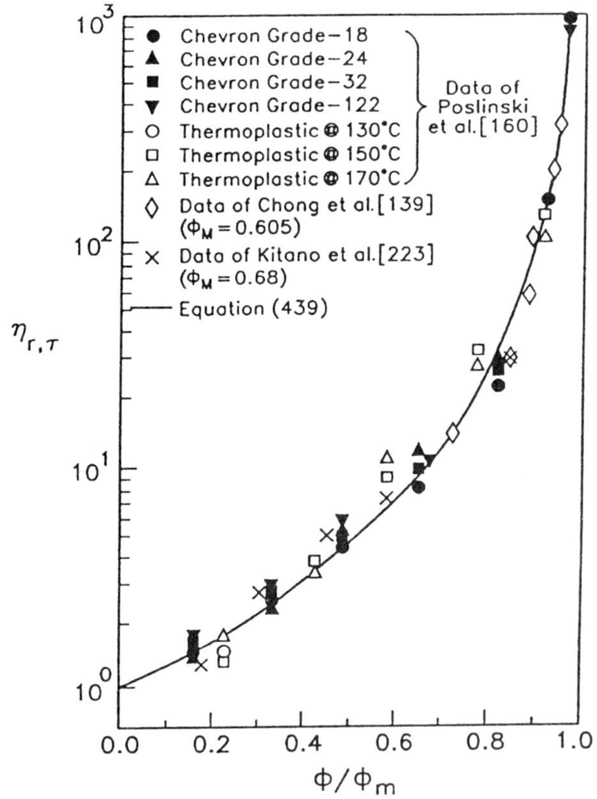

Figure 88. Comparison of experimental data with the empirical equation of Maron and Pierce. (From Ref. 160.)

Poslinski et al. [160] have recently proposed empirical correlations to correlate the viscoelastic properties of dispersions in polymeric media. They found that the relative first normal stress coefficient (Ψ_{1r}) can be correlated as follows:

$$\Psi_{1r} = \left(1 - \frac{\phi}{\phi_m}\right)^{-2} \tag{440}$$

where Ψ_{1r} is defined as the ratio of the first normal stress coefficient of the dispersion (Ψ_1) to that of the dispersion medium (Ψ_{1c}) at the same shear rate. Note that the first normal stress coefficient is defined as $(\tau_{11} - \tau_{22})/\dot{\gamma}^2$. Figure 89 compares Eq. (440) with the experimental data of Poslinski et al. [160]. As can be seen, the equation describes the experimental data reasonably well.

The relative dynamic viscosity η'_r, defined as the ratio of the dynamic viscosity of the dispersion (η') to that of the dispersion medium (η'_c) at the same frequency, was successfully correlated by the following equation:

$$\eta'_R = \frac{\eta'|_\omega}{\eta'_c|_\omega} = \left(1 + \frac{0.75\phi/\phi_m}{1 - \phi/\phi_m}\right)^2 \tag{441}$$

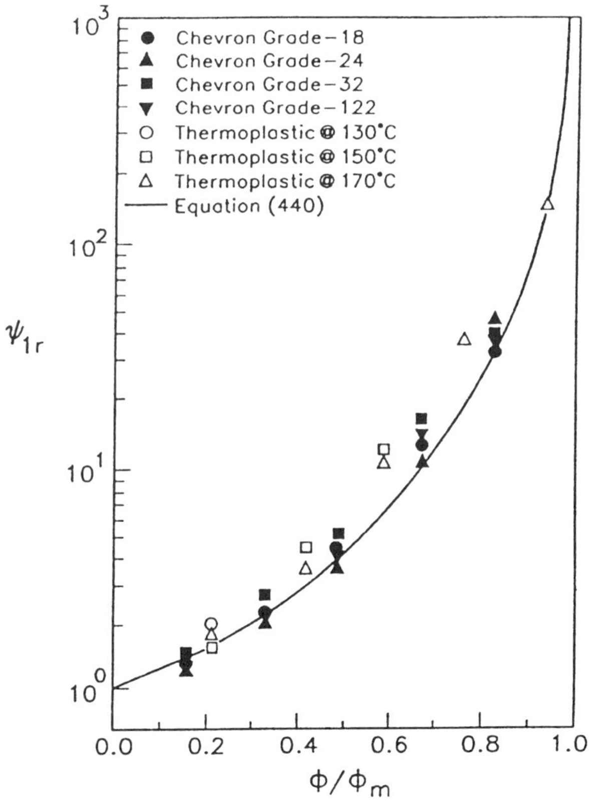

Figure 89. Plot of relative first normal stress coefficient Ψ_{1r} as a function of ϕ/ϕ_m. (From Ref. 160.)

Figure 90 shows the experimental data of Poslinksi et al. for η'_R. Clearly, Eq. (441) describes the experimental data adequately. Note further that the relative loss modulus G''_R is the same as η'_R; that is,

$$G''_R = \frac{G''|_\omega}{G''_C|_\omega} = \frac{\omega\eta'|_\omega}{\omega\eta'_C|_\omega} = \eta'_R \tag{442}$$

Therefore, the relative loss modulus data could be adequately described by Eq. (441). Finally, the relative storage modulus G'_R was correlated by the following equation:

$$G'_R = \frac{G'|_\omega}{G'_C|_\omega} = \left(1 - \frac{\phi}{\phi_m}\right)^{-2} \tag{443}$$

where G' is the storage modulus of the dispersion at frequency ω and G'_C is the storage modulus of the dispersion medium at the same frequency. The comparison between Eq. (443) and the experimental data of Poslinski et al. [160] is shown in Fig. 91. The equation describes the experimental data very well.

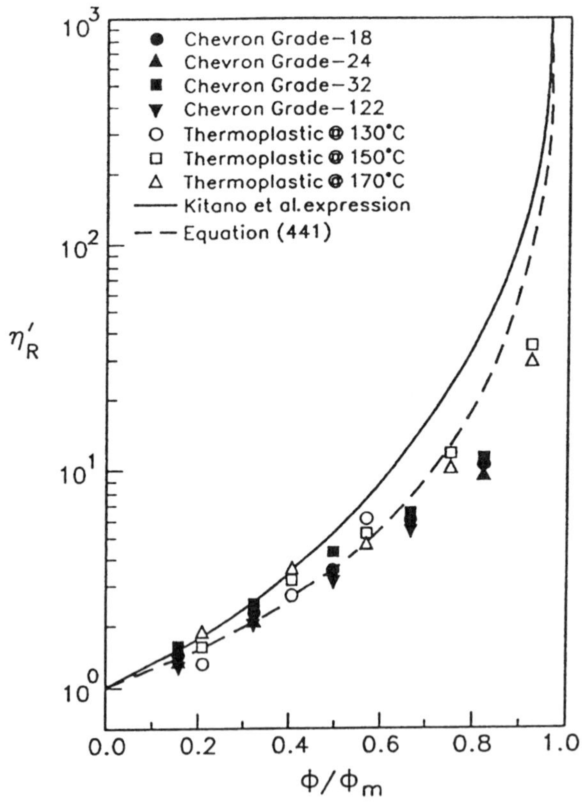

Figure 90. Plot of relative dynamic viscosity η'_R as a function of ϕ/ϕ_m. (From Ref. 160.)

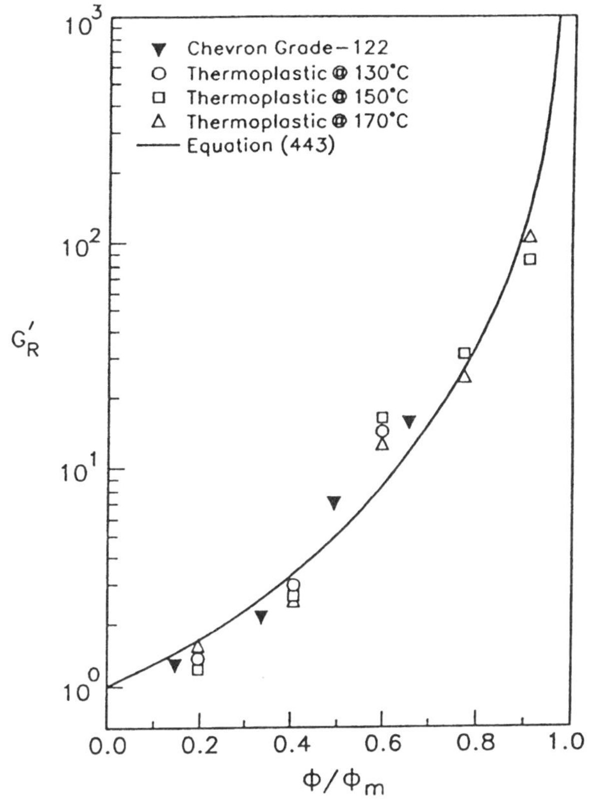

Figure 91. Plot of relative storage modulus G'_R as a function of ϕ/ϕ_m. (From Ref. 160.)

B. Stable (Unflocculated) Emulsions of Deformable Particles

While extensive literature is available on the flow behavior of dispersions of rigid particles in non-Newtonian polymeric media, only a few papers have been published on the rheological behavior of emulsions of deformable particles [52,75,76,217,218].

Figure 92 shows the apparent viscosity as a function of shear stress for oil-in-water emulsions prepared from 1.1 wt % Polyox WSR-301 (high-molecular-weight polyethylene oxide resin) solution as the dispersion medium. For oil droplet concentrations up to 60 vol %, the flow curves for the emulsions are similar to that of the dispersion medium alone (without oil). The flow curves exhibit two distinct regions: the lower Newtonian region and the shear-thinning region. With the increase in oil droplet concentration, flow curves shift upward to higher viscosities. The flow curve for the 70% oil-in-water emulsions is different from other flow curves in two respects. First, the increase in viscosity from 60% O/W to 70% O/W emulsion is more dramatic. Second, the shear-thinning region is extended over a wider range of shear stress. Clearly, these observations are indicative of particle–particle interactions, which become important at high concentrations.

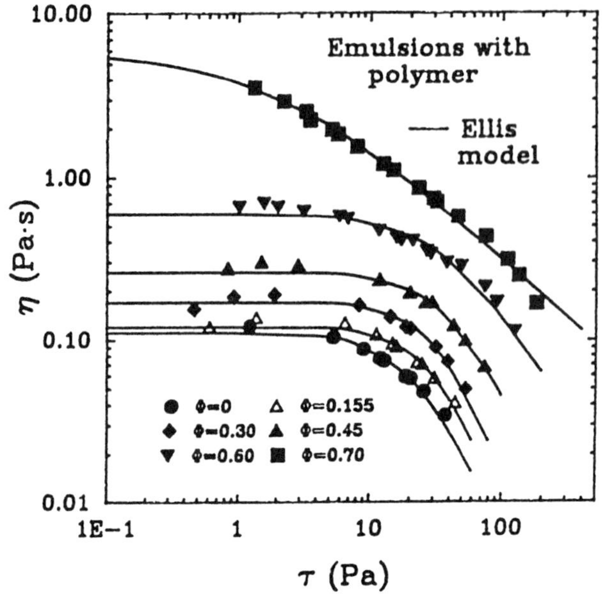

Figure 92. Apparent viscosity as a function of shear stress for oil-in-water emulsions prepared from 1.1 wt % Polyox WSR-301 solution as the dispersion medium. ●, $\phi = 0$; △, $\phi = 0.155$; ◆, $\phi = 0.30$; ▲, $\phi = 0.45$; ▼, $\phi = 0.60$; ■, $\phi = 0.70$.

Figure 93 shows the flow curves for water-in-oil emulsions prepared from a non-Newtonian polymeric suspending medium (1.98 wt % polyisobutylene solution in oil). The flow curves for the emulsions are similar to that of the dispersion medium up to about 40% by volume water concentration. At higher water droplet concentrations, the viscosity–shear stress curves bend upward, indicating the presence of a yield stress. The presence of a yield stress is probably due to particle–particle interactions, which become important at high water concentrations.

An interesting point to note about polymer-thickened emulsions is that their relative viscosities are generally significantly lower than those of the unthickened emulsions. For example, Fig. 94 compares the zero-shear-rate relative viscosity data for polymer-thickened emulsions and unthickened emulsions. The relative viscosities of the thickened emulsions are significantly lower. We believe that the main cause for the observed lower relative viscosities for polymer-thickened emulsions is the increase in the viscosity ratio, η_c/η_d, where η_c is the dispersion-medium viscosity and η_d is the dispersed-phase viscosity. Since we are dealing with fluid particles as the dispersed phase, there is a strong possibility of internal circulation of fluid within the particles, which reduces the distortion of flow patterns around the particles. The internal circulation effect becomes stronger with the increase in the viscosity ratio η_c/η_d, and, therefore, the relative viscosity decreases.

Figure 95 gives the plots of storage and loss moduli, in the linear viscoelastic region, for oil-in-water emulsions prepared from 1 wt % xanthan gum solution.

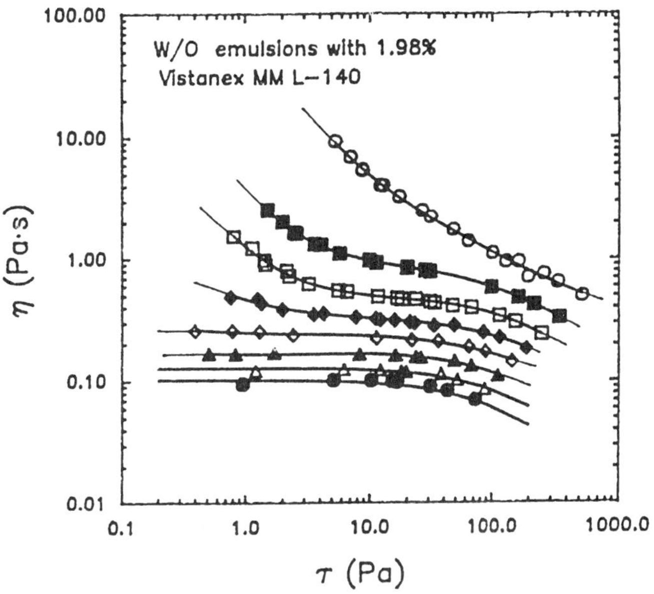

Figure 93. Flow curves for water-in-oil emulsions prepared from non-Newtonian polymeric suspending medium (1.98 wt % polyisobutylene in oil). ●, $\phi = 0.0$; △, $\phi = 0.1$; ▲, $\phi = 0.2$; ◇, $\phi = 0.3$; ◆, $\phi = 0.4$; □, $\phi = 0.5$; ■, $\phi = 0.6$; ○, $\phi = 0.7$.

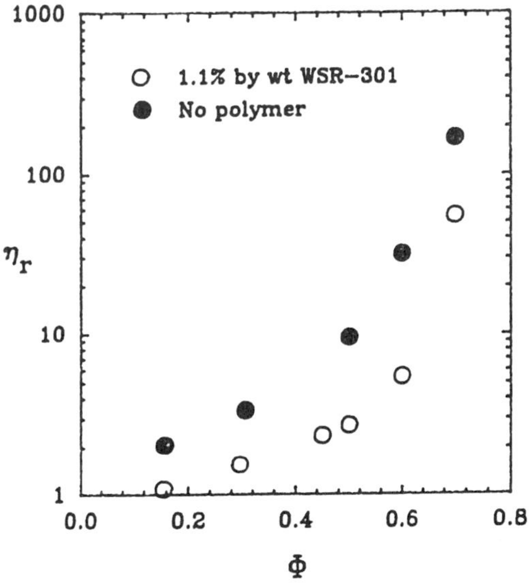

Figure 94. Comparison of zero-shear relative vicosity data for polymer-thickened emulsions and unthickened emulsions.

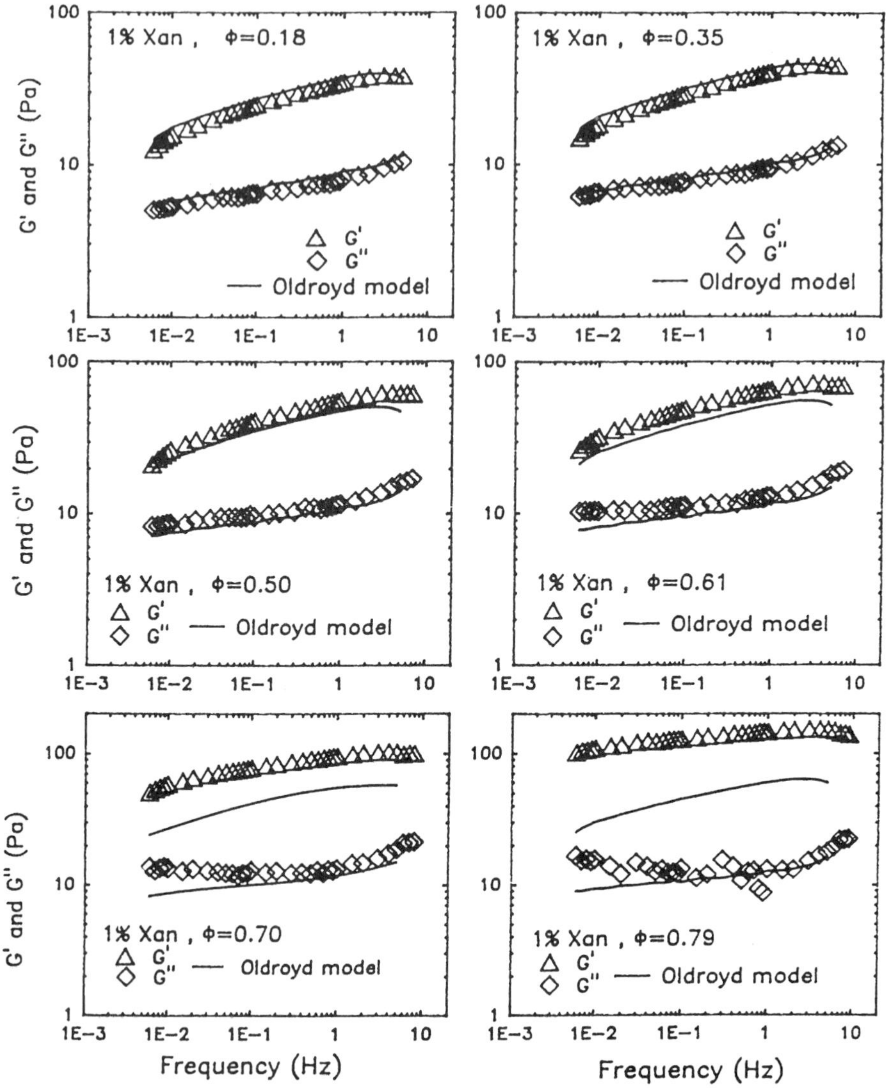

Figure 95. Plots of storage and loss moduli for oil-in-water emulsions prepared from 1 wt % xanthan gum solution.

For comparison, the values of the moduli predicted from the Oldroyd model (Eq. 361) are also shown. The important points to be noted from this figure are as follows: (1) The storage modulus G' for various differently concentrated oil-in-water emulsions falls above the loss modulus (G'') over the entire frequency range, indicating that the emulsions prepared from 1 wt % xanthan gum solution are predominantly elastic in nature. (2) The Oldroyd model gives a good prediction of G' and G'' over the entire frequency range for emulsions with dispersed-phase concentration $(\phi) \leq 0.61$. (3) At very high values of ϕ (≥ 0.70), the Oldroyd model predicts the values of the loss modulus reasonably well, but the storage modulus predicted is significantly lower than the experimental values. The difference between the predicted and experimental values increases with an increase in ϕ.

The high values of the storage modulus exhibited by xanthan-thickened oil-in-water emulsions are due to two reasons. First, the dispersion medium of the emulsions (1 wt % xanthan gum solution) itself is predominantly elastic. Second, an additional mechanism for elasticity is present in emulsions, that is, elasticity of shape.

The complex viscosities $|\eta^*|$ of oil-in-water emulsions prepared from 1 wt % xanthum gum solution as the dispersion medium are compared with the steady-shear viscosities in Fig. 96. The complex viscosities fall well above the steady-shear viscosities for various differently concentrated emulsions. That is, the Cox and Merz rule [90] is not obeyed by the emulsions. One major reason for this deviation is that the dispersion medium of these emulsions itself does not obey the Cox–Merz rule [76].

The dynamic viscosities η' of the xanthan-thickened emulsions are compared with the steady-shear viscosities in Fig. 97. The dynamic viscosities appear to agree quite well with the steady-shear viscosities, especially at low frequencies.

C. Unstable (Flocculated) Emulsions

The presence of polymer in the dispersion medium can induce flocculation of particles (droplets). There are two major mechanisms by which polymers can induce flocculation, referred to as "depletion flocculation" and "bridging flocculation." While depletion flocculation occurs when the polymer molecules present in the dispersion medium are of the "free" nonadsorbing type, bridging flocculation occurs only when the polymer molecules exhibit the tendency to adsorb onto the particles. In general, bridging flocculation occurs when the same polymer molecule adsorbs simultaneously onto two or more particles and hence forms "bridges" or "links" between the particles.

1. Rheological Behavior of Depletion-Flocculated Dispersions

Tadros and co-workers [127–130] have published several papers investigating the rheological behavior of depletion-flocculated dispersions. Their findings can be summarized as follows:

1. Below a certain critical value of the volume fraction of free polymer (ϕ_p^+), the depletion effect is insignificant and the dispersions (moderately concentrated) are generally Newtonian in nature.

2. Above ϕ_p^+, depletion flocculation of dispersed particles becomes important. Depletion-flocculated dispersions are generally non-Newtonian. The rheological parameters—such as Bingham yield stress (τ_0), complex modulus (G^*), storage modulus (G'), loss modulus (G''), and high-frequency shear modulus (G_∞)—all increase with the increase in polymer concentration.

3. The critical flocculation concentration (ϕ_p^+) decreases with an increase in the molecular weight of the free polymer.

4. ϕ_p^+ decreases with an increase in particle size. In other words, larger particles require a smaller concentration of polymer to undergo depletion flocculation.

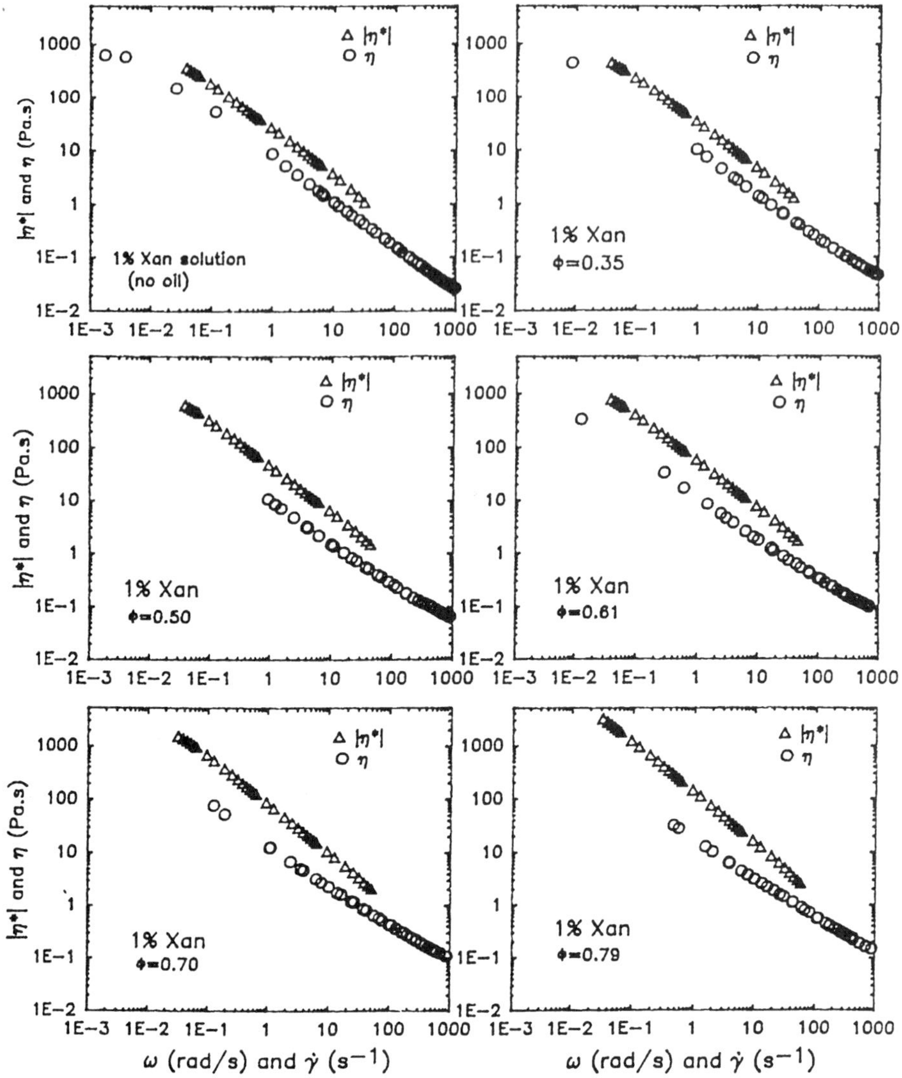

Figure 96. Comparison of complex viscosity data with steady-shear viscosity data for oil-in-water emulsions prepared from 1 wt % xanthan gum solution.

5. ϕ_p^+ is generally independent of the volume fraction of the dispersed phase (ϕ) when either the osmotic pressure of the polymer solution is small or the ratio of polymer coil diameter to particle diameter is small (< 0.12). Otherwise, ϕ_p^+ may decrease with an increase in ϕ.

6. The rheological parameters ($\tau_0, G^*, G', G'', G_\infty$) generally increase with an increase in ϕ_p, ϕ, and molecular weight of the free polymer. Note that an increase in these rheological parameters implies that either the number of particle–particle bonds in the dispersion increases or the depth of the attractive potential energy well increases. For example, an increase in particle volume fraction ϕ results in an increase in the number of particle–particle contacts,

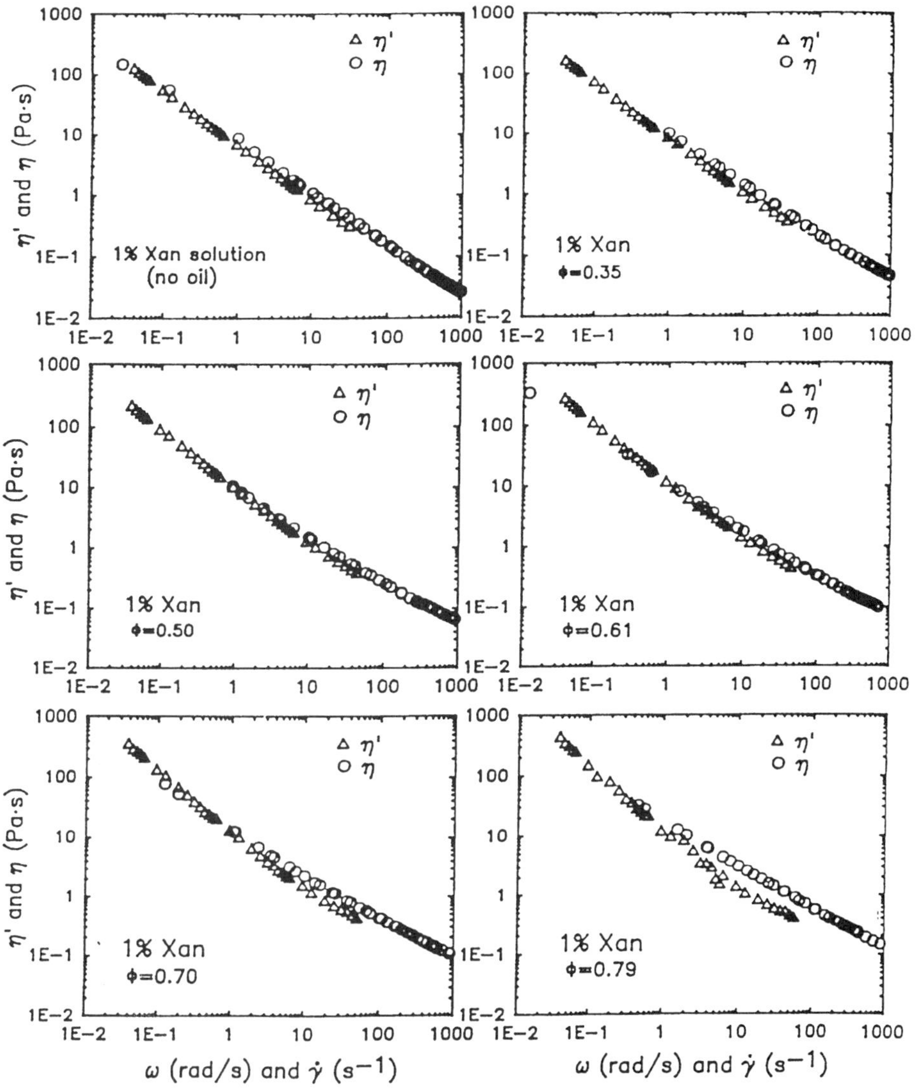

Figure 97. Comparison of dynamic viscosity data with steady-shear viscosity data for oil-in-water emulsions prepared from 1 wt % xanthan gum solution.

whereas an increase in ϕ_p or molecular weight results in an increase in the depth of attractive potential energy well.

Based on a large amount of experimental data, Tadros and co-workers proposed the following empirical correlation for the yield stress in depletion-flocculated dispersions:

$$\tau_0 = K\phi^{2.8}(\phi_p - \phi_p^+) \tag{444}$$

where K is claimed to be a constant. Note that K cannot be a constant; it must depend on particle size. The yield stress of depletion-flocculated dispersions is found to increase very significantly with a decrease in particle size, whereas Eq. (444) (if K is assumed to be a constant) indicates that τ_0 decreases with a decrease in particle size (ϕ_p^+ increases with a decrease in particle size). Tadros and co-workers also proposed the following relation for Bingham yield stress in flocculated dispersions:

$$\tau_0 = NE_{sep} \tag{445}$$

where N is the total number of particle–particle bonds per unit volume of a flocculated dispersion and E_{sep} is the energy needed to break the particle–particle bond. The total number of particle–particle bonds per unit volume of the dispersion is given by

$$N = \frac{1}{2}\left(\frac{3\phi n}{4\pi R^3}\right) \tag{446}$$

where n is the coordination number in a floc, i.e., the number of particle–particle contacts per particle. From Eqs. (445) and (446):

$$\tau_0 = \frac{1}{2}\left(\frac{3\phi n}{4\pi R^3}\right)E_{sep} \tag{447}$$

This equation may be used to calculate E_{sep} from the yield-stress data. The values of E_{sep} so calculated are generally found to be comparable to the free energy of depletion (G_{dep}) calculated from the models of Asakura and Oosawa [125,126] and Fleer et al. [226].

2. Rheological Behavior of Dispersions Flocculated by Polymer Bridging

Otsubo and co-workers [227–237] have published a number of interesting articles on the rheological behavior of dispersions flocculated by polymer bridging. The bridging flocculation of particles is influenced by several factors, such as particle volume fraction, particle size, solubility of polymer in the dispersion medium, molecular weight of polymer, polymer concentration, affinity of polymer for the particle surface, and presence of surfactant.

Figure 98 shows the flow curves for various differently concentrated dispersions of silica particles having a diameter of 20 nm. The continuous phase of these dispersions is a mixture of glycerine (25%) and water (75%), and it contains 1 wt % of polyacrylamide (PAAm). At low concentration of particles (< 3 wt %), the flow behavior of the dispersion is normal in that there are no discontinuities in the flow curve and that the flow behavior is time-independent. Also, the shape of the flow curves at low concentrations is similar to that of the dispersion medium. However, at high particle concentrations (≥ 4 wt %), the flow curves exhibit discontinuities. If one starts with a low shear rate, the viscosity of the dispersion decreases when the shear rate is increased, up to a certain critical value ($\dot{\gamma}_c$). As long as the shear rate is below $\dot{\gamma}_c$, the flow curve obtained is the same in the increasing or decreasing direction of shear rate. That is, below $\dot{\gamma}_c$, there is no hysteresis or time-dependent effect; however, at $\dot{\gamma} = \dot{\gamma}_c$ the viscosity begins to rise with shearing time, and when steady state is

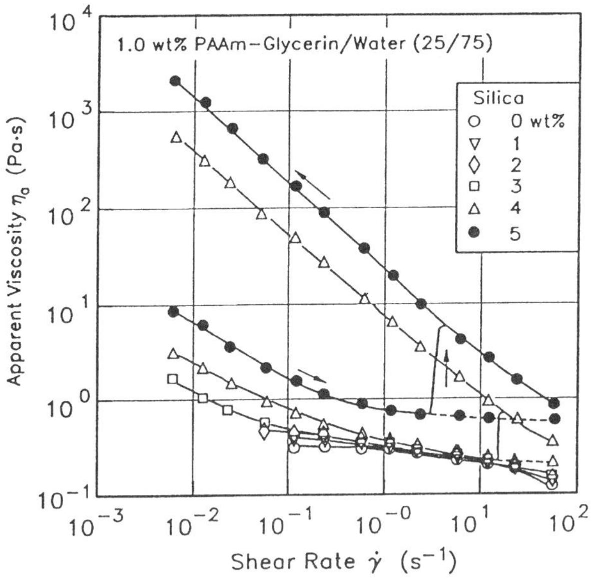

Figure 98. Flow curves for various differently concentrated dispersions of silica particles in 1 wt % polyacrylamide solution. (From Ref. 227.)

reached, the new value of viscosity at $\dot{\gamma} = \dot{\gamma}_c$ is much higher than the starting value (see Fig. 98). Furthermore, once an abrupt increase in the viscosity occurs, the dispersion never returns to the initial flow curve, even when the shear rate is decreased. Also, if the shear rate is increased beyond $\dot{\gamma}_c$ without providing enough time for viscosity buildup to take place, the dashed flow curve in Fig. 99 is obtained. But at shear rates higher than $\dot{\gamma}_c$, the time required for viscosity buildup to reach steady state is less. This can be seen clearly in Fig. 100, where viscosity buildup at different shear rates is shown for 5 wt % dispersion (the critical shear rate $\dot{\gamma}_c$ in this case is $2.4\,\text{sec}^{-1}$).

The equilibrium value of the storage modulus for the dispersions sheared above $\dot{\gamma}_c$ is observed to be very high. Also, at any given shear rate above $\dot{\gamma}_c$, the storage modulus, like viscosity, increases with shearing time. Figure 101 shows how the storage modulus increases with the shearing time for a 5 wt % silica dispersion. The dispersion not only shows an increase in G' with shearing time, but it also exhibits a plateau at low frequencies. Such a plateau is normally observed in dispersions when the dispersed particles form a network structure.

The effect of dispersed-phase concentration on the equilibrium value of the storage modulus is shown in Fig. 102. At low particle concentrations ($\leq$ 4 wt %), the storage modulus of the dispersions is nearly the same as that of the dispersion medium. However, at high particle concentrations, the storage modulus is very high and exhibits a plateau indicative of flocculation of particles.

The observed irreversible increase in the viscosity and storage modulus of the dispersions beyond $\dot{\gamma}_c$ is believed to be due to shear-induced irreversible bridging of particles. Since the dispersion medium is a good solvent for PAAm, the polymer molecule adsorbed on any given particle would have extended loops.

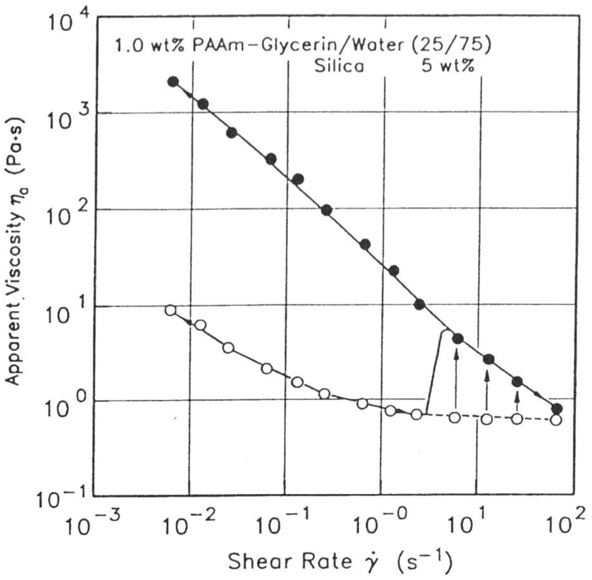

Figure 99. Flow curve for 5 wt % silica dispersion in 1 wt % polyacrylamide solution. (From Ref. 227.)

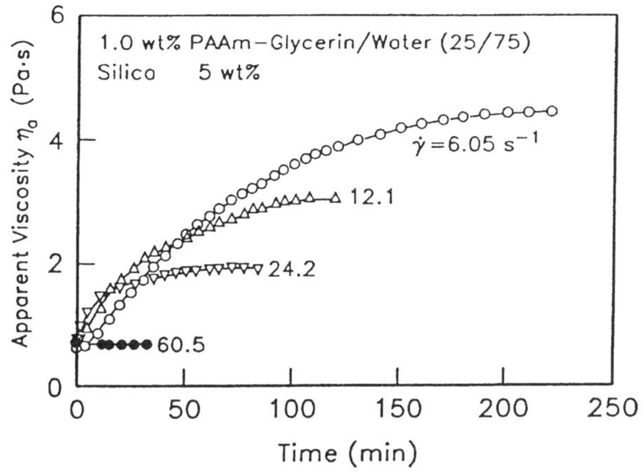

Figure 100. Viscosity buildup with time at different shear rates. (From Ref. 228.)

If under the influence of shear, two particles are brought closer together to a distance that is smaller than the extended loop length, the bridging of particles by polymer molecules will take place. It appears that PAAm molecules adsorb onto silica particles quite strongly, and the process of bridging is irreversible. The buildup of viscosity and storage modulus at shear rates beyond $\dot{\gamma}_c$ is then due to progressive growth of flocculated structure.

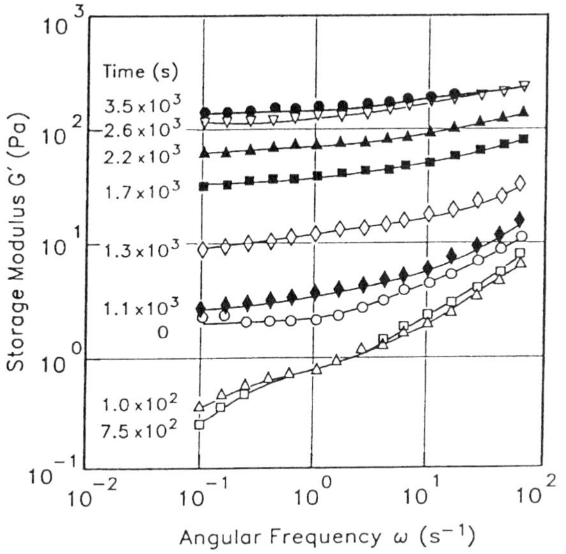

Figure 101. Storage modulus as a function of frequency after different durations of shearing. (From Ref. 231.)

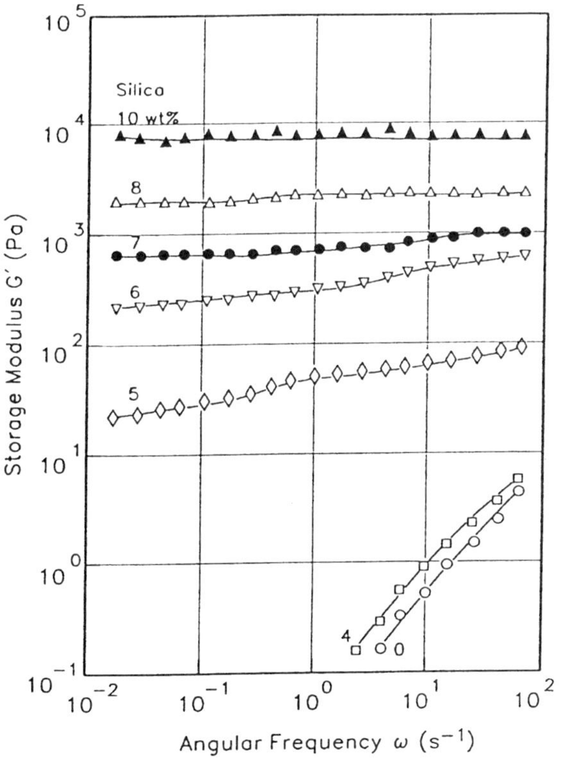

Figure 102. Equilibrium value of storage modulus as a function of frequency at different values of dispersed-phase concentration. (From Ref. 237.)

Otsubo and co-workers also observed a significant effect of aging on the rheological behavior of silica dispersions flocculated by polymer bridging. While the freshly prepared (or less aged) dispersion gave flow curves like those shown in Fig. 98, the aged dispersions (stored for more than 10 days) did not exhibit any discontinuities in the flow curves. However, the aged dispersions exhibited high viscosity and storage modulus almost equal to the equilibrium values for fresh dispersions above the critical shear rate. This indicates that the bridging flocculation process occurs even in a quiescent state; and if the final equilibrium state is reached before the rheological measurements are made, the dispersions would exhibit high equilibrium values of viscosity and storage modulus without any time dependency (and without any discontinuities in the flow curves).

Figure 103 shows the effect of particle size on the viscosity of 5 wt % silica dispersions in 0.5 wt % PAAm solution. The flow curve for the dispersion of large particles (40 nm) is very similar to that of the dispersion medium, except that the viscosity is slightly higher. This indicates that 40-nm particles remain in a well-dispersed state without any significant polymer bridging. However, the dispersions of smaller particles (10 nm and 20 nm) exhibit very high viscosities. Also, the dispersion of 20-nm silica particles exhibits a discontinuity in the flow curve above a certain critical shear rate. This clearly indicates that the small particles undergo bridging flocculation. We do not observe any discontinuity in the flow curve for the dispersion of 10-nm particles because the critical shear rate in this case seems to be so small that bridging flocculation is almost completed during the sample preparation. Thus, the flocculating power of polymer molecules increases with a decrease in particle size. It appears that small par-

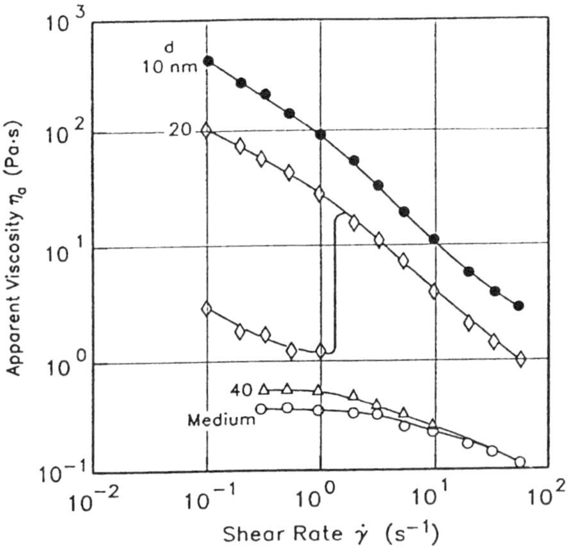

Figure 103. Flow curves for dispersions of silica particles with different diameters in 0.5 wt % polyacrylamide solution. (From Ref. 230.)

ticles are easily flocculated because one polymer chain can adsorb onto many particles and link the particles together.

The flocculating power of polymer molecules was also found to vary with the solubility of polymer in the dispersion medium and with the molecular weight of the polymer. When the solubility of the polymer in the dispersion medium was enhanced, the flocculating power of the polymer molecules increased because in a good solvent the adsorbed polymer molecules have extended loops that increase the chances for bridge formation between the particles. With the increase in molecular weight of the polymer, the flocculating power first increased and then decreased; that is, there exists an optimum molecular weight at which the bridging flocculation of particles is maximum [230].

The concentration of polymer in the dispersion medium can have a significant effect on the flocculation behavior of dispersions. Figure 104 shows the influence of polymer concentration on the rheological behavior of 5 wt % silica dispersion. The viscosity of the dispersion increases with an increase in polymer concentration. However, an important point to note is that the critical shear rate, where irreversible increase in the viscosity occurs, decreases with an increase in polymer concentration.

All the dispersions just discussed consisted of silica as the dispersed particles and polyacrylamide (PAAm) as the polymer. In general, the bridging flocculation in such systems is believed to be irreversible, in that the bridges, once formed, do not easily break down by Brownian motion in a quiescent state. This is mainly due to the fact that the polymer molecule tends to attach to the particle surface

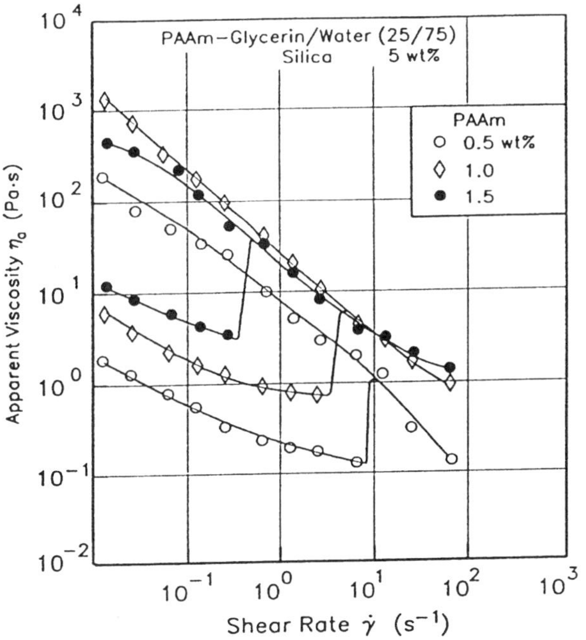

Figure 104. Effect of polymer concentration on the rheological behavior of 5 wt % silica dispersion. (From Ref. 228.)

at many sites; i.e., the polymer molecule undergoes multipoint attachment. However, desorption of polymer molecules is certainly possible in a shear field.

If the polymer molecule does not have a strong affinity for the particle surface or if the polymer chain is not attached to the surface at several points, the polymer bridges formed between the particles are expected to be reversible, in the sense that they are constantly formed, broken down, and reformed by Brownian motion in a quiescent state. The rheological behavior of such dispersions is expected to be different.

Figure 105 shows the flow curves for various differently concentrated dispersions of polymeric particles (styrene-methylacrylate copolymer) having a diameter of 80 nm [235]. The continuous medium of the dispersions is an aqueous solution of 1 wt % polyacrylic acid (PAA). At low concentration of particles (<7.5 vol %), the dispersion is nearly Newtonian. However, at 10 vol % and above, the dispersions exhibit more complicated behavior: At low shear rates, the viscosity is nearly constant, indicating Newtonian behavior; at moderate shear rates, there is a steep rise in the viscosity, indicating dilatant behavior; and at high shear rates, shear-thinning behavior is observed.

The effect of polymer (PAA) concentration on the rheological behavior of 15 vol % dispersion is shown in Fig. 106. At low concentrations of polymer (≤ 0.4 wt %), the flow behavior is nearly Newtonian. At higher polymer concentrations, the flow behavior is similar to that observed for concentrated dispersions in Fig. 105; that is, at low shear rate the behavior is Newtonian, and at moderate shear rates the dispersion becomes dilatant.

According to Otsubo [235,237], the polymer molecules in these dispersions do not have a strong affinity for the particle surface. Consequently, only reversible bridging takes place between the particles; that is, the polymer bridges

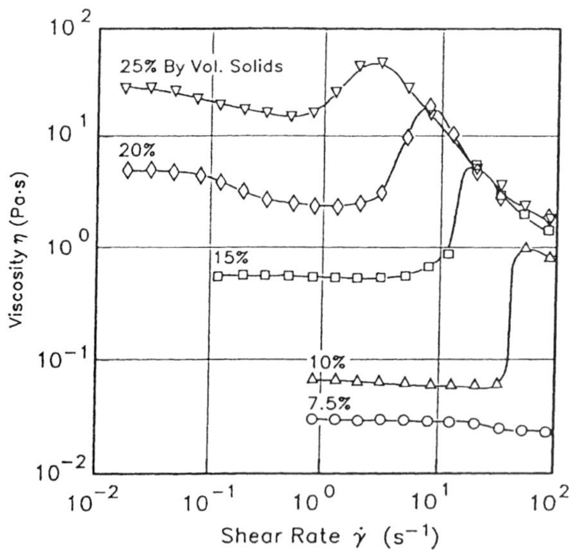

Figure 105. Flow curves for various differently concentrated dispersions of 80-nm-diameter particles in an aqueous solution of 1 wt % polyacrylic acid. (From Ref. 235.)

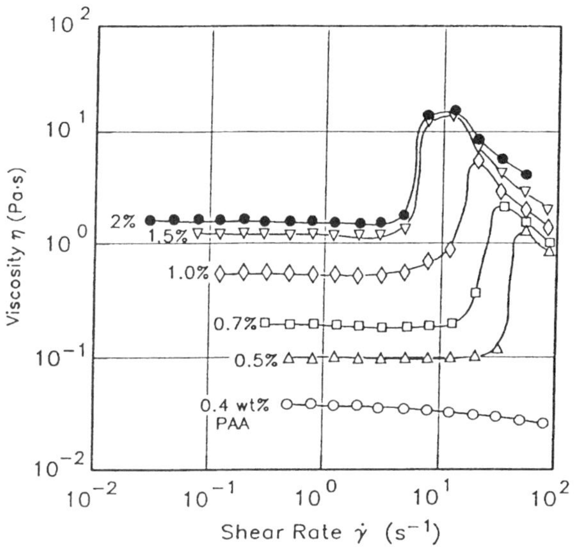

Figure 106. Effect of polymer concentration on the rheological behavior of 15 vol % dispersion. (From Ref. 235.)

between the particles undergo constant formation, breakdown, and reformation by Brownian motion. In a quiescent state, as well as in a low-shear-rate region, an equilibrium exists between bridge formation and bridge breakdown processes; therefore, Newtonian behavior is observed. The shear-thickening observed at moderate shear rates is due to the rapid extension of polymer coils (before the desorption). When the polymer bridges are rapidly extended, the restoring force causes an increase in flow resistance. The shear-thinning behavior observed at very high shear rates is due to the excessive breakdown of polymer bridges.

Finally, Fig. 107 shows the effect of added surfactant (Triton X-100) on the flow of 15 vol % dispersion in 2 wt % PAA solution. With an increase in surfactant concentration, the Newtonian viscosity at low shear rates decreases; at 2 wt % surfactant concentration, the dilatant behavior in the dispersion disappears. These observations clearly indicate that the addition of surfactant reduces bridging flocculation of particles.

VII. Rheology of Emulsions of Immiscible Polymeric Melts

Physical blending of two immiscible polymeric melts to generate new performance materials is of growing industrial importance [42–63]. A large variety of new products with desirable properties can be created simply by the physical blending of polymer melts without the need for synthesizing new chemical structures. The properties of blends of incompatible (immiscible) polymeric melts are a strong function of the morphology of the blend, i.e., particle size and shape and the spatial distribution of the phases. Other factors influencing the rheological behavior of an incompatible polymer blend are interfacial tension between the

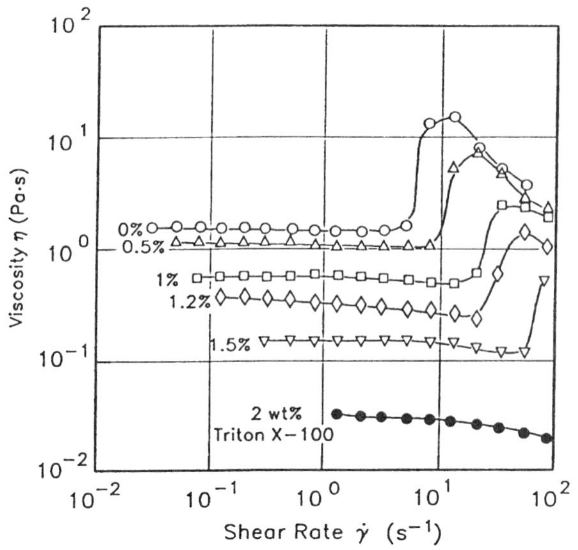

Figure 107. Effect of added surfactant (Triton X-100) on the flow of 15 vol % dispersion in 2 wt % polyacrylic acid. (From Ref. 235.)

two immiscible polymers and the rheological properties of the individual components.

A. Morphology of Immiscible Polymer Blends

Unlike ordinary emulsions, blends of two immiscible polymer melts exhibit a diversity of morphologies. The primary factors determining the morphology of the blends are:

1. Interfacial tension between the phases
2. The viscosity ratio of constituent polymers
3. Processing conditions (e.g., the deformation field in a mixing device)
4. The volume fractions of constituent polymers

The magnitude of the interfacial tension between the two immiscible polymers has a strong influence on the degree of dispersion during mixing. If the interfacial tension is high, it is difficult to achieve a fine dispersion of phases. However, the interfacial tension can be controlled by adding compatibilizers (interfacial agents) to the blends. The addition of compatibilizers not only affects the interfacial tension between the phases, but it also helps to stabilize the dispersed phase (droplets) once it is formed during the mixing. Thus, compatibilizers promote the formation of relatively finer dispersions. However, compatibilizers used in polymer blends are not as effective in reducing the interfacial tension—and hence in improving the stability of the dispersed phase—as are surfactants in the case of ordinary emulsions. This is mainly because the molecular diffusion rates of compatibilizers in polymeric blends are several orders of magnitude smaller than the diffusion rates of surfactants in emulsions.

The viscosity ratio of the two immiscible polymer melts has a strong influence on the morphology of the blend. When the dispersed phase has a much lower viscosity than that of the continuous phase, the dispersed phase may take the form of elongated fibrils or platelets with varying aspect ratios, i.e, a fibrous structure is favored. In the reverse case, when dispersed phase has a higher viscosity than that of the continuous phase, discrete spherical droplets of dispersed phase are favored.

Another important factor influencing the morphology of polymer blends is the deformation field generated by the blending device. For example, uniaxial elongational flows, such as extrusion through a die, favor the formation of a fibrilous dispersed phase, whereas biaxial stretching, such as blow molding, favors the formation of platelets. Figure 108 shows some examples of the possible blend morphologies [44].

The volume fractions of the constituent polymers in the blend also play a major role in determining the morphology of the blend. In general, the major component, i.e., the high-volume-fraction component, tends to form the continuous phase. Also, with the increase in dispersed-phase concentration, the blend morphology undergoes a significant change. At low dispersed-phase concentrations, the dispersed-phase units are discrete and separate. But with an increase in dispersed-phase concentration, a co-continuous structure, sometimes called "dual phase continuity," of two polymers is formed. With a further increase in the concentration of the dispersed phase, "phase inversion" occurs; i.e., the original dispersed phase now becomes the continuous phase, and the original continuous phase now becomes the dispersed phase.

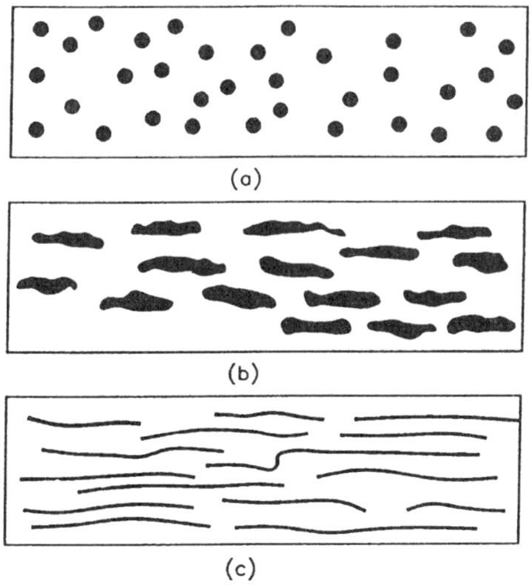

Figure 108. Possible blend morphologies. (From Ref. 44.)

In summary, the morphology of a blend of two immiscible polymer melts is not an inherent characteristic of the polymer pairs involved. It is the outcome of the complex interactions between the various factors just discussed.

1. *Phase Inversion Concentration*

Several expressions have been proposed in the literature to predict the phase inversion point in blends of two immiscible polymers. Here we will present the expressions proposed by Paul and Barlow [238], Metelkin and Blekht [239], and Utracki [45].

Paul and Barlow [238] proposed the following equation for predicting the phase inversion of polymeric blends:

$$\left(\frac{\eta_1}{\eta_2}\right)\left(\frac{\phi_{2\mathrm{I}}}{\phi_{1\mathrm{I}}}\right) = 1 \tag{448}$$

where η_1 and η_2 are the viscosities of polymers 1 and 2 and $\phi_{1\mathrm{I}}$ and $\phi_{2\mathrm{I}}$ are the volume fractions of polymers 1 and 2 at the phase inversion point. Since

$$\phi_{1\mathrm{I}} = 1 - \phi_{2\mathrm{I}} \tag{449}$$

the Paul and Barlow criterion could be rewritten as

$$\phi_{2\mathrm{I}} = \frac{1}{1+\lambda} \tag{450}$$

where $\lambda = \eta_1/\eta_2$. When $\phi_2 < \phi_{2\mathrm{I}}$, phase 1 is likely to form the continuous phase and phase 2 the dispersed phase. When $\phi_2 > \phi_{2\mathrm{I}}$, the reverse situation is likely to occur; i.e., phase 2 is expected to be the continuous phase and phase 1 to be the dispersed phase.

Metelkin and Blekht [239] derived the following criterion for the phase inversion point:

$$\phi_{2\mathrm{I}} = \frac{1}{1+\lambda \mathrm{F}(\lambda)} \tag{451}$$

where

$$\mathrm{F}(\lambda) = 1 + 2.25 \log\lambda + 1.81(\log\lambda)^2 \tag{452}$$

Utracki [45] recently proposed another expression for predicting the phase inversion point:

$$\frac{0.84 - \phi_{2\mathrm{I}}}{\phi_{2\mathrm{I}} - 0.16} = \lambda^{0.626} \tag{453}$$

In the range $0.1 \le \lambda \le 10$, Eq. (453) could be approximated as

$$\phi_{2\mathrm{I}} = \frac{1 - 0.5263 \log\lambda}{2} \tag{454}$$

Figure 109 shows the experimental data for phase inversion concentration $\phi_{2\mathrm{I}}$ as a function of λ for a large number (18) of thermoplastics blends [45]. For comparison, the Paul and Barlow equation as well as Utracki's equation are plotted. Clearly, the equations proposed by Utracki predict well the phase inversion concentration.

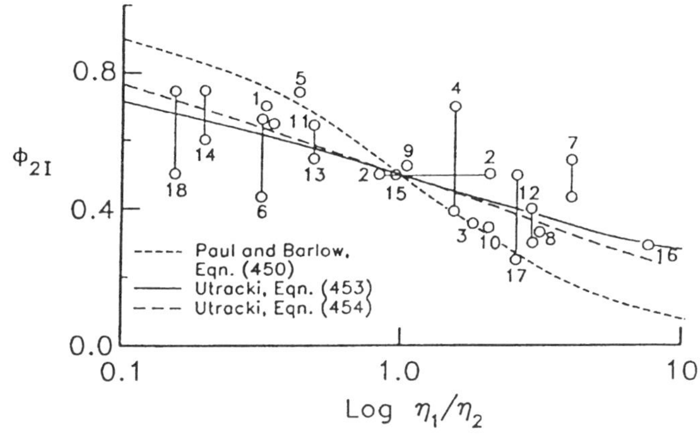

Figure 109. Experimental data for phase inversion concentration ϕ_{2I} as a function of viscosity ratio λ for 18 different thermoplastics blends. (From Ref. 45.)

B. Viscosity of Immiscible Polymeric Blends

The viscosity of polymeric blends is influenced by the following factors:

1. The viscosity of individual polymeric components
2. The interfacial interaction between the two polymeric phases
3. The blend morphology
4. The volume fractions of components (i.e., the composition)

As an example, Fig. 110 shows the viscosity versus shear rate curves for polystyrene (PS)/butadiene (BR) blends at various blend compositions [47]. The

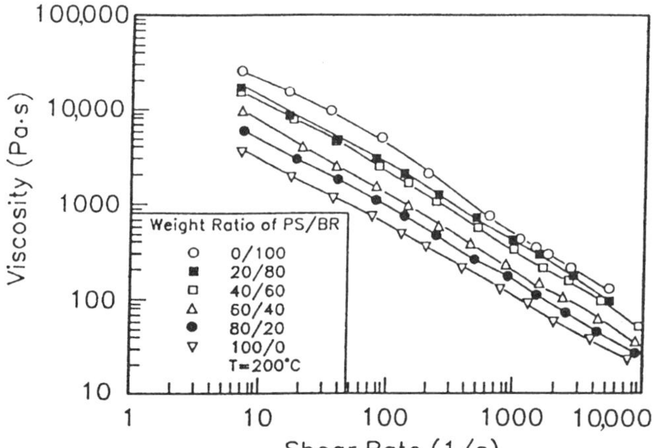

Figure 110. Viscosity versus shear rate curves for polystyrene (PS)/butadiene (BR) blends at various blend compositions. (From Ref. 47.)

pure components and the blends exhibit shear-thinning. The corresponding viscosity versus blend composition (wt % BR) curves are shown in Fig. 111. For comparison, the following mixing rule is also plotted:

$$\eta = x_1\eta_1 + x_2\eta_2 \tag{455}$$

where η is blend viscosity, x_1 and x_2 are weight fractions of components 1 and 2, and η_1 and η_2 are viscosities of components 1 and 2. The polymer blends of PS/BR exhibit negative deviation from the simple additivity rule. One possible reason for this is the "slippage" between the two phases. The slippage effect (also referred to as the "interlayer slip effect") between the phases becomes important when the phases are dissimilar in structure and there is little affinity, and hence little interfacial interaction, between them [45,240]. Polymeric blends where the phases have high affinity for each other, and hence large interfacial interaction, generally exhibit positive deviation from the additivity rule because of the absence of interlayer slip. Figure 112 shows viscosity versus blend composition for blends of polyoxymethylene (POM) and copolyamide (CPA) [46,241]. These blends exhibit positive deviation from the additivity rule.

According to Utracki [49], polymeric blends could be classified into three broad groups, depending on the nature of their deviation from the log additivity rule:

$$\log \eta = \phi_1 \log \eta_1 + \phi_2 \log \eta_2 \tag{456}$$

Here, ϕ_1 and ϕ_2 are volume fractions of components 1 and 2. The three groups are:

1. Blends that exhibit a negative deviation from the log additivity rule (Eq. 456) (referred to as "NDB")

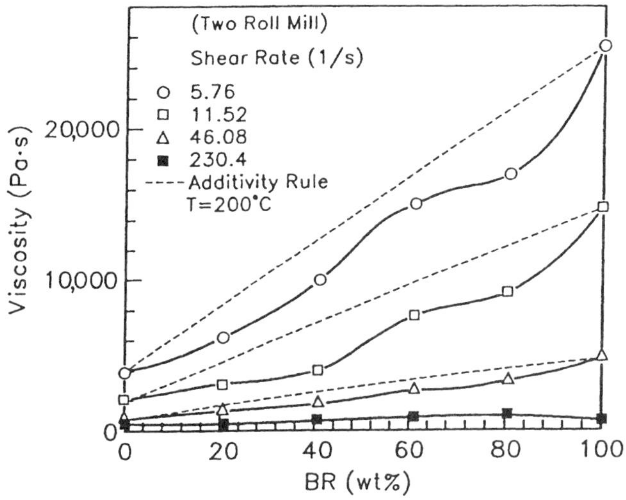

Figure 111. Viscosity data for polystyrene (PS)/butadiene (BR) blends as a function of butadiene (BR) content. (From Ref. 47.)

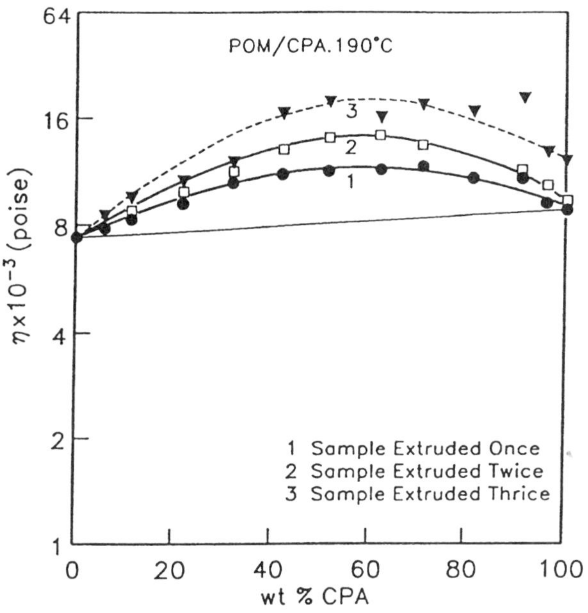

Figure 112. Viscosity versus blend composition for blends of polyoxymethylene (POM) and copolyamide (CPA). (From Refs. 46 and 241.)

2. Blends that exhibit a positive deviation from the log additivity rule (referred to as "PDB")
3. Blends that exhibit both positive and negative deviations from the log additivity rule (referred to as "PNDB")

When there is a lack of adhesion between the segregated domains of the two polymeric phases so that interlayer slip is significant, the blend generally exhibits NDB behavior. This usually occurs when the two polymers are quite dissimilar in structure and there is little interfacial interaction between the two phases.

Polymer blends exhibit PDB behavior when the interfacial interaction between the two phases is large and there is little or no interlayer slip.

When the interfacial interaction between the two phases changes with the composition (the volume fractions of the components) of the blend, blends may exhibit PNDB behavior. An example of PNDB is polybutadiene/*cis*-polyisoprene blend (see Fig. 113) [49,242].

However, note that the most frequently encountered blends exhibit PDB behavior [243]. According to Utracki [45], about 60% of the blends reported in the literature are PDB type. It is interesting to note that ordinary emulsions also exhibit PDB behavior, particularly if the viscosities of oil and water are not drastically different. Figure 114 shows the viscosity/concentration behavior of ordinary emulsions. When water is dispersed in oil, water-in-oil emulsions having viscosities greater than that of oil are formed. With an increase in the dispersed-phase (water) concentration, the viscosity of the water-in-oil emulsion increases.

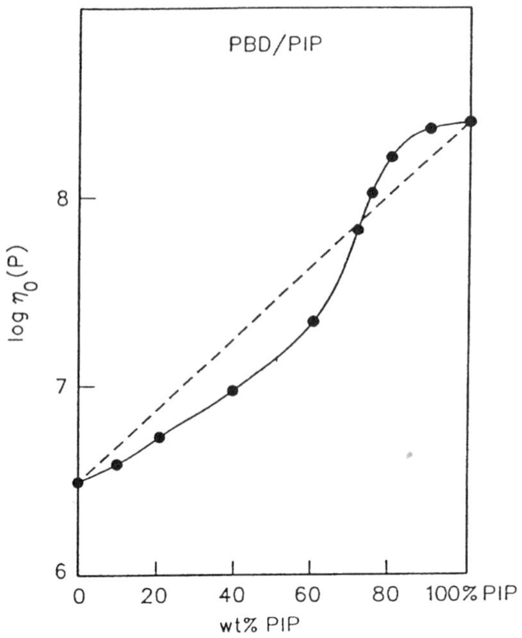

Figure 113. Polybutadiene (PBD)/*cis*-polyisoprene (PIP) blend as an example of PNDB. (From Refs. 49 and 242.)

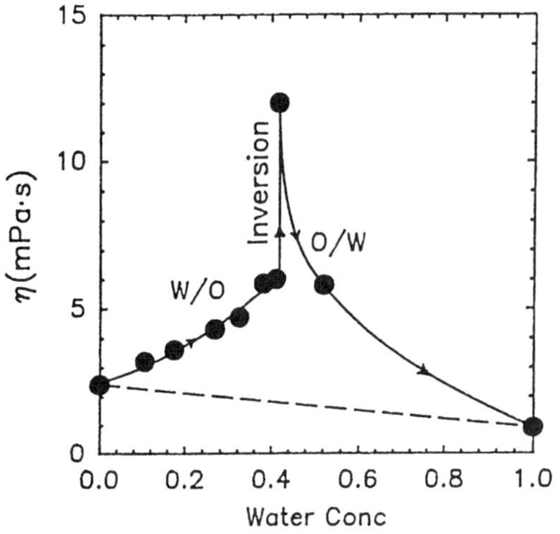

Figure 114. Viscosity/concentration behavior of ordinary emulsions.

However, at some critical value of water concentration, the water-in-oil emulsion inverts to an oil-in-water emulsion and there is a sudden increase in the viscosity. With a further increase in water concentration, the emulsion remains oil-in-water type and the viscosity decreases because of the dilution effect.

Finally, it is important to point out that because of the non-Newtonian nature of blends, the blend behavior may change from one class (say, PDB) to another (say, NDB) with an increase in shear stress. Thus, blend classification based on the log additivity rule is dependent not only on the nature of the polymer pair but also on the flow conditions.

1. Viscosity/Concentration Equations for Polymer Blends

In the absence of interlayer slip, the viscosity/concentration behavior of well-dispersed polymeric blends is very similar to that of ordinary emulsions. Thus, all the equations discussed for ordinary emulsions could be applied to such blends. However, according to Utracki [244], the equations of Simha [146] and Mooney [152] are most successful in describing the zero-shear relative viscosity (η_r) of polymeric blends with negligible slip. The Simha equation may be written as

$$\eta_r = 1 + [\eta]\mathrm{I}(\lambda)\phi \tag{457}$$

where $[\eta]$ is the intrinsic viscosity of emulsion, $\mathrm{I}(\lambda)$ is as given by Eq. (182), and λ is as given by

$$\lambda = \left[2\left(\frac{\phi_m}{\phi}\right)^{1/3} - 1\right]^{-1} \tag{458}$$

This expression for λ is valid at high values of ϕ.

The Mooney equation is given by

$$\ln \eta_r = \frac{[\eta]\phi}{1 - \phi/\phi_m} \tag{459}$$

The maximum packing-volume fraction (ϕ_m) in these Simha and Mooney equations should be treated as an adjustable parameter. If no interfacial agent is present in the blend, the intrinsic viscosity is given by Taylor's equation:

$$[\eta] = 2.5\left[\frac{1 + 0.4(\eta_c/\eta_d)}{1 + \eta_c/\eta_d}\right] \tag{460}$$

where η_c is the continuous-phase viscosity and η_d is the dispersed-phase viscosity. If interfacial agent is present on the droplet surface, internal circulation within the droplets is inhibited to some extent and the intrinsic viscosity is given by Oldroyd's equation [209]:

$$[\eta] = 2.5\left[\frac{1 + 0.4(\eta_c/\eta_d) + (0.4/R\eta_d)(2\eta_{si} + 3\eta_{ei})}{1 + (\eta_c/\eta_d) + (0.4/R\eta_d)(2\eta_{si} + 3\eta_{ei})}\right] \tag{461}$$

where η_{si} is shear viscosity of the interfacial film, η_{ei} is the extensional viscosity of the film, and R is the droplet radius.

It is interesting to note that Pal [75,218] recently proposed the same Mooney equation (Eq. 459 with $[\eta]$ given by either Eq. 460 or Eq. 461) for polymer-thickened oil-in-water and water-in-oil emulsions.

When interlayer slip is present between the two phases of the blend, the viscous behavior of the blend cannot be described adequately by emulsion-type equations. For blends with interlayer slip, Lin [240] developed the following equation for the blend viscosity:

$$\log \eta = -\log\,(1 + \beta\sqrt{\phi_1\phi_2}) - \log\left(\frac{\phi_1}{\eta_1} + \frac{\phi_2}{\eta_2}\right) \tag{462}$$

where the factor β takes into account the interlayer-slip effect. Utracki [45] recently modified Eq. (462) to

$$\log \eta = -\log\left[(1 + \beta\sqrt{\phi_1\phi_2})\left(\frac{\phi_1}{\eta_1} + \frac{\phi_2}{\eta_2}\right)\right] + \mathrm{A}\left[1 - \frac{(\phi_1 - \phi_{1\mathrm{I}})^2}{(\phi_1\phi_{2\mathrm{I}}^2 + \phi_2\phi_{1\mathrm{I}}^2)}\right] \tag{463}$$

In this equation, A and β are adjustable parameters, and the phase-inversion concentration $\phi_{1\mathrm{I}}$ and $\phi_{2\mathrm{I}}$ $(= 1 - \phi_{1\mathrm{I}})$ are to be calculated from Eq. (454).

Equation (463) is capable of predicting NDB, PDB, and PNDB behaviors by adjusting the parameters A and β. Figure 115 shows different behaviors of blend predicted from Eq. (463) when $\eta_1 = 10\eta_2$. Utracki was able to describe the viscosity/concentration behavior for over 30 immiscible polymer blends using this equation. Based on the experimental data for these blends, the following general observations were noted: (1) With an increase in temperature, the interlayer-slip parameter β increases; (2) compatibilization usually reduces β; (3) β increases with the increase in shear stress; (4) the parameter A in Eq. (463) decreases with temperature and with the shear stress.

C. Viscoelastic Properties of Immiscible Polymeric Blends

In recent years, a large number of papers [51–60] have been published on the viscoelastic properties of blends of two immiscible polymeric melts. But, for the most part, these studies are restricted to linear viscoelastic behavior.

The linear viscoelastic behavior of blends of immiscible polymeric melts is generally interpreted in terms of the models originally developed for ordinary

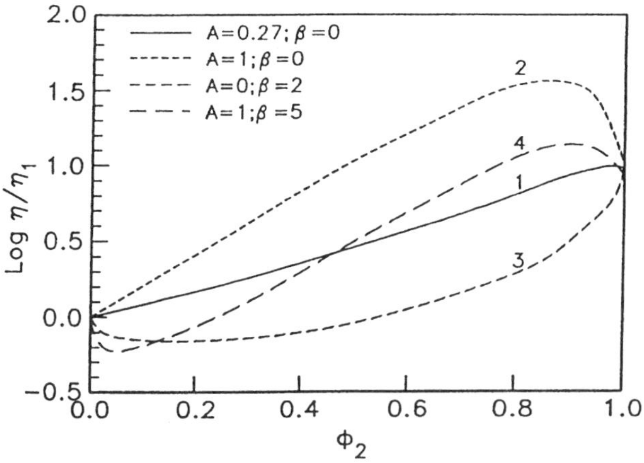

Figure 115. Different behaviors of blend predicted from Utracki Eq. (463) when $\eta_1 = 10\eta_2$. (From Ref. 45.)

emulsions (such as the Choi and Schowalter model [213]). When one polymeric melt is dispersed into another immiscible polymeric melt, the resulting dispersion (emulsion) exhibits viscoelastic properties because of two mechanisms: (1) The components (i.e., the two polymers) of the blend are viscoelastic in nature; (2) the dispersed polymeric droplets possess elasticity of shape due to interfacial tension.

Figure 116 shows the dynamic moduli data for blends of polydimethylsiloxane (PDMS) and polyoxyethylene (POE) as a function of frequency [56]. The volume fraction ϕ of the dispersed phase (POE) in Fig. 116a is 0.1; in Fig. 116b, $\phi = 0.2$. The average value of the droplet size of the dispersed phase is 30 μm, and the interfacial tension between the two phases is 10.4 mN/m. For comparison, the dynamic moduli data for the pure continuous phase (PDMS) are also shown in the figures. The solid curves are predicted from the Choi–Schowalter model (Eqs. 374 and 375) using the zero-shear viscosities of the two phases. At high frequencies, the dynamic moduli of the blends are similar to those of the matrix (continuous phase). But at low frequencies, the storage modulus of the blends deviates remarkably from that of the matrix. The values of G'_{blend} in the low-frequency range are much higher than those of the matrix. Furthermore, the data for G'_{blend} show a plateau and appear to follow the Choi–Schowalter model at low frequencies. Clearly, the increase in the storage modulus for the blend (with respect to the matrix) at low frequencies is due to the elasticity of the shape of the dispersed droplets. This becomes more obvious when the rheological data are plotted in terms of Cole–Cole diagrams [58]. As an example, Fig. 117 shows the Cole–Cole diagrams for PDMS/POE blends for two values of ϕ: $\phi = 0.10$ and $\phi = 0.20$ [58]. For comparison, the Cole–Cole diagram for the pure matrix (PDMS) is also shown. Clearly, two frequency domains are present. The elastic behavior in the low-frequency domain is due to interfacial-tension effects (elasticity of shape of droplets), whereas in the

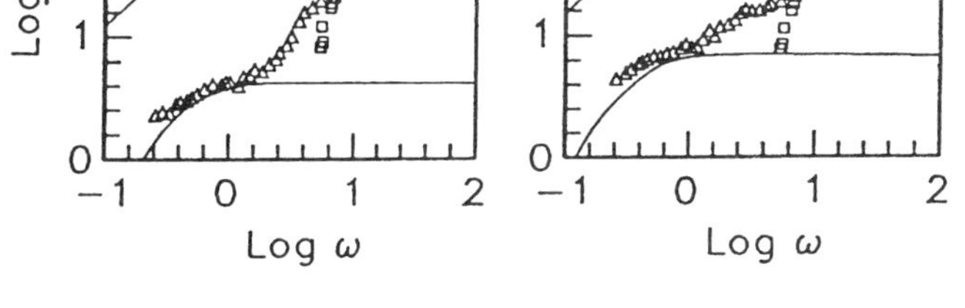

Figure 116. Dynamic moduli data for blends of polydimethyl siloxane (PDMS) and polyoxyethylene (POE) as a function of frequency. (From Ref. 56.)

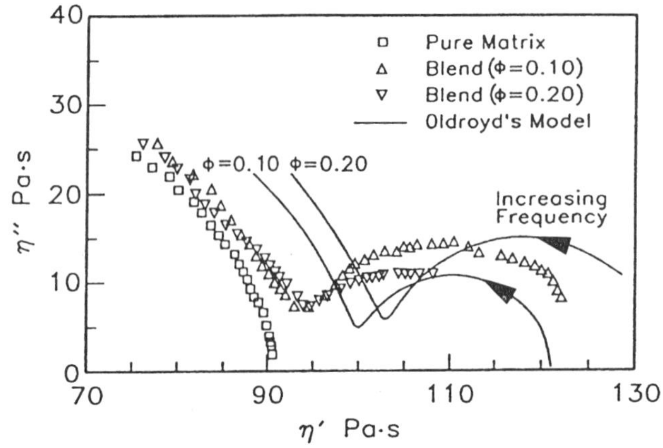

Figure 117. Cole–Cole diagrams for PDMS/POE blends for two values of ϕ. (From Ref. 58.)

high-frequency domain the elastic behavior is due to the elasticity of individual polymer components.

Note further from Fig. 116 that at high frequencies, the experimental data for the blends (particularly the storage modulus data) do not follow the Choi–Schowalter model. This is because that model is valid only for emulsions of two Newtonian liquids, whereas the blends of PDMS/POE consist of a viscoelastic matrix (PDMS). Thus, the viscoelastic behavior of blends of immiscible polymeric melts can be described by emulsion models (of Newtonian phases) only at low frequencies.

Figure 118 shows the effect of droplet size on the dynamic moduli of blends, as predicted from the Choi–Schowalter model (hence valid only at low frequencies) [53]. With an increase in droplet size (keeping all other parameters constant), the plateau value of the storage modulus decreases. Also, the frequency at which the terminal zone (where $G' \propto \omega^2$) begins is shifted to lower values with an increase in droplet size. This is because the relaxation time of droplets increases with an increase in droplet size. Droplet size appears to have little effect on the loss modulus (G'').

Figure 119 shows the effect of viscosity ratio k $(= \eta_d/\eta_c)$ on the viscoelastic behavior of blends [53]. The viscosity ratio k is varied by varying the dispersed-phase viscosity η_d. All other factors $(\eta_c, \phi, R, \sigma)$ are kept constant. While the plateau value of the storage modulus decreases (when $k > 1$), the loss modulus increases with an increase in k.

Both G' (in the plateau zone) and G'' increase when ϕ or η_c is increased. With increase in interfacial tension σ, G' in the plateau zone increases and the droplet relaxation time decreases. Interfacial tension has a negligible effect on G''.

These observations are based on the Choi–Schowalter model, which is applicable to blends only at low frequencies. At high frequencies, viscoelasticity of individual phases (components) becomes important.

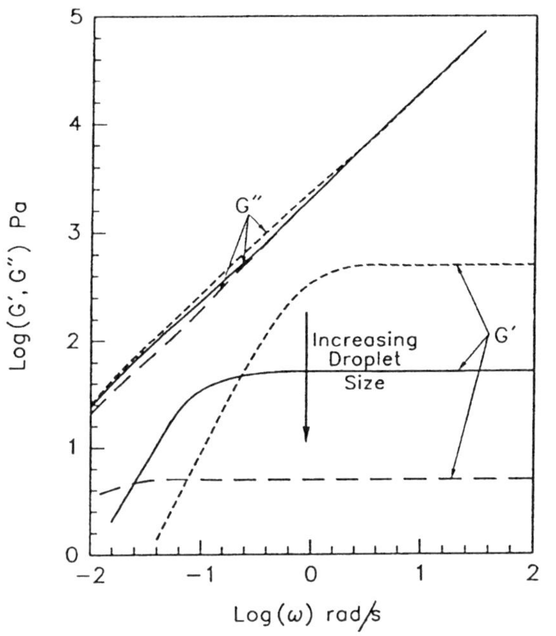

Figure 118. Effect of droplet size on the dynamic moduli of blends, as predicted from the Choi–Schowalter model. (From Ref. 53.)

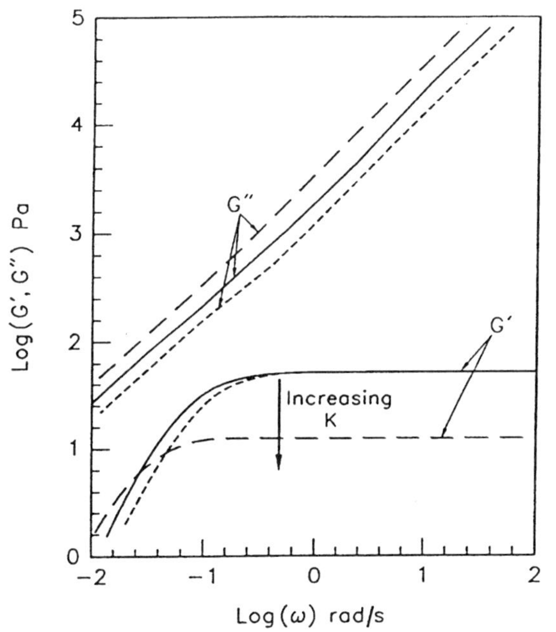

Figure 119. Effect of viscosity ratio k $(= \eta_d/\eta_c)$ on the viscoelastic behavior of blends. (From Ref. 53.)

Recently, Gramespacher and Meissner [54] proposed new relations for the dynamic moduli of blends of immiscible polymeric melts. According to these authors, the complex modulus of a blend (G^*_{blend}) can be written as

$$G^*_{blend} = G^*_{components} + G^*_{interface} \tag{464}$$

where $G^*_{components}$ is the contribution due to viscoelastic properties of the components and $G^*_{interface}$ is the contribution due to the elasticity of the shape of the dispersed droplets (i.e., interfacial-tension effect). A simple mixing rule was used to determine $G^*_{components}$:

$$G^*_{components} = \phi G^*_d + (1-\phi) G^*_c \tag{465}$$

For $G^*_{interface}$, the emulsion model of Choi and Schowalter was used:

$$G^*_{interface} = G'_{interface} + jG''_{interface} \tag{466}$$

$$G'_{interface} = \frac{\eta_0 \omega^2 (h_1 - h_2)}{1 + \omega^2 h_1} = \left(\frac{\eta_0}{h_1}\right)\left(1 - \frac{h_2}{h_1}\right)\left(\frac{\omega^2 h_1^2}{1 + \omega^2 h_1^2}\right) \tag{467}$$

$$G''_{interface} = \frac{\eta_0 \omega (1 + \omega^2 h_1 h_2)}{1 + \omega^2 h_1^2} = \left(\frac{\eta_0}{h_1}\right)\left(1 - \frac{h_2}{h_1}\right)\left(\frac{\omega h_1}{1 + \omega^2 h_1^2}\right) + \omega \eta_0 \left(\frac{h_2}{h_1}\right) \tag{468}$$

where h_1 is the relaxation time and h_2 is the retardation time for the droplets. The formulae for h_1, h_2, and η_0 are given by Eqs. (371)–(373). Note that in these equations, the viscosity ratio k ($= \eta_d/\eta_c$) and the continuous-phase viscosity (η_c) are to be evaluated at zero shear rate.

From these equations for $G^*_{components}$ and $G^*_{interface}$, the following results are obtained for the dynamic moduli of polymeric blends:

$$G'_{blend} = \phi G'_d + (1-\phi) G'_c + \left(\frac{\eta_0}{h_1}\right)\left(1 - \frac{h_2}{h_1}\right)\left(\frac{\omega^2 h_1^2}{1 + \omega^2 h_1^2}\right) \tag{469}$$

$$\begin{aligned} G''_{blend} &= \phi G''_d + (1-\phi) G''_c + \left(\frac{\eta_0}{h_1}\right)\left(1 - \frac{h_2}{h_1}\right)\left(\frac{\omega h_1}{1 + \omega^2 h_1^2}\right) + \omega \eta_0 \left(\frac{h_2}{h_1}\right) \\ &\simeq \phi G''_d + (1-\phi) G''_c + \left(\frac{\eta_0}{h_1}\right)\left(1 - \frac{h_2}{h_1}\right)\left(\frac{\omega h_1}{1 + \omega^2 h_1^2}\right) \end{aligned} \tag{470}$$

Equations (469) and (470) were able to describe adequately the dynamic moduli data for polystyrene (PS)/polymethylmethacrylate (PMMA) blends over a broad frequency range [54].

Gramespacher and Meissner [54] also analyzed the relaxation-time spectra for blends. The relaxation-time spectrum $H(\tau)$ is related to the dynamic moduli as follows:

$$G'(\omega) = \int_{-\infty}^{+\infty} H(\tau) \left(\frac{\omega^2 \tau^2}{1 + \omega^2 \tau^2}\right) d \ln \tau \tag{471}$$

$$G''(\omega) = \int_{-\infty}^{+\infty} H(\tau)\left(\frac{\omega\tau}{1+\omega^2\tau^2}\right) d\ln\tau \tag{472}$$

The weighted relaxation spectrum $\tau H(\tau)$ (where τ is relaxation time) as a function of $\log \tau$ can be calculated from Eqs. (471) and (472) and the dynamic moduli data by using a modified Tikhonov regularization method [54].

The plots of weighted relaxation-time spectra $\tau H(\tau)$ for pure components (PS and PMMA, respectively) are shown in Fig. 120. The pure components give a single peak. In the case of blends, however, there are two major peaks: one due to the distribution of relaxation times for pure components and one due to the distribution of relaxation times for dispersed droplets. A typical weighted relaxation spectrum for blends is given in Fig. 121. Note that the second peak (indicated by the arrow) is found to occur at $\tau = h_1$, where h_1 is the relaxation time for droplets given by the Choi–Schowalter model (Eq. 371). Gramespacher and Meissner [54] further calculated the interfacial tension (σ) for blends from Eq. (371) using the relaxation spectra. The calculated value was found to be in good agreement with the measured value.

However, the major drawback of the Gramespacher–Meissner model is that it is not based on any theoretical analysis. It is, therefore, difficult to judge its generality.

Palierne [51] recently developed a theory for the linear viscoelastic behavior of concentrated emulsions of incompatible viscoelastic liquids. Although the theory developed by Palierne takes into account the most general behavior of the interface (where interfacial tension is dependent on local shear deformation and on the variation of area during deformation of droplets), here we will restrict ourselves to situations where the interfacial tension between the matrix

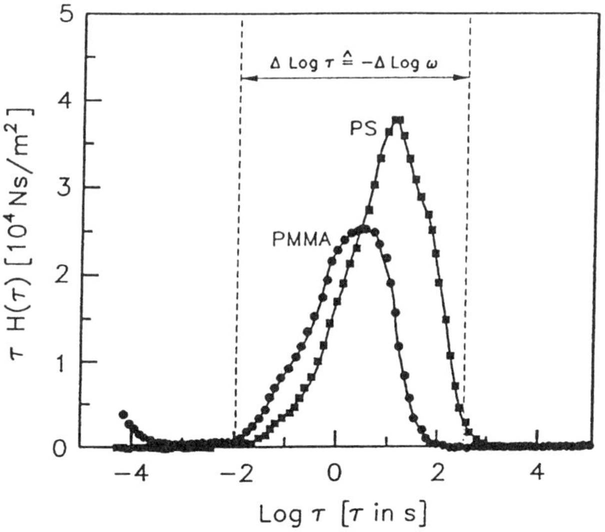

Figure 120. Plots of weighted relaxation-time spectra $\tau H(\tau)$ for polystyrene (PS) and polymethylmethacrylate (PMMA), respectively. (From Ref. 54.)

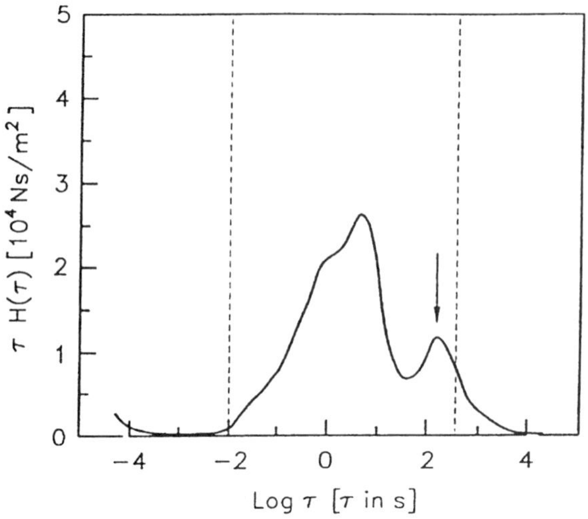

Figure 121. Plot of weighted relaxation-time spectrum for a blend of polystyrene/polymethylmethacrylate. (From Ref. 54.)

and the dispersed phase is constant. Under such conditions, Palierne's theory leads to the following expression for the complex modulus of emulsions:

$$G^* = G_c^* \left[\frac{1 + 3\sum_i \phi_i H_i}{1 - 2\sum_i \phi_i H_i} \right] \tag{473}$$

where G_c^* is the complex modulus for the matrix (continuous phase), ϕ_i is the volume fraction of inclusions (droplets) of radius R_i, and H_i is given as

$$H_i = \frac{(4\sigma/R_i)(2G_c^* + 5G_d^*) + (G_d^* - G_c^*)(16G_c^* + 19G_d^*)}{(40\sigma/R_i)(G_c^* + G_d^*) + (2G_d^* + 3G_c^*)(16G_c^* + 19G_d^*)} \tag{474}$$

Note that for uniform-size droplets, Eq. (473) reduces to Oldroyd's model (Eq. 361):

$$G^* = G_c^* \left(\frac{1 + 3\phi H}{1 - 2\phi H} \right) \tag{361}$$

For emulsions of two immiscible Newtonian liquids with uniform droplet size, Eq. (361) gives

$$G^* = j\omega\eta_0 \left(\frac{1 + j\omega\lambda_2}{1 + j\omega\lambda_1} \right) = \frac{\eta_0\omega^2}{1 + \omega^2\lambda_1^2} (\lambda_1 - \lambda_2) + j\,\frac{\eta_0\omega}{1 + \omega^2\lambda_1^2} (1 + \omega^2\lambda_1\lambda_2) \tag{475}$$

where η_0 is the zero-shear viscosity of the emulsion, λ_1 is relaxation time, and λ_2 is retardation time. The expression for η_0, λ_1, and λ_2 are:

$$\eta_0 = \eta_c \left[\frac{10(k + 1) + 3\phi(5k + 2)}{10(k + 1) - 2\phi(5k + 2)} \right] \tag{476}$$

$$\lambda_1 = \left(\frac{R\eta_c}{4\sigma}\right)\left\{\frac{(19k+16)[2k+3-2\phi(k-1)]}{10(k+1)-2\phi(5k+2)}\right\} \tag{477}$$

$$\lambda_2 = \left(\frac{R\eta_c}{4\sigma}\right)\left\{\frac{19k+16)[2k+3+3\phi(k-1)]}{10(k+1)+3\phi(5k+2)}\right\} \tag{478}$$

For *dilute* emulsions, Eqs. (476)–(478) reduce to the corresponding expressions given in Eqs. (357)–(360).

The viscoelastic behavior of emulsions or blends of polymeric melts becomes more complicated when the components are not Newtonian. For example, consider the case where the components are viscoelastic liquids, which follow the Maxwell model [59]:

$$G_c^* = \frac{j\omega\eta_c}{1+j\omega\lambda_c} \tag{479}$$

$$G_d^* = \frac{j\omega\eta_d}{1+j\omega\lambda_d} \tag{480}$$

where η_c and η_d are zero-shear viscosities of continuous and dispersed phase liquids, respectively, and λ_c and λ_d are the corresponding relaxation times. The dynamic moduli (G' and G'') for the emulsion can now be calculated from Eq. (361) using Eqs. (479) and (480).

As an example, Fig. 122 shows the behavior of the dynamic moduli of emulsion calculated from Eq. (361) under the following conditions [59]: $\lambda_c = \lambda_d = 0.1$ sec, $\eta_c = \eta_d = 10^6$ Pa·sec, $R = 1\,\mu m$, $\sigma = 10$ mN/m, and $\phi = 0.20$. For comparison, the dynamic moduli for emulsions of Newtonian liquids (with

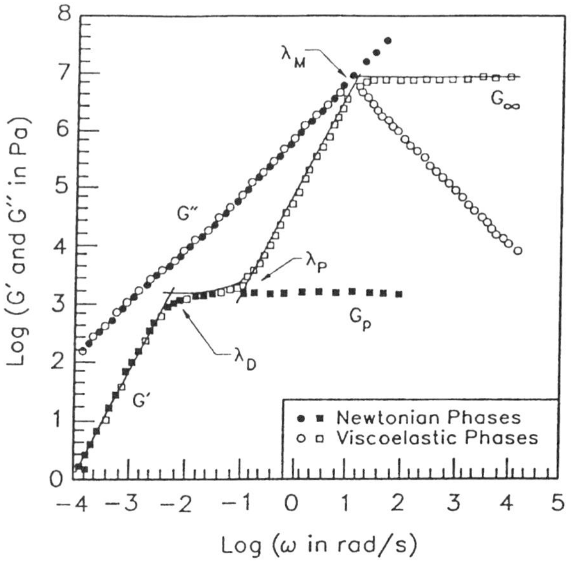

Figure 122. Behavior of dynamic moduli for emulsions. (From Ref. 59.)

the same parameters except that $\lambda_c = \lambda_d = 0$) are also plotted. From the G′ versus frequency plot, four frequency zones and three characteristic relaxation times ($\lambda_D, \lambda_p, \lambda_m$) can be identified. At frequencies greater than $1/\lambda_m$, G′ exhibits a plateau. This zone, where G′ is constant, is referred to as the "first plateau zone." In the frequency range $(1/\lambda_p)$–$(1/\lambda_m)$, the behavior of dynamic moduli may be described approximately as $G' \propto \omega^2$ and $G'' \propto \omega$. However, in the frequency range $(1/\lambda_D)$–$(1/\lambda_p)$, another plateau zone for G′ is observed. This zone is referred to as the "secondary plateau zone." At frequencies less than $1/\lambda_D$, we have the final terminal zone of the emulsion, where the dynamic moduli may be described as $G' \propto \omega^2$ and $G'' \propto \omega$.

Upon comparison of the dynamic moduli data for emulsions of viscoelastic liquids with the corresponding data for emulsions of Newtonian liquids (see Fig. 122), it is clear that λ_D is approximately equal to the relaxation time for emulsions of Newtonian liquids:

$$\lambda_D \simeq \lambda_1 = \left(\frac{R\eta_c}{4\sigma}\right) \frac{(19k + 16)[2k + 3 - 2\phi(k - 1)]}{[10(k + 1) - 2\phi(5k + 2)]} \tag{481}$$

Also,

$$G_p \simeq \lim_{\omega \to \infty} G'_{\text{emulsion of Newtonian liquids}} = 20\left(\frac{\sigma}{R}\right)\phi \frac{1}{[2k + 3 - 2\phi(k - 1)]^2} \tag{482}$$

Figure 123 shows the effect of the viscosity ratio k on the dynamic moduli of emulsion. The viscosity ratio is varied by varying η_d. All other factors

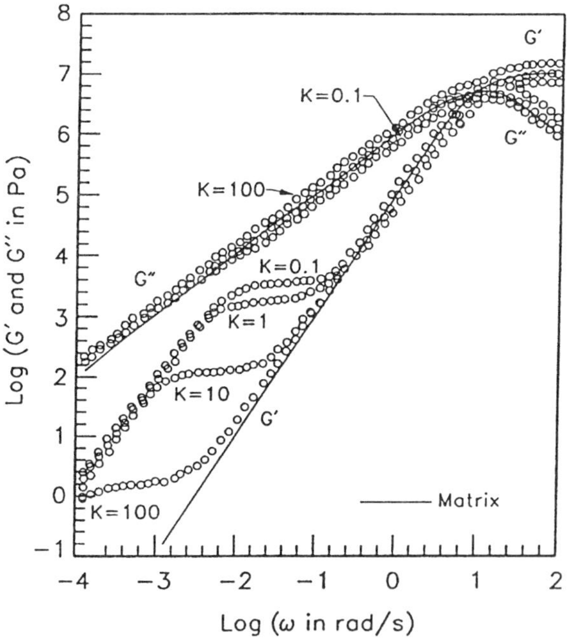

Figure 123. Effect of viscosity ratio k ($= \eta_d/\eta_c$) on the dynamic moduli of emulsion. (From Ref. 59.)

(η_c, λ_c, λ_d, R, σ, and ϕ) are kept constant. With an increase in k, G_p decreases, as expected from Eq. (482). Also, both λ_D and λ_p increase with increasing k. This implies that measurement of the storage modulus in the secondary plateau zone becomes more difficult at high values of k (high-viscosity droplets) because G_p is low and the secondary plateau is shifted to low frequencies.

The effect of the relaxation-time ratio for pure phases (λ_d/λ_c) is shown in Fig. 124. This ratio is varied by varying the relaxation time of the matrix and keeping all other factors constant. By decreasing the ratio λ_d/λ_c, λ_p and λ_D become closer; below a certain value of λ_d/λ_c, the secondary plateau becomes ill defined. This effect is more clear on a Cole–Cole diagram. The Cole–Cole diagram in Fig. 125 shows two frequency domains. The low-frequency domain is due to the relaxation of droplets, whereas the high-frequency domain is due to the relaxation of pure phases. The two domains are well separated only at high values of λ_d/λ_c.

The effect of σ/R, on the dynamic moduli of emulsion is shown in Fig. 126. With an increase in σ/R, G_p increases and λ_D decreases as expected from Eqs. (481) and (482).

The volume fraction of the dispersed phase can also have a significant influence on the dynamic moduli of emulsion. Figure 127 shows the effect of ϕ, keeping all other variables constant. With an increase in ϕ, the value of the storage modulus in the secondary plateau increases. This is in agreement with Eq. (482).

So far we have assumed that the emulsion consists of droplets of uniform size. However, in most practical situations the droplets have a size distribution.

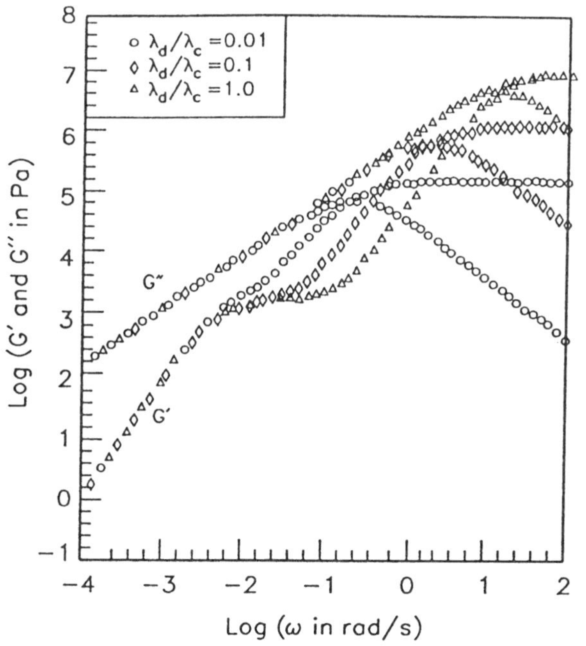

Figure 124. Effect of relaxation-time ratio for pure phases (λ_d/λ_c) on the dynamic moduli of emulsion. (From Ref. 59.)

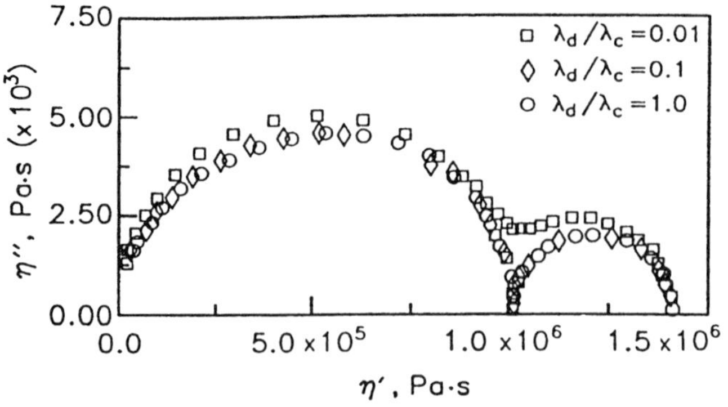

Figure 125. Effect of relaxation-time ratio for pure phases (λ_d/λ_c) on the Cole–Cole diagram of emulsion. (From Ref. 59.)

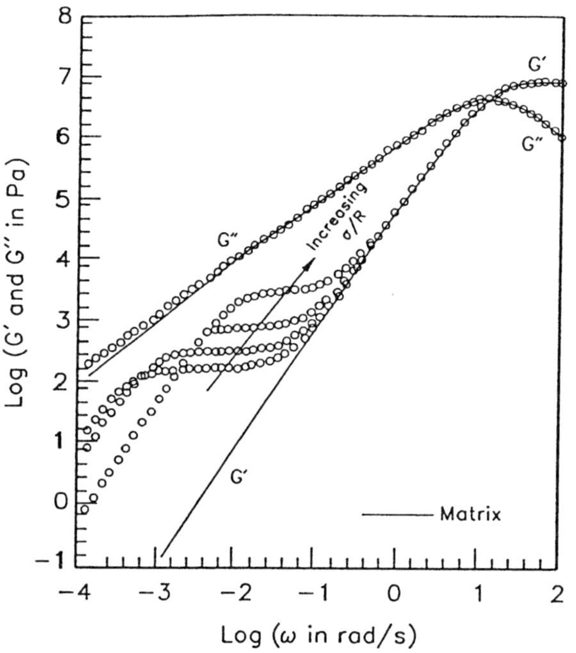

Figure 126. Effect of σ/R on the dynamic moduli of emulsion. (From Ref. 59.)

According to Graebling et al. [59], the effect of droplet size distribution can be taken into account by using a volume average particle radius ($\bar{R}_V$) in the Oldroyd model. $\bar{R}_V$ is defined as

$$\bar{R}_V = \sum \frac{\phi_i R_i}{\phi} = \frac{\sum N_i R_i^4}{\sum N_i R_i^3} \tag{483}$$

where N_i is the number of droplets with radius R_i.

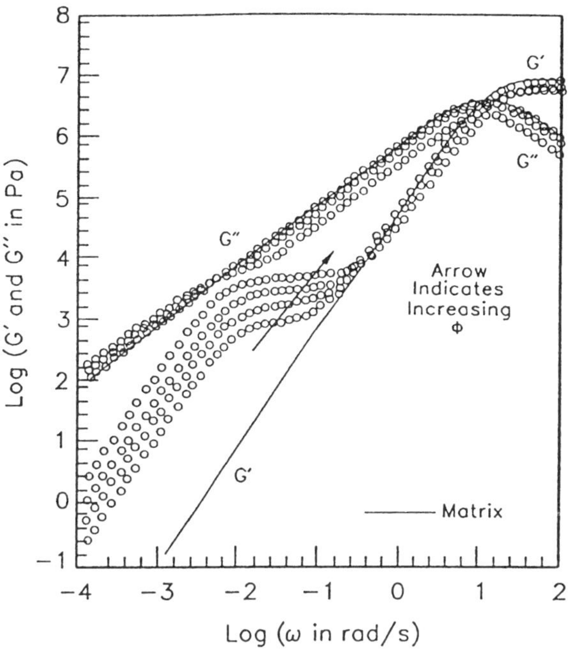

Figure 127. Effect of dispersed-phase concentration (ϕ) on the dynamic moduli of emulsion. (From Ref. 59.)

Finally, note that the Oldroyd (or Palierne) model does not account for aggregation of droplets. If the droplets or particles exhibit the tendency to flocculate, deviation of dynamic moduli from the Oldroyd model is expected. Recently, Bousmina and Muller [60] confirmed this for rubber-toughened PMMA blends (i.e., blends of rubber particles and PMMA matrix). As an example, Fig. 128 compares experimental data with the Oldroyd model (solid curves) for rubber-toughened PMMA blends at various concentrations of rubber. While at low rubber concentrations (<10 wt %) the model gives good predictions, there is a significant deviation at high concentrations, where aggregation of rubber particles is expected.

VIII. Concluding Remarks

The presence of polymers can have a significant effect on the stability and rheology of emulsions. In this chapter, a thorough review of the rheological behavior of emulsions containing polymeric liquids is given. Some basic aspects about polymers and stability of disperse systems are also reviewed. From the review presented here, it is clear that the existing literature, for the most part, is restricted to solids-in-liquid suspensions. For example, very few rheological studies have been conducted on depletion and bridging effects in emulsions. Thus, more experimental work needs to be done.

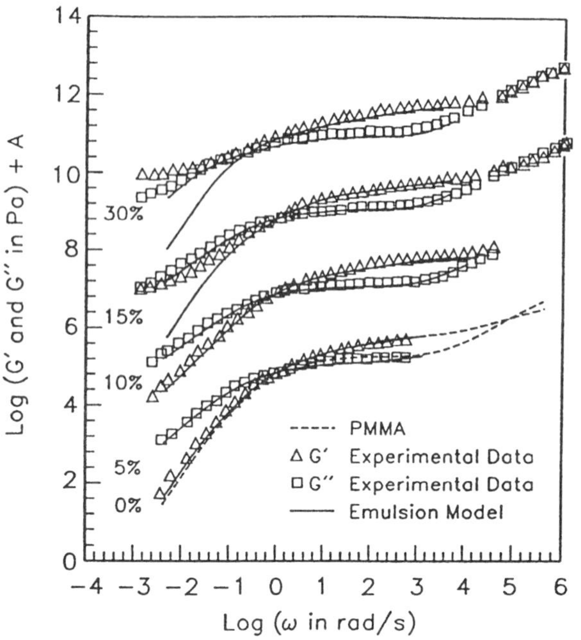

Figure 128. Comparison between experimental data and the Oldroyd model (solid curves) for rubber-toughened PMMA blends at various concentrations of rubber. (From Ref. 60.)

Acknowledgments

Financial support from the Natural Sciences and Engineering Research Council (NSERC) of Canada is gratefully acknowledged.

References

1. H. Bennett, J. L. Bishop, and M. F. Wulfinghoff, *Practical Emulsions—Applications*, Chemical, New York, 1968.
2. K. J. Lissant, *Emulsions and Emulsion Technology*, Dekker, New York, 1974.
3. A. L. Smith, *Theory and Practice of Emulsion Technology*, Academic Press, London, 1976.
4. S. Friberg, *Food Emulsions*, Dekker, New York, 1976.
5. P. Becher, *Emulsions: Theory and Practice*, Krieger, Malabar, Fla., 1977, Ch. 8.
6. S. Torrey, *Emulsions and Emulsifier Applications—Recent Developments*, Noyes Data Corp., Park Ridge, N.J., 1984.
7. P. Becher, ed., *Encyclopedia of Emulsion Technology*, Vol. 2, Dekker, New York, 1986.
8. M. El-Nakaly and D. Cornell, *Microemulsions and Emulsions in Foods*, ACS Symposium Series 448, ACS, Washington, D.C., 1991.
9. A. Jerome Ganz, in *Food Colloids* (H. D. Graham, ed.), AVI Publishing, Westport, Conn., 1977. Ch. 9.

10. Aqualon Company, *Aqualon Cellulose Gum*, Bulletin No. 250-10C Rev. 1-88 10M IP, Aqualon Co., Wilmington, Del., 1988.
11. Aqualon Company, *Culminal*, Bulletin No. 22.001-E2, Aqualon Co., Wilmington, Del., 1988.
12. J. M. V. Blanshard and J. R. Mitchell, *Polysaccharides in Food*, Butterworths, London, 1979.
13. E. Dickinson and G. Stainsby, *Colloids in Food*, Applied Science, London, 1982.
14. P. Kovacs and K. S. Kang, in *Food Colloids* (H. D. Graham, ed.), AVI Publishing, Westport, Conn., 1977, Ch. 13.
15. D. J. Pettit, in *Polysaccharides in Food* (J. M. V. Blanshard and J. R. Mitchell, eds.), Butterworths, London, 1979, Ch. 17.
16. M. Glicksman, in *Polysaccharides in Food* (J. M. V. Blanshard and J. R. Mitchell, eds.,), Butterworths, London, 1979, Ch. 12.
17. W. H. McNeely and K. S. Kang, in *Industrial Gums* (R. L. Whistler and J. N. BeMiller, eds.), Academic Press, New York, 1973, Ch. XXI.
18. J. K. Baird, in *Encyclopedia of Polymer Science and Engineering*, Vol. 17, (J. L. Kroschwitz, ed.), Wiley, New York, 1986, pp. 901–918.
19. W. A. Meer, in *Food Colloids* (H. D. Graham, ed.), AVI Publishing, Westport, Conn., 1977, Ch. 14.
20. E. Dickinson and V. B. Galazka, in *Food Polymers, Gels, and Colloids* (E. Dickinson, ed.), Royal Society of Chemistry, Cambridge, 1991, pp. 494–497.
21. M. M. Robins, in *Microemulsions and Emulsions in Foods* (M. El-Nakaly and D. Cornell, eds.), ACS Symposium Series 448, ACS, Washsington, D.C., 1991, Ch. 17.
22. E. Dickinson, M. I. Goller, and D. J. Wedlock, *Colloids Surfaces 75*:195 (1993).
23. H. Luyten, M. Jonkman, W. Kloek, and T. van Vliet, in *Food Colloids and Polymers: Stability and Mechanical Properties* (E. Dickinson and P. Walstra, eds.), Royal Society of Chemistry, Cambridge, 1993, pp. 224–234.
24. S. J. Gouldby, P. A. Gunning, D. J. Hibberd, and M. M. Robins, in *Food Polymers, Gels, and Colloids* (E. Dickinson, ed.), Royal Society of Chemistry, Cambridge, 1991, pp. 494–497.
25. R. Y. Lockhead, J. A. Davidson, and G. M. Thomas, in *Polymers in Aqueous Media* (J. E. Glass, ed.), Advances in Chemistry Series 223, ACS, Washington, D.C., 1989, pp. 113–147.
26. O. E. Araujo, *J. Pharm. Sci. 56*:1141 (1967).
27. L. R. Stanislav and J. K. Sheets, *Food Prod. Develop. 5*:52 (1971).
28. American Cyanamid Company, *CYANAMER Polyacrylamides*, Bulletin No. 2-1214 5K 8/82 PRT-21D, American Cyanamid Co., Chemical Products Div., Process Chemicals Dept., Wayne, N.J., 1982.
29. D. Reichman and N. Garti, in *Food Polymers, Gels, and Colloids* (E. Dickinson, ed.), Royal Society of Chemistry, Cambridge, 1991, pp. 549–556.
30. P. Sherman, in *Encyclopedia of Emulsion Technology*, Vol. 1, (P. Becher, ed.), Dekker, New York, 1983, pp. 427–429.
31. G. C. March and D. H. Napper, *J. Colloid & Int. Sci. 61*:383 (1977).
32. A. Lips, I. J. Campbell, and E. G. Pelan, in *Food Polymers, Gels, and Colloids* (E. Dickinson, ed.), Royal Society of Chemistry, Cambridge, 1991, pp. 1–21.
33. J. M. G. Lankveld and J. Lyklema, *J. Colloid & Int. Sci. 41*:454 (1972); 41:466 (1972); *41*:475 (1972).
34. D. H. Napper, *Polymeric Stabilization of Colloidal Dispersions*, Academic Press, London, 1983.
35. T. F. Tadros and B. Vincent, in *Encyclopedia of Emulsion Technology*, Vol. 1, (P. Becher, ed.), Dekker, New York, 1983, pp. 232–285.
36. T. W. Schwartz, *Amer. Perfum. Cosmet. 77*:85 (1962).

37. E. E. Isaacs and R. S. Chow, in *Emulsions—Fundamentals and Applications in the Petroleum Industry* (L. L. Schramm, ed.), Advances in Chemistry Series 231, ACS, Washington, D.C., 1992, Ch. 2.
38. R. Grace, in *Emulsions—Fundamentals and Applications in the Petroleum Industry* (L. L. Schramm, ed.), Advances in Chemistry Series 231, ACS, Washington, D.C., 1992, Ch. 9.
39. G. Leopold, in *Emulsions—Fundamentals and Applications in the Petroleum Industry*, Advances in Chemistry Series, ACS, Washington, D.C., 1992, Ch. 10.
40. V. B. Menon and D. T. Wasan, in *Encyclopedia of Emulsion Technology*, Vol. 2, (P. Becher, ed.), Dekker, New York, 1985, pp. 20–27.
41. J. Gregory, in *Solid/Liquid Dispersions* (T. F. Tadros, ed.), Academic Press, London, 1987, pp. 163–181.
42. D. R. Paul, in *Multicomponent Polymer Materials* (D. R. Paul and L. H. Sperling, eds.), Advances in Chemistry Series 211, ACS, Washington, D.C., 1986, pp. 3–19.
43. D. R. Paul and J. W. Barlow, in *Multiphase Polymers* (S. L. Cooper and G. M. Estes, eds.), Advances in Chemistry Series 176, ACS, Washington, D.C., 1979, pp. 315–335.
44. D. R. Paul, J. W. Barlow, and H. Keskkula, in *Encyclopedia of Polymer Science and Engineering*, Vol. 12, (J. L. Droschwitz, ed.), Wiley, New York, 1986, pp. 399–461.
45. L. A. Utracki, *J. Rheol. 35*:1615 (1991).
46. L. A. Utracki and M. R. Kamal, *Poly. Eng. and Sci. 22*:96 (1982).
47. L. Chu, G. Guo, W. Chin, and H. Tseng, *J. Appl. Poly. Sci. 49*:1791 (1993).
48. R. Fayt, R. Jerome, and P. Teyssie, *Poly. Eng. and Sci. 27*:328 (1987).
49. L. A. Utracki, *Poly. Eng. and Sci. 23*:602 (1983).
50. M. Heino, P. T. Hietaoja, T. P. Vainio, and J. V. Seppala, *J. Appl. Poly. Sci. 51*:259 (1994).
51. J. F. Palierne, *Rheol. Acta. 29*:204 (1990).
52. C. D. Han and R. G. King, *J. Rheol. 24*:213 (1980).
53. P. Scholz, D. Froelich, and R. Muller, *J. Rheol. 33*:481 (1989).
54. H. Gramespacher and J. Meissner, *J. Rheol. 36*:1127 (1992).
55. J. G. M. van Gisbergen and H. E. H. Meijer, *J. Rheol. 35*:63 (1991).
56. D. Graebling, D. Froelich, and R. Muller, *J. Rheol. 33*:1283 (1989).
57. D. Graebling and R. Muller, *Colloids Surfaces 55*:89 (1991).
58. D. Graebling and R. Muller, *J. Rheol. 34*:193 (1990).
59. D. Graebling, R. Muller, and J. F. Palierne, *Macromolecules 26*:320 (1993).
60. M. Bousmina and R. Muller, *J. Rheol. 37*:663 (1993).
61. S. Wu, *Poly. Sci. and Eng. 27*:335 (1987).
62. C. D. Han, *Rheology in Polymer Processing*, Academic Press, New York, 1976.
63. C. D. Han, *Polymer Blends and Composites in Multiphase Systems*, Advances in Chemistry Series 206, ACS, Washington, D.C., 1984.
64. C. Tanford, *Physical Chemistry of Macromolecules*, Wiley, New York, 1965.
65. F. W. Billmeyer, *Textbook of Polymer Science*, Interscience, New York, 1965.
66. F. Rodriguez, *Principles of Polymer Systems*, McGraw-Hill, New York, 1970.
67. R. G. Mortimer, *Physical Chemistry*, Benjamin/Cummings, Redwood City, Calif., 1993.
68. A. Einstein, *Ann. Physik 19*:289 (1906).
69. A. Einstein, *Ann. Physik 34*:591 (1911).
70. M. L. Huggins, *J. ACS 64*:2716 (1942).
71. E. O. Kraemer, *Ind. Eng. Chem. 30*:1200 (1938).
72. R. B. Bird, R. C. Armstrong, and O. Hassager, *Dynamics of Polymeric Liquids*, Vol. 1, Wiley, New York, 1987.
73. T. Kotaka, H. Suzuki, and H. Inagaki, *J. Chem. Phys. 45*:2770 (1966).

74. J. Ferguson and Z. Kemblowski, *Applied Fluid Rheology*, Elsevier Applied Science, London, 1991.
75. R. Pal, *J. Rheol. 36(7)*:1245 (1992).
76. R. Pal, *AIChE J.*, in press (1995).
77. H. A. Barnes, J. F. Hutton, and K. Walters, *An Introduction to Rheology*, Elsevier Science, B.V., Amsterdam, Netherlands, 1989.
78. W. Ostwald, *Kolloid-Z. 36*:99 (1925).
79. A. de Waele, *Oil Color Chem. Assoc. J. 6*:33 (1923).
80. M. M. Cross, *J. Colloid Sci. 20*:417 (1965).
81. Bohlin VOR Rheometer Reference User Manual, Bohlin Reologi AB 22370 Lund, Sweden, 1991.
82. P. J. Carreau, *Trans. Soc. Rheol. 16*:99 (1972).
83. R. V. Williamson, *Ind. Eng. Chem. 21*:1108 (1929).
84. M. Reiner, *Deformation, Strain and Flow*, H. K. Lewis, London, 1960, p. 275.
85. A. W. Sisko, *Ind. Eng. Chem. 50*:1789 (1958).
86. E. C. Bingham, *Fluidity and Plasticity*, McGraw-Hill, New York, 1992.
87. H. Herschel and R. Bulkley, *Proc. Am. Soc. Test. Mater. 26*:621 (1926).
88. W. Parzonka and J. Vocadlo, *Rheol. Acta 7*:260 (1968).
89. N. Casson, in *Rheology of Disperse Systems* (C. C. Mill, ed.), Pergamon Press, London, 1959.
90. W. P. Cox and E. H. Merz, *J. Polymer Sci. 28*:619 (1958).
91. H. M. Laun, *J. Rheol. 30*:459 (1986).
92. H. A. Kramers, *Atti Congr. Int. Fisici Como. 2*:545 (1927).
93. R. de L. Kronig, *J. Opt. Soc. Am. 12*:547 (1926).
94. E. D. Goddard, *Colloids Surfaces 19*:255 (1986).
95. E. D. Goddard, *Colloids Surfaces 19*:301 (1986).
96. R. Nagarajan, *Colloids Surfaces 13*:1 (1985).
97. M. N. Jones and A. Brass, in *Food Polymers, Gels, and Colloids* (E. Dickinson, ed.), Royal Society of Chemistry, Cambridge, 1991.
98. A. P. Mast, R. K. Prud'homme, and J. E. Glass, *Langmuir 9*:708 (1993).
99. B. Lindman, A. Carlsson, S. Gerdes, G. Karlstrom, L. Piculell, K. Thalberg, and K. Zhang, in *Food Colloids and Polymers* (E. Dickinson and P. Walstra, eds.), Royal Society of Chemistry, Cambridge, 1993.
100. P. L. Dubin and D. Davis, *Colloids Surfaces 13*:113 (1985).
101. G. B. Arnold and M. M. Breuer, *Colloids Surfaces 13*:103 (1985).
102. B. M. Moudgil and P. Somasundaran, *Colloids Surfaces 13*:87 (1985).
103. P. Somasundaran and J. Cleverdon, *Colloids Surfaces 13*:73 (1985).
104. P. S. Leung and E. D. Goddard, *Colloids Surfaces 13*:47 (1985).
105. B. Cabane and R. Duplessix, *Colloids Surfaces 13*:19 (1985).
106. K. P. Ananthapadmanabhan, P. S. Leung, and E. D. Goddard, *Colloids Surfaces 13*:63 (1985).
107. M. Barck and P. Stenius, *Colloids Surfaces 89*:59 (1994).
108. T. Sato and R. Ruch, *Stabilization of Colloidal Dispersions by Polymer Adsorption*, Dekker, New York, 1980.
109. R. Perkel and R. Ullman, *J. Polym. Sic. 54*:127 (1961).
110. I. Langmuir, *Science 88*:450 (1938).
111. T. F. Tadros and B. Vincent, in *Encyclopedia of Emulsion Technology*, Vol. 1, (P. Becher, ed.), Dekker, New York, 1983. p. 41.
112. R. R. Stromberg, D. J. Tutas, and E. Passaglia, *J. Phy. Chem. 69*:3955 (1965).
113. G. W. Castellan, *Physical Chemistry*, 3d ed., Benjamin/Cummings, Menlo Park, Calif., 1983, pp. 936–937.

114. T. F. Tadros, in *Solid/Liquid Dispersions* (T. F. Tadros, ed.), Academic Press, London, 1987, Ch. 13.
115. M. J. Garvey, T. F. Tadros, and B. Vincent, *J. Colloid Int. Sci. 49*:57 (1974).
116. M. J. Garvey, T. F. Tadros, and B. Vincent, *J. Colloid Int. Sci. 55*:440 (1976).
116a. T. F. Tadros and B. Vincent, in *Encyclopedia of Emulsion Technology*, Vol. 1, (P. Becher, ed.), Dekker, New York, 1983, Ch. 7.
116b. P. Walstra, this volume, Ch. 1.
117. B. H. Bijsterbosch, in *Solid/Liquid Dispersions* (T. F. Tadros, ed.), Academic Press, London, 1987, Ch. 4.
118. J. Visser, *Adv. Coll. Interface Sci. 3*:331 (1972).
119. V. T. Crowl, in *Pigments—An Introduction to Their Physical Chemistry* (D. Patterson, ed.), Elsevier, New York, 1967, Ch. 10.
120. M. J. Vold, *J. Colloid Sci. 16*:1 (1961).
121. R. H. Ottewill, in *Nonionic Surfactants* (M. J. Schick, ed.), Dekker, New York, 1967, Ch. 9.
121a. P. Becher, *J. Colloid Int. Sci. 42*:645 (1973).
122. E. W. Fischer, *Kolloid-Z. 160*:120 (1958).
123. E. L. Mackor, *J. Colloid Int. Sci. 6*:492 (1951).
124. V. K. La Mer, *Discuss. Faraday Soc. 42*:248 (1966).
125. S. Asakura and F. Oosawa, *J. Chem. Phys. 22*:1255 (1954).
126. S. Asakura and F. Oosawa, *J. Polym. Sci. 33*:183 (1958).
127. T. F. Tadros and A. Zsednai, *Colloids Surfaces 43*:105 (1990).
128. W. Liang, T. F. Tadros, and P. F. Luckham, *J. Colloid Int. Sci. 155*:156 (1993).
129. W. Liang, T. F. Tadros, and P. F. Luckham, *J. Colloid Int. Sci. 158*:152 (1993).
130. W. Liang, T. F. Tadros, and P. F. Luckham, *J. Colloid Int. Sci. 160*:183 (1993).
130a. P. Sherman, in *Encyclopedia of Emulsion Technology*, Vol. 1 (P. Becher, ed.), Dekker, New York, 1983, Ch. 7.
130b. B. W. Davis, in *Encyclopedia of Emulsion Technology*, Vol. 3 (P. Becher, ed.), Dekker, New York, 1988, Ch. 8.
131. I. M. Krieger, *Adv. Colloid Int. Sci. 3*:111 (1972).
132. B. E. Rodriguez, E. W. Kaler, and M. S. Wolfe, *Langmuir 8*:2382 (1992).
133. W. B. Russel, *J. Rheol. 24(3)*:287 (1980).
134. S. J. Willey and C. W. Macosko, *J. Rheol. 22(5)*:525 (1978).
135. R. Pal, *AIChE J. 39*:1754 (1993).
136. D. A. R. Jones, B. Leary, and D. V. Boger, *J. Colloid Int. Sci. 147(2)*:479 (1991).
137. R. L. Hoffman, *Adv. Colloid Int. Sci. 17*:161 (1982).
138. H. A. Barnes, *J. Rheol. 33(2)*:329 (1989).
139. J. S. Chong, E. B. Christiansen, and A. D. Baer, *J. Appl. Poly. Sci. 15*:2007 (1971).
140. G. K. Batchelor, *J. Fluid Mech. 41(3)*:545 (1970).
141. G. K. Batchelor and J. T. Green, *J. Fluid Mech. 56(3)*:401 (1972).
142. G. K. Batchelor, *J. Fluid Mech. 83(1)*:97 (1977).
143. W. B. Russel, D. A. Saville, and W. R. Schowalter, *Colloidal Dispersions*, Cambridge University Press, Cambridge, 1989.
144. D. J. Jeffrey and A. Acrivos, *AIChE J. 22(3)*:417 (1976).
145. A. T. J. M. Woutersen and C. G. de Kruif, *J. Rheol. 37(4)*:681 (1993).
146. R. Simha, *J. Appl. Phys. 23*:1020 (1952).
147. J. Happel, *J. Appl. Phys. 28*:1288 (1957).
148. N. A. Frankel and A. Acrivos, *Chem. Eng. Sci. 22*:847 (1967).
149. S. Arrhenius, *Z. Physik. Chem. 1*:285 (1887).
150. R. Roscoe, *Br. J. Appl. Phys. 3*:267 (1952).
151. H. C. Brinkman, *J. Chem. Phys. 20(4)*:571 (1952).
152. M. Mooney, *J. Colloid Sci. 6*:162 (1951).

153. I. M. Krieger and T. J. Dougherty, *Trans. Soc. Rheol. 3*:137 (1959).
154. R. Pal and E. Rhodes, *J. Colloid Int. Sci. 107(2)*:301 (1985).
155. R. Pal and E. Rhodes, *J. Rheol. 33(7)*:1021 (1989).
156. H. A. Barnes, *Colloids Surfaces A91*:89 (1994).
157. C. R. Wildemuth and M. C. Williams, *Rheol. Acta 23*:627 (1984).
158. R. D. Sudduth, *J. Appl. Polym. Sic. 48*:37 (1993).
159. R. N. Gupta and S. G. Seshadri, *J. Rheol. 30*:503 (1986).
160. A. J. Poslinski, M. E. Ryan, R. K. Gupta, S. G. Seshadri, and F. J. Frechette, *J. Rheol. 32(7)*:703 (1988).
161. A. J. Poslinski, M. E. Ryan, R. K. Gupta, S. G. Seshadri, and F. J. Frechette, *J. Rheol. 32(8)*:751 (1988).
162. C. Castillo and M. C. Williams, *Chem. Eng. Commun. 3*:529 (1979).
163. D. G. Thomas, *J. Colloid Sci. 20*:267 (1965).
164. I. M. Krieger and M. Eguiluz, *Trans. Soc. Rheol. 20(1)*:29 (1976).
165. W. H. Boersma, J. Laven, and H. N. Stein, *J. Colloid Int. Sci. 149(1)*:10 (1992).
166. R. L. Hoffman, *Trans. Soc. Rheol. 16*:155 (1972).
167. R. L. Hoffman, *J. Colloid Int. Sci. 46*:491 (1974).
168. R. L. Hoffman, *J. Rheol. 36(5)*:947 (1992).
169. W. H. Boersma, J. Laven, and H. N. Stein, *AIChE J. 36(3)*:321 (1990).
170. W. Liang, T. F. Tadros, and P. F. Luckham, *J. Colloid Int. Sci. 153(1)*:131 (1992).
171. B. A. Costello, P. F. Luckham, and T. F. Tadros, *J. Colloid Int. Sci. 152(1)*:237 (1992).
172. J. W. Goodwin, in *Colloid Science 2*, Royal Society of Chemistry specialist periodical report, London, 1975, Ch. 7.
173. I. G. Watterson and L. R. White, *J. Chem. Soc. Faraday Trans. 2. 77*:115 (1971).
174. W. G. Russel, *J. Fluid Mech. 85(2)*:209 (1978).
175. W. B. Russel, *J. Colloid Int. Sci. 55(3)*:590 (1976).
176. F. S. Chan, J. Blachford, and D. A. I. Goring, *J. Colloid Int. Sci. 22*:378 (1966).
177. J. W. Goodwin, T. Gregory, and J. A. Stile, *Adv. Colloid Int. Sci. 17*:185 (1982).
178. J. W. Goodwin, in *Solid/Liquid Dispersions* (T. F. Tadros, ed.), Academic Press, London, 1987, Ch. 10.
179. B. E. Conway and A. Dobry-Duclaux, in *Rheology*, Vol. 3 (F. R. Eirich, ed.), Academic Press, New York, 1960, p. 83.
180. M. von Smoluchowski, *Kolloid-Z. 18*:194 (1916).
181. B. Krasny-Ergen, *Kolloid-Z. 74*:172 (1936).
182. F. Booth, *Proc. Roy. Soc. A203*:533 (1950).
183. M. van der Waarden, *J. Colloid Sci. 9*:215 (1954).
184. R. J. Hunter, *Adv. Colloid Int. Sci. 17*:197 (1982).
185. C. F. Goodeve, *Trans. Faraday Soc. 35*:342 (1939).
186. T. Gillespie, *J. Colloid Sci. 15*:219 (1960).
187. T. Gillespie, *J. Polym. Sci. 46*:383 (1960).
188. T. Gillespie and R. M. Wiley, *J. Phys. Chem. 66*:1077 (1962).
189. R. S. Bradley, *Phil. Mag. 13*:853 (1932).
190. H. C. Hamaker, *Physica 4*:1058 (1937).
191. B. A. Firth and R. J. Hunter, *J. Colloid Int. Sci. 57*:248 (1976).
192. B. A. Firth, *J. Colloid Int. Sci. 57*:257 (1976).
193. B. A. Firth and R. J. Hunter, *J. Colloid Int. Sci. 57(2)*:266 (1976).
194. T. G. M. van de Ven and R. J. Hunter, *Rheol. Acta. 16*:534 (1977).
195. R. J. Hunter and J. Frayne, *J. Colloid. Int. Sci. 71(1)*:30 (1979).
196. R. J. Hunter and J. Frayne, *J. Colloid Int. Sci. 76(1)*:107 (1980).
197. N. Ekdawi and R. J. Hunter, *J. Colloid Int. Sci. 94(2)*:355 (1983).
198. M. van den Tempel, *J. Colloid Sci. 16*:284 (1961).

199. P. Sherman, in *Proceedings of the Fifth International Congress on Rheology*, Vol. 2 (S. Onogi, ed.), University of Tokyo Press, Tokyo, and University Park Press, Manchester, Eng., 1970, pp. 327–338.
200. V. K. Strenge and H. Sonntag, *Colloid Polym. Sci. 252*:133 (1974).
201. J. Mewis and A. J. B. Spaull, *Adv. in Colloid. Int. Sci. 6*:173 (1976).
202. T. G. M. van de Ven and R. J. Hunter, *J. Colloid Int. Sci. 68(1)*:135 (1979).
203. R. Buscall, J. W. Goodwin, M. W. Hawkins, and R. H. Ottewill, *J. Chem. Soc. Faraday Trans. 1 78*:2873 (1982).
204. R. Buscall, J. W. Goodwin, M. W. Hawkins, and R. H. Ottewill, *J. Chem. Soc. Faraday Trans. 1 78*:2889 (1982).
205. J. W. Goodwin, T. Gregory, and J. A. Stile, *Adv. Colloid Int. Sci. 17*:185 (1982).
206. L. G. B. Bremer and T. van Vliet, *Rheol. Acta. 30*:98 (1991).
207. V. Chaplain, P. Mills, and M. Djabourov, *Colloid Polym. Sci. 272*:991 (1994).
208. G. I. Taylor, *Proc. Roy. Soc. A. 138*:41 (1932).
209. J. G. Oldroyd, *Proc. Roy. Soc. A. 218*:122 (1953).
210. J. G. Oldroyd, in *Rheology of Disperse Systems* (C. C. Mill, ed.), Pergamon Press, London, 1959, Ch. 1.
211. W. R. Schowalter, C. E. Chaffey, and H. Brenner, *J. Colloid Int. Sci. 26*:152 (1968).
212. N. A. Frankel and A. Acrivos, *J. Fluid Mech. 44*:65 (1970).
213. S. J. Choi and W. R. Schowalter, *The Physics of Fluids 18(4)*:420 (1975).
214. L. Nicodemo, L. Nicolais, and R. F. Landel, *Chem. Eng. Sci. 29*:729 (1974).
215. H. Tanaka and J. L. White, *Polymer Eng. Sci. 20(14)*:949 (1980).
216. D. Chan and R. L. Powell, *J. Non-Newtonian Fluid Mech. 15*:165 (1984).
217. R. Pal, *Chem. Eng. Comm. 111*:45 (1992).
218. R. Pal, *J. Appl. Polym. Sci. 49*:65 (1993).
219. J. Laven. and H. N. Stein, *J. Rheol. 35(8)*:1523 (1991).
220. G. J. Jarzebski, *Rheol. Acta 20*:280 (1981).
221. H. Tanaka and J. L. White, *J. Non-Newtonian Fluid Mech. 7*:333 (1980).
222. Y. Kawase and J. J. Ulbrecht, *Chem. Eng. Commun. 20*:127 (1983).
223. T. Kitano, T. Kataoka, and T. Shirota, *Rheol. Acta. 20*:207 (1981).
224. S. H. Maron and P. E. Pierce, *J. Colloid Sci. 11*:80 (1956).
225. A. B. Metzner, *J. Rheol. 29(6)*:739 (1985).
226. G. J. Fleer, J. H. M. H. Scheutjens, and B. Vincent, *ACS Symp. Ser. 240*:245 (1984).
227. Y. Otsubo and K. Umeya, *J. Colloid Int. Sci. 95(1)*:279 (1983).
228. Y. Otsubo and K. Umeya, *J. Rheol. 28(2)*:95 (1984).
229. Y. Otsubo, *J. Colloid Int. Sci. 112(2)*:380 (1986).
230. Y. Otsubo and K. Watanabe, *J. Colloid Int. Sci. 122(2)*:346 (1988).
231. Y. Otsubo and K. Watanabe, *J. Colloid Int. Sci. 133(2)*:491 (1989).
232. Y. Otsubo, *Langmuir 6*:114 (1990).
233. Y. Otsubo and K. Watanabe, *Colloids Surfaces 50*:341 (1990).
234. Y. Otsubo, *J. Colloid Int. Sci. 153(2)*:584 (1992).
235. Y. Otsubo, *Langmuir 8*:2336 (1992).
236. Y. Otsubo, *J. Rheol. 37(5)*:799 (1993).
237. Y. Otsubo, *Adv. Colloid Int. Sci. 53*:1 (1994).
238. D. R. Paul and J. W. Barlow, *J. Macromol. Sci., Rev. Macromol. Chem. C18*:109 (1980).
239. V. I. Metelkin and V. S. Blekht, *Kolloid Zh. 46*:476 (1984).
240. C. C. Lin, *Polym. J. 11(3)*:185 (1979).
241. M. V. Tsebrenko, A. I. Benzar, A. V. Yudin, and G. V. Vinogradov, *Vysokomol Soed A21*:830 (1979).

242. V. N. Kuleznev, O. L. Melnikova, V. D. Klykova, V. P. Skvortsov, and V. S. Gludhovskoi, *Kolloid Zh.* *37*:273 (1975).
243. L. A. Utracki, *Polymer Alloys and Blends*, Hanser, Munich, 1989.
244. L. A. Utracki, in *Current Topics in Polymer Science* (R. M. Ottenbrite, L. A. Utracki, and S. Inoue, eds.), Hanser, Munich, 1987, Ch. 5.1.

4
Repulsive Forces in Aqueous Solubility, Micelle Stability, Microemulsion Formation, and Phase Separation

CAREL J. VAN OSS / Department of Microbiology, School of Medicine and Biomedical Sciences, State University of New York at Buffalo, Buffalo, New York

I. Introduction

Repulsive forces caused by electrical double-layer (EL) interactions between electrically charged biopolymers, micelles, droplets, or particles have long been known to exist and to play an important role in colloidal stability [1–3]. The classical DLVO theory, named after Derjaguin and Landau [4] and Verwey and Overbeek [2,3], holds that particle stability should prevail when the electrostatic repulsion between suspended particles is larger than their van der Waals attraction. However, electrically neutral, hydrophilic particles can also form stable suspensions in aqueous media. It became clear about a decade ago that this type of stability is caused by net repulsive Lewis acid–base (AB) interactions [5–7]. Repulsive AB interactions were subsequently shown to be the driving force for: polymer solubility in aqueous media [8], micelle stability [9], aqueous polymer phase separation [10], and stability of clay particle suspensions [11].

Repulsive AB interactions were also proposed as the principal driving force for microemulsion formation [6]; this hypothesis, however, is analogous to earlier suggestions that negative interfacial tensions are at the origin of microemulsion formation [12–15]. Although phase separation in apolar systems does not give rise to the formation of more than two phases [16], in polar systems separation into as many phases as there are different solutes is theoretically possible [10] and practically feasible [17], as long as each of the solutes has a different free energy of repulsion [18]. Thus, if in emulsions or microemulsions, particles or droplets of different surface energies coexist (e.g., droplets of similar surface energies per unit surface area but of different sizes), multiple phase separations can occur within a major phase compartment. With the help of schlieren optics, the existence of such multiple phase systems has been demonstrated [19–21].

II. Apolar and Polar Interfacial Interactions in Aqueous Media

A. Apolar Interactions

Surface tensions of condensed-phase materials i are composed of apolar, or Lifshitz–van der Waals (LW), and polar, or Lewis acid–base (AB), components:

$$\gamma_i = \gamma_i^{LW} + \gamma_i^{AB} \tag{1}$$

The same holds for interfacial tensions, $\gamma_{ij} = \gamma_{iw}$. The apolar component of the interfacial tension, γ_{ij}^{LW}, obeys the Good–Girifalco–Fowkes combining rule [22,23],

$$\gamma_{ij}^{LW} = \left(\sqrt{\gamma_i^{LW}} - \sqrt{\gamma_j^{LW}}\right)^2 \tag{2}$$

with an error that is smaller than 2% [24].

The free energy of cohesion of (condensed-phase) material i is:

$$\Delta G_{ii}^{coh} = -2\gamma_i \tag{3}$$

and the interfacial (IF) free energy of adhesion between molecules or particles of material i immersed in liquid j is:

$$\Delta G_{iji}^{IF} = -2\gamma_{ij} \tag{4}$$

where, in analogy with Eq. (1),

$$\Delta G^{IF} = \Delta G^{LW} + \Delta G^{AB} \tag{5}$$

which is valid with subscripts ii, iji, and ikj. In the last case, according to the Dupré equation,

$$\Delta G_{ikj}^{IF} = \gamma_{ij} - \gamma_{ik} - \gamma_{jk} \tag{6}$$

where k represents the liquid in which the interaction occurs. Thus, for the interaction between molecules or particles i and j in water w:

$$\Delta G_{iwj}^{IF} = \gamma_{ij} - \gamma_{iw} - \gamma_{jw} \tag{6a}$$

which, for apolar reactions (cf. Eq. 2), can be expressed as

$$\begin{aligned}\Delta G_{iwj}^{LW} &= 2\left(\sqrt{\gamma_i^{LW}\gamma_w^{LW}} + \sqrt{\gamma_j^{LW}\gamma_w^{LW}} - \sqrt{\gamma_i^{LW}\gamma_j^{LW}} - \gamma_w^{LW}\right) \\ &= 2\left(\sqrt{\gamma_i^{LW}} - \sqrt{\gamma_w^{LW}}\right)\left(\sqrt{\gamma_w^{LW}} - \sqrt{\gamma_j^{LW}}\right)\end{aligned} \tag{7}$$

B. Polar Interactions

The γ_i^{AB} component of material i (cf. Eq. 1) is composed of two parameters: $\gamma_i^{\oplus}$, the electron-acceptor parameter, and $\gamma_i^{\ominus}$, the electron-donor parameter [6,7]:

$$\gamma_i^{AB} = 2\sqrt{\gamma_i^{\oplus}\gamma_i^{\ominus}} \tag{8}$$

As a consequence of the dual nature of γ_i^{AB} [25], the combining rule for the polar component of the interfacial tension between materials i and j is quite different from Eq. (2), which defined γ_{ij}^{LW}:

$$\begin{aligned}\gamma_{ij}^{AB} &= 2\left(\sqrt{\gamma_i^{\oplus}\gamma_i^{\ominus}} + \sqrt{\gamma_j^{\oplus}\gamma_j^{\ominus}} - \sqrt{\gamma_i^{\oplus}\gamma_j^{\ominus}} - \sqrt{\gamma_i^{\ominus}\gamma_j^{\oplus}}\right) \\ &= 2\left(\sqrt{\gamma_i^{\oplus}} - \sqrt{\gamma_j^{\oplus}}\right)\left(\sqrt{\gamma_i^{\ominus}} - \sqrt{\gamma_j^{\ominus}}\right)\end{aligned} \tag{9}$$

The most important difference is that γ_{ij}^{LW} always has a positive value (cf. Eq. 2), whereas γ_{ij}^{AB} can be either positive or negative. Thus:

$$\begin{aligned}\gamma_{ij} = \gamma_{ij}^{LW} + \gamma_{ij}^{AB} &= \left(\sqrt{\gamma_i^{LW}} - \sqrt{\gamma_j^{LW}}\right)^2 \\ &+ 2\left(\sqrt{\gamma_i^{\oplus}\gamma_i^{\ominus}} + \sqrt{\gamma_j^{\oplus}\gamma_j^{\ominus}} - \sqrt{\gamma_i^{\oplus}\gamma_j^{\ominus}} - \sqrt{\gamma_i^{\ominus}\gamma_j^{\oplus}}\right)\end{aligned} \tag{10}$$

can also be either positive or negative.

Using the Dupré equation (Eq. 6a), the polar free energy of interaction between materials i and j immersed in water is [6,7]:

$$\Delta G_{\mathrm{iwj}}^{\mathrm{AB}} = 2\left[\sqrt{\gamma_{\mathrm{w}}^{\oplus}}\left(\sqrt{\gamma_{\mathrm{i}}^{\ominus}} + \sqrt{\gamma_{\mathrm{j}}^{\ominus}} - \sqrt{\gamma_{\mathrm{w}}^{\ominus}}\right) + \sqrt{\gamma_{\mathrm{w}}^{\ominus}}\left(\sqrt{\gamma_{\mathrm{i}}^{\oplus}} + \sqrt{\gamma_{\mathrm{j}}^{\oplus}} - \sqrt{\gamma_{\mathrm{w}}^{\oplus}}\right) - \sqrt{\gamma_{\mathrm{i}}^{\oplus}\gamma_{\mathrm{j}}^{\ominus}} - \sqrt{\gamma_{\mathrm{i}}^{\ominus}\gamma_{\mathrm{j}}^{\oplus}}\right] \tag{11}$$

Thus, following Eq. (5) and combining Eqs. (17) and (11), the total interfacial (IF) free energy of interaction between materials i and j immersed in water is expressed as

$$\begin{aligned}\Delta G_{\mathrm{iwj}}^{\mathrm{IF}} = {} & 2\left(\sqrt{\gamma_{\mathrm{i}}^{\mathrm{LW}}} - \sqrt{\gamma_{\mathrm{w}}^{\mathrm{LW}}}\right)\left(\sqrt{\gamma_{\mathrm{w}}^{\mathrm{LW}}} - \sqrt{\gamma_{\mathrm{j}}^{\mathrm{LW}}}\right) \\ & + 2\left[\sqrt{\gamma_{\mathrm{w}}^{\oplus}}\left(\sqrt{\gamma_{\mathrm{i}}^{\ominus}} + \sqrt{\gamma_{\mathrm{j}}^{\ominus}} - \sqrt{\gamma_{\mathrm{w}}^{\ominus}}\right) + \sqrt{\gamma_{\mathrm{w}}^{\ominus}}\left(\sqrt{\gamma_{\mathrm{w}}^{\oplus}} + \sqrt{\gamma_{\mathrm{j}}^{\oplus}} - \sqrt{\gamma_{\mathrm{w}}^{\oplus}}\right) \right. \\ & \left. - \sqrt{\gamma_{\mathrm{i}}^{\oplus}\gamma_{\mathrm{j}}^{\ominus}} - \sqrt{\gamma_{\mathrm{i}}^{\ominus}\gamma_{\mathrm{j}}^{\oplus}}\right]\end{aligned} \tag{12}$$

Also, following Eq. (4), the total interfacial (IF) free energy of interaction between molecules or particles i immersed in water (combining Eqs. 2 and 10 and multiplying by -2):

$$\Delta G_{\mathrm{iwi}}^{\mathrm{IF}} = -2\left(\sqrt{\gamma_{\mathrm{i}}^{\mathrm{LW}}} - \sqrt{\gamma_{\mathrm{w}}^{\mathrm{LW}}}\right)^2 - 4\left(\sqrt{\gamma_{\mathrm{i}}^{\oplus}\gamma_{\mathrm{i}}^{\ominus}} + \sqrt{\gamma_{\mathrm{w}}^{\oplus}\gamma_{\mathrm{w}}^{\ominus}} - \sqrt{\gamma_{\mathrm{i}}^{\oplus}\gamma_{\mathrm{w}}^{\ominus}} - \sqrt{\gamma_{\mathrm{i}}^{\ominus}\gamma_{\mathrm{w}}^{\oplus}}\right) \tag{13}$$

C. Measurement of Apolar and Polar Surface Properties

For any solid material S, the values of $\gamma_{\mathrm{S}}^{\mathrm{LW}}$, $\gamma_{\mathrm{S}}^{\oplus}$, and $\gamma_{\mathrm{S}}^{\ominus}$ can be measured by contact angle (θ) determination with a liquid L of known $\gamma_{\mathrm{L}}^{\mathrm{LW}}$, $\gamma_{\mathrm{L}}^{\oplus}$, and $\gamma_{\mathrm{L}}^{\ominus}$, using [6,7]

$$1 + \cos\theta = 2\left(\sqrt{\gamma_{\mathrm{S}}^{\mathrm{LW}}\gamma_{\mathrm{L}}^{\mathrm{LW}}} + \sqrt{\gamma_{\mathrm{S}}^{\oplus}\gamma_{\mathrm{L}}^{\ominus}} + \sqrt{\gamma_{\mathrm{S}}^{\ominus}\gamma_{\mathrm{L}}^{\oplus}}\right) \tag{14}$$

Because in the above case (Eq. 14), there are three unknowns ($\gamma_{\mathrm{S}}^{\mathrm{LW}}$, $\gamma_{\mathrm{S}}^{\oplus}$, and $\gamma_{\mathrm{S}}^{\ominus}$), Eq. (14) must be used (at least) three times, with liquids of known $\gamma_{\mathrm{L}}^{\mathrm{LW}}$, $\gamma_{\mathrm{L}}^{\oplus}$, and $\gamma_{\mathrm{L}}^{\ominus}$ values, of which two must be polar. Furthermore, it is necessary that $\gamma_{\mathrm{L}} > \gamma_{\mathrm{S}}$, to obtain nonzero contact angles. When the γ_{L} components and parameters are the unknowns, the same procedure can be used, employing surfaces with known $\gamma_{\mathrm{S}}^{\mathrm{LW}}$, $\gamma_{\mathrm{S}}^{\oplus}$, and $\gamma_{\mathrm{S}}^{\ominus}$ values (see, e.g., van Oss et al. [25a]). A list of the $\gamma_{\mathrm{L}}^{\mathrm{LW}}$, $\gamma_{\mathrm{L}}^{\oplus}$, and $\gamma_{\mathrm{L}}^{\ominus}$ values of some of the more useful high-energy contact angle liquids may be found in Refs. 26 and 50.

III. Electrical Double-Layer Interactions

The total interaction energies of molecules, droplets, or particles suspended in a liquid comprises not only ΔG^{IF} (Eqs. 12 and 13), for in those cases where such molecules, droplets, or particles have a significant electrical surface potential, the free energy of electrical double-layer interactions must also be taken into account for short-range as well as for long-range interactions:

$$\Delta G^{TOT} = \Delta G^{IF} + \Delta G^{EL} \tag{15}$$

The value of ΔG^{EL} is determined from the (ψ_0) potential at the potential-determining surface of the molecules, droplets, or particles in question [3,4], which is a function of the ζ-potential, determined at the slipping plane of these entities, via electrokinetic measurements (usually electrophoresis); see, e.g., Refs. 6 and 27.

IV. Interfacial Tensions and Interfacial Free Energies

A. Attractive and Repulsive Interfacial Free Energies

1. LW Energies

Whilst LW interfacial tensions cannot be negative (cf. Eq. 2), so that LW repulsive forces between similar entities immersed in a liquid (Eq. 4) cannot be repulsive, LW repulsions *can* occur between two different entities immersed in a liquid (Eq. 7), as long as $\gamma_i^{LW} > \gamma_{liquid}^{LW} > \gamma_j^{LW}$ or $\gamma_i^{LW} < \gamma_{liquid}^{LW} < \gamma_j^{LW}$ [16]. Thus even in water (with $\gamma_w^{LW} = 21.8\,\mathrm{mJ/m^2}$), LW repulsions can occur, e.g., between an alkane smaller than octane and a higher alkane or other organic liquid with $\gamma^{LW} \gtrsim 22\,\mathrm{mJ/m^2}$. However, whether repulsive or attractive, the LW interaction energies between low-energy compounds immersed in water are usually fairly close to zero because of the closeness of their γ^{LW} values to the $21.8\,\mathrm{mJ/m^2}$ of water. Thus, in aqueous media the LW interaction energies between similar (ΔG_{iwi}^{LW}: Eq. 4) or different (ΔG_{iwj}^{LW}: Eq. 7) low-energy compounds is rarely more than about $|1\,\mathrm{mJ/m^2}|$ and is usually much less. In comparison with the ΔG_{iwi}^{AB} (see upcoming Eq. 16) and the ΔG_{iwj}^{AB} (Eq. 11) attractions (or hydrophobic attractions), which can be as significant as $-102\,\mathrm{mJ/m^2}$, these LW interactions are virtually negligible.

2. AB (Hydrophobic) Attractions

The hydrophobic free energy of attraction between two similar low-energy entities immersed in water is $-2\gamma_{ij}^{AB}$ (Eqs. 4 and 9):

$$\Delta G_{iwi}^{AB} = -4\left(\sqrt{\gamma_i^{\oplus}\gamma_i^{\ominus}} + \sqrt{\gamma_w^{\oplus}\gamma_w^{\ominus}} - \sqrt{\gamma_i^{\oplus}\gamma_w^{\ominus}} - \sqrt{\gamma_i^{\ominus}\gamma_w^{\oplus}}\right) \tag{16}$$

The value of ΔG_{iwi}^{AB} in the case of apolar moieties reduces to $\Delta G_{iwi}^{AB} = -4\sqrt{\gamma_w^{\oplus}\gamma_w^{\ominus}}$, because when i is apolar, $\sqrt{\gamma_i^{\oplus}\gamma_i^{\ominus}}$, $\sqrt{\gamma_i^{\oplus}\gamma_w^{\ominus}}$, and $\sqrt{\gamma_i^{\ominus}\gamma_w^{\oplus}}$ (Eq. 16) are all zero. The value of $\sqrt{\gamma_w^{\oplus}\gamma_w^{\ominus}}$ is $25.5\,\mathrm{mJ/m^2}$ at 20°C [6], so the hydrophobic attraction energy amounts to $-102\mathrm{mJ/m^2}$ [28]. Thus the general statement can be made that the hydrophobic attraction is driven solely by the AB component of the free energy of cohesion of the water molecules, i.e., *by the hydrogen-bonding energy of cohesion of water*.

3. AB (Hydrophilic) Repulsions

To begin with, it should be emphasized that even when a hydrophilic repulsion (or "hydration pressure") between hydrophilic entities immersed or dissolved in water prevails, *the hydrogen-bonding energy of cohesion of water has not disappeared.* It is still very much there, but the hydrophilic repulsion then is simply stronger than the large background hydrophobic attraction. Therefore, net repulsion energies between hydrophilic entities in water tend to be smaller than the hydrophobic attraction between apolar moieties immersed in water. Nonetheless, for very hydrophilic compounds, the net hydrophilic repulsive energy can be considerable. For instance, for poly(ethylene oxide) (PEO), $\Delta G_{iwi}^{IF} = +52.5\,mJ/m^2$ [29], which is slightly more than half of the underlying hydrophobic attraction [26] of $-102.0\,mJ/m^2$ that first had to be surmounted. In addition, an LW attraction amounting to $\Delta G_{iwi}^{LW} = -7.1\,mJ/m^2$ must be overcome. For PEO, the value of the repulsive ΔG_{iwi}^{AB} is $+161.6\,mJ/m^2$ at 20°C, but its final net repulsive ΔG_{iwi}^{IF}, as just mentioned, is only $+52.5\,mJ/m^2$.

B. Range of Interfacial and EL Interactions

The free energy of interfacial interactions between equal spheres of radius R is proportional to R for LW, AB, and EL interactions. LW interactions decay with distance ℓ as ℓ^{-1}.

Acid–base interactions decay with distance ℓ as $\exp[(\ell - \ell_0)/\lambda]$, where ℓ_0 is the minimum equilibrium distance at "contact" ($\ell_0 \approx 0.157\,nm$ [7]) and λ is the decay length of water. For hydrophilic repulsions, λ is of the order of 1.0 nm [30]; the best value we obtained experimentally was 0.6 nm [26,31]. For *hydrophobic attractions*, however, λ, which may still be of the order of 1.0 nm at fairly close range, may be much greater at longer ranges; values of 13 nm and more have been reported [26,30,32]; see however more reliable results that were reported recently [73]. The range of EL interactions varies with the ionic strength; the decay length (or Debye length), $1/\kappa$, is inversely proportional to the square root of the ionic strength [3,4]. ΔG^{EL} decays with ℓ as $\ln\,[1 + \exp(-\kappa\ell)]$. In practice, LW forces can be measurable up to ℓ of approximately 5–10 nm, AB forces can be measurable up to $\ell \approx 5\,nm$ (repulsive) and up to $\ell \approx 10\,nm$ or more (attractive), and EL forces can be measurable up to ℓ of approximately 3–6 nm at high ionic strengths ($\Gamma/2 \geq 0.1$) and to ℓ of approximately 10–20 nm or more at low ionic strengths ($\Gamma/2 < 0.01$). For a number of equations describing decay with distance, see Appendix A of this chapter.

C. Hydrophobicity and Hydrophilicity

The degree of hydrophobicity or hydrophilicity of compounds has hitherto usually been expressed via the hydrophile–lipophile balance (HLB) [33]. It now seems possible to express the degree of hydrophobicity or hydrophilicity of compounds or materials in a more rigorously quantitative manner, in S.I. units, by using the surface thermodynamic approach, outlined earlier.

The boundary between hydrophobicity and hydrophilicity is best placed at the point where the difference between the apolar attraction and the polar repulsion between molecules or particles of material (i) immersed in water (w) is exactly equal to the cohesive polar attraction between the water molecules. Under these conditions, the interfacial free energy of interaction between particles i immersed in water is $\Delta G_{iwi}^{IF} = 0$. When ΔG_{iwi}^{IF} is positive, the interaction of the material with water dominates and the surface of the material is hydrophilic; when ΔG_{iwi}^{IF} is negative, the polar cohesive attraction between the water molecules dominates and the material is hydrophobic. Thus, the sign of ΔG_{iwi}^{IF} defines the *nature* of the surface (hydrophilic or hydrophobic), and the *magnitude* of ΔG_{iwi}^{IF} serves as the quantitative measure of its hydrophilicity or hydrophobicity [34,35]: The more positive ΔG_{iwi}^{IF} is, the more hydrophilic are molecules i; and the more negative ΔG_{iwi}^{IF} is, the more hydrophobic are molecules i. As discussed earlier, polar (AB) interactions usually are the dominant contributors to ΔG_{iwi}, by as much as 95% or more. However, in many cases, ΔG_{iwi}^{LW} must also be taken into account, cf. Eq. (13). One of the most hydrophobic "materials" we know is vacuum, or air, with $\Delta G_{iwi}^{IF} = -147.6\,\text{mJ/m}^2$ at 20°C. And one of the most hydrophilic materials is PEO, with $\Delta G_{iwi}^{IF} = +52.5\,\text{mJ/m}^2$ (see Sec. IV.B.3).

In some cases it can be more advantageous to express ΔG_{iwi}^{IF} in terms of kT instead of millijoules per square meter (mJ/m^2), where k is Boltzmann's constant ($=1.38\times10^{-23}\,\text{Jdeg}^{-1}$) and T the absolute temperature in degrees K; see the next section.

D. The Contactable Surface Area Between Particles and/or Solutes

When it is expedient to express interfacial interaction energies between droplets, particles, or macromolecules ΔG_{iwi}^{IF} in energy units per particle or molecule pair, i.e., in kT units, it is necessary first to ascertain the contactable surface area Sc between these entities:

$$\frac{\Delta G_{iwi}\,(\text{in mJ/m}^2) \times \text{Sc}\,(\text{in nm}^2)}{4.04} = \Delta G_{iwi}\,(\text{in kT}) \tag{17}$$

(Eq. 17 is valid at 20°C.) The Sc values of, e.g., alkanes can be estimated from adsorption cross sections of adsorbed molecules [9,36], or more accurately from aqueous solubilities [52]. For octane this yields a value of $\text{Sc} = 0.52\,\text{nm}^2$. For alkanes or alkyl groups higher as well as lower than the octyl group, the Sc values may be taken to be roughly proportional to the number of carbon atoms, relative to the eight carbon atoms of octane.

In the repulsive mode, molecules—and especially macromolecules—do not tend to contact each other over as large a surface as possible (as, e.g., alkanes do), but rather they approach each other via the smallest possible Sc. Thus, linear polymers, when repelling each other, would strive to oppose each other at right angles. The Sc for PEO strands has been estimated at $0.26 \times 0.375 = 0.0975\,\text{nm}^2$ [9,29]. At $\Delta G_{iwi} = +52.5\,\text{mJ/m}^2$ for PEO, this yields $+52.5 \times 0.0975/4.04 = +1.27\,\text{kT}$ for the interaction between two dissolved unattached PEO strands. For PEO strands that are part of nonionic surfactant molecules, somewhat larger surface areas are induced to approach each other rather closely, yielding ΔG_{iwi} values for PEO chains from $+2.76\,\text{kT}$ $(\text{EO})_6$ to $+4.11\,\text{kT}$ for $(\text{EO})_{21}$ [9].

E. Interaction Energies Between Spherical Particles or Droplets

For spherical particles or droplets, see the equations in Appendix A. The interaction energies ΔG_{iwi} between spherical particles, which are expressed in joules, should be divided by 4.04×10^{-21} (at 20°C) to obtain the interaction energies in kT. The interaction energy between a flat surface and a spherical particle or droplet of radius R is equal to twice the interaction energy between two spherical particles or droplets of radius R.

V. Stability and Solubility

A. Similarities Between Stability and Solubility

Except for the fact that there usually (but not always) is a difference in size, there is no other intrinsic difference between the physical conditions governing the stability of particles in aqueous suspensions and the solubility of solutes in water. Both particles and solutes may be:

1. *Homogeneous*, e.g., polystyrene latex particles, oil droplets, polysaccharides, or small solute molecules
2. *Heterogeneous*, e.g., blood cells, "sterically" stabilized latex particles, droplets surrounded by surfactant molecules
3. *Spherical or almost spherical*, e.g., latex particles, some micelles, osmotically swollen cells, staphylococci, globular proteins (such as hemoglobin)
4. *Nonspherical*, e.g., erythrocytes, clay particles, Bacilli, most proteins, polysaccharides
5. *Amphipathic*, e.g., kaolinite clay particles, surfactant molecules, lipopolysaccharides, RNA, DNA, phospholipids

B. Stability of Droplet and Particle Suspensions

Good particle stability and pronounced polymer solubility both require mutual repulsion between particles or polymer molecules, i.e., $\Delta G_{iwi} > 0$. Even a slight attractive tendency would already seriously destabilize fairly small particles or droplets. Assuming electrically neutral particles with a negligible LW interaction when immersed in water with a (slightly hydrophobic) ΔG_{iwi}^{AB} of only $-1\,\text{mJ/m}^2$, this translates, for $\lambda = 0.6$ nm and $R = 10$ nm, into a value for ΔG_{iwi} of -4.7 kT, denoting incipient instability. At $R = 20$ nm, $\Delta G_{iwi} \approx -9.3$ kT, indicating greater instability; at $R = 100$ nm, $\Delta G_{iwi} = -46.7$ kT, corresponding to total and practically irreversible flocculation. Thus, for only very slightly hydrophobic particles just to attain metastability at best, particle or droplet sizes are needed that are not much greater than about 5 nm (radius), and for significant stability a positive value of ΔG_{iwi} is essential.

For even mildly hydrophobic particles or droplets that are larger than a very tiny 10 nm (radius), stability can only be attained by coating their surfaces with a more hydrophilic material. Table 1 shows the factors determining the stability of polystyrene latex particles of $R = 110$ nm in terms of the ΔG values at closest

Table 1. Surface Potentials, Surface Free Energies,[a] and Interfacial Tensions of Polystyrene (PST) Latex Particles of 220-nm Diameter, Uncoated or Coated with Various Materials

	ψ_0 mV	ΔG_{iwi}^{LW} kT	ΔG_{iwi}^{AB} kT	ΔG_{iwi}^{EL} kT	ΔG_{iwi}^{TOT} kT	γ_{iwi}^{i} mJ/m^2
Uncoated PST	0[b]	−88.2[e]	−4148.0[e]	0	−4236.2	43.7
SDS-coated PST	−129[c]	−39.5[f]	+1796.3[f]	+2013.6	+3770.4	−16.0
Tween 80-coated PST	−13.1[d]	−95.8[g]	+3058.9[g]	+20.8	+2983.9	−26.2
IgG-coated PST	−5.4[d]	−38.9[h]	+1636.1[h]	+3.7	+1601.8	−13.9

[a]ΔG values at close range.
[b]Clean, uncoated PST is here assumed to have no surface charge.
[c]From Refs. 37 and 39.
[d]From Ref. 40.
[e]From Ref. 6.
[f]From Ref. 39.
[g]From Ref. 29.
[h]From Ref. 34.
[i]Eq. (10).
(SDS=sodium dodecyl sulfate; IgG=immunoglobulin-G).

approach ($\ell = \ell_0$; see Appendix A). Electrophoretic mobilities have been measured with these particles coated with sodium dodecyl sulfate (SDS) [37], Tween 80, and human immunoglobulin G (IgG) [38]. From these mobilities the ζ-potentials, and thus also the ψ_0-potentials and hence the ΔG_{iwi}^{EL} values, could be obtained (Table 1). The ΔG_{iwi}^{LW} and ΔG_{iwi}^{AB} values were obtained from surface tension data on polysytrne (PST) and polyethylene oxide (PEO) [5] and human immunoglobulin G (IgG) [34]. For the hydrophilic moiety of the nonionic surfactant Tween 80, the data on PEO were utilized. It is clear from Table 1 that pure, bare PST particles (presumably without any significant surface potential) would attract each other strongly, which causes their suspension to be totally unstable (cf. the strongly negative ΔG_{iwi}^{TOT}). On the other hand, coating of the PST particles, with either a strongly anionic surfactant (SDS), or with a nonionic surfactant (Tween 80) or a virtually nonionic protein (IgG at pH 7), endows their suspension with solid stability (cf. the strongly positive ΔG_{iwi}^{TOT} values).

The stabilization of hydrophobic particles or droplets by means of ionic (e.g., anionic) surfactants is believed to be a consequence of the distal (negative) electric potential conferred by the surfactant. It can be seen from Table 1 that, e.g., in SDS-stabilized PST latex particles, the negative ψ_0-potential here is responsible for about 58% of the total repulsive energy at close range; 42% of the repulsive contribution is due to AB forces. However, in the cases of particles stabilized by essentially noncharged surfactants (e.g., Tween 80) or polymers (IgG), a sizable net Lewis acid–base repulsive energy ($\Delta G_{iwi}^{AB} \gg 0$) is virtually the only driving force conferring stability to the particles (Table 1). Thus, in the case of nonelectrically charged particles immersed in water, it is largely their surface electron donicity [6] that gives them stability through net repulsive hydrogen-bonding interactions [6,7,26,50].

Because the SDS type of particle or droplet stabilization appeared traditionally to be clearly understandable as feasible only through EL repulsion, the successful stabilization of particle and droplet suspensions by means of nonionic surfactants or polymers [49] was a source of puzzlement to many in the early 1960s. Well before the importance of long-range repulsive hydration forces [41–43] and their underlying net repulsive Lewis acid–base forces [5–6] was realized, the concept of "steric" stabilization became popular as an interpretation of the phemonenon; see, e.g, the monograph by Napper [44]. However, the "steric" stabilization concept has not proven to be a simple, or at least a lucid, theoretical construct that readily allows a quantitative prediction of stability or nonstability based on easily measurable surface properties [45].

It has been known for more than half a century [46] that the pertinent surface properties of a particle are solely those of the molecules situated at, or attached to, its surface. This applies not only to the electrokinetic properties of particles and other surfaces [46], but also to their surface and interfacial tension components and parameters [26,34; see also 47,48, 50]. Thus, when water-soluble surfactants or polymers are appended to the surface of a particle or other solid, they simply confer their own surface properties to that particle or solid. When given surfactant or polymer molecules mutually repel each other, through EL and/or AB forces, then upon attachment to particles or other solid surfaces, these particles or solid surfaces will similarly repel each other. With the possible exception of a (usually fairly minor) chain elasticity contribution [49], there is nothing steric about this mechanism [45].

C. Solubility of Macromolecules and Other Solutes

When an attraction exists between polymer molecules—i.e., when $\Delta G_{iwi} < 0$— the large value of Sc, which is an inherent property of large molecules, gives rise to a large negative value of ΔG_{iwi} in units of kT (cf. Eq. 17), thus leading to insolubility. The solubility s of macrosolutes [26] as well as of microsolutes can be quantitatively expressed [8,26] as

$$\Delta G_{iwi} \cdot Sc = kT \cdot \ln s \tag{18}$$

There are three parameters in Eq. (18) (ΔG_{iwi}, Sc and s): knowledge of any two of these allows one to obtain the third one. For instance, for a symmetrical, nonamphipathic small organic solute such as benzene, one only needs to know its interfacial tension with water, γ_{iw} (which is $35.0\,mJ/m^2$ [22], so that $\Delta G_{iwi} = -70\ mJ/m^2$) and its solubility, s ($s = 1.75\ g/L = 22.436\ mM = 4.038 \times 10^{-4}$ mole fraction,* at 20°C). Using Eq. (18), one finds $Sc = 0.459\ nm^2$, or 45.9 Å^2, which is close to the Sc value one would estimate from the structure of benzene. For large water-soluble molecules, such as poly(ethylene oxide) (PEO) and Destran, in which cases $\Delta G_{iwi} > 0$, Eq. (18) can also serve to predict the solubility [8].

*It is necessary to express the solubility s (in the context of Eq. 18), as mole fractions instead of moles/L. In aqueous media, a concentration of 1 mole/L equals 1/55.6, or 0.018 mole fraction.

However, with amphipathic molecules, i.e., molecules comprising hydrophilic as well as hydrophobic moieties (e.g., alcohols, surfactants), the precise value of ΔG_{iwi} (Eq. (18)) is not as readily assessed. Even when the interfacial tension with water γ_{iw} of such an amphipathic compound can be measured, the measured result then reflects only (or mainly) the interfacial tension with water of the most polar (hydrophilic) moiety. The γ_{iw} thus obtained would then give rise (Eqs. 4 and 18) to an aqueous solubility that is much greater than the solubility that is found experimentally [50,52]. The same difficulty exists when the solubility of an amphipathic compound is calculated by Eq. (18) and γ components and parameters obtained via contact angle measurements and Eq. (14), because with amphipathic compounds one measures contact angles *either* at its hydrophobic *or* at its hydrophilic moieties, thus leading to too small *or* too large a predicted solubility. The solution to this dilemma is to determine the γ_{iw} values of the hydrophilic (1) and the hydrophobic (2) moieties of such compounds separately and subsequently to add the ΔG_{1W1} and the ΔG_{2W2} values (expressed in units of kT) thus obtained, to obtain ΔG_{iwi}:

$$\Delta G_{1W1} + \Delta G_{2W2} = \Delta G_{iwi} \tag{19}$$

resulting in the correct magnitude for the solubility when using Eq. (18).

D. Critical Micelle Concentration of Surfactants

Using Eqs. (18) and (19) for calculating the critical micelle concentration (cmc) of nonionic surfactants (assuming cmc = s; cf. [53]), it proved possible to obtain values that are within ±9% of the reported experimental cmc values [54] for the following compounds: C_8EO_6, $C_{10}EO_6$, $C_{12}EO_6$, $C_{14}EO_6$, $C_{16}EO_6$, $C_{16}EO_{12}$, $C_{16}EO_{15}$, $C_{16}EO_{21}$, C_8-phenyl-EO_6, C_8-phenyl-EO_9 and C_8-phenyl-EO_{10} [37].* Only for C_6EO_6 a calculated cmc was found that was 15% higher than the experimental cmc [9].

In the case of ionic surfactants, not only the hydrophilic (ΔG_{1W1}) and the hydrophobic (ΔG_{2W2}) are to be considered separately (Eq. 19), but, in addition, ΔG_{iwi}^{EL} must be taken into account. In doing so, the cmc of sodium dodecylsulfate (SDS) was calculated, at ionic strengths 0.10, 0.03, and 0.01. The calculated cmc corresponded to the experimentally found cmc values [55] to within 1% [39].

Once the limit of solubility of the combined hydrophobic and hydrophilic chains is reached, micelle formation sets in [9,53]. By sheltering the water-insoluble hydrophobic chains in the interior of the micelle, and by orienting the soluble hydrophilic moieties outward, micelle formation allows the solubilization (or stabilization) of the multimolecular micellar complexes. For instance, in the case of nonionic PEO-type surfactant micelles, with a radius $R \approx 20$ nm, $\Delta G_{iwi}^{AB} \approx +556$ kT (cf. Table 1), which confers pronounced solubility (or stability) on the micellar units.

---

*Here the kT values used in Refs. 9 and 39 should be adjusted by -4 kT (because of the conversion of moles/L to mole fraction; see the preceding footnote), entailing an increase in the estimated Sc values by about 0.161 nm^2 for the alkyl groups and by 0.235 nm^2 for the phenyl moiety.

VI. Repulsive Interfacial Free Energies in Aqueous Phase Separation

When aqueous solutions of two different charged or neutral polymers (or other solutes) are mixed together, phase separation can occur. (For a classification of the different modes of phase separation in liquids, see Appendix B.)

The phenomenon of phase separation in aqueous solutions of two polymers was first described by Beijerinck [56,57] almost a century ago and was later intensively investigated by Bungenberg de Jong [58,59], who called it "coacervation" (from the Latin *coacervo*, or heaping together), culminating in a number of chapters in the second volume of Kruyt's *Colloid Science* [58]. Beijerinck [56,57] described the phase separation, or coacervation, between water-soluble polymers dissolved in water, such as gelatin and agar, and gelatin and soluble starch. More recently, Albertsson [17] pioneered the use of other water-soluble systems, starting in the 1950s. Of these, the dextran-polyethylene oxide system is typical and is now also the best-known example. Nevertheless, as Albertsson still stated in the most recent edition of his work [17]: "The mechanism governing (aqueous phase) partition is largely unknown." With dextran and polyethylene glycol in water, phase separation occurs when the two polymers are present at concentrations of above $\approx$8% (w/v) each or $\approx$8% total polymer concentration [17]. Liftshitz–van der Waals repulsions, which are the driving force for phase separation in totally apolar systems of polymer solutions [16], cannot exist between dextran and polyethylene oxide (PEO) dissolved or suspended in water, because the van der Waals–Lifshitz surface tension (γ^{LW}) of water is only $21.8\,mJ/m^2$, while the γ^{LW} of both dextran (DEX) and PEO is of the order of $42–43\,mJ/m^2$, so nonhydrated DEX and PEO would attract each other (in water) with an apolar energy of about $7\,mJ/m^2$. Second, DEX is one of the biopolymers endowed with a remarkably low surface charge [60], so in systems containing DEX as one of the repelling partners, electrostatic repulsion can be completely ruled out.

Using Eq. (12) and the data from Ref. 6, the total interaction energy ΔG_{132}^{TOT} between DEX and PEO immersed in water is found to equal $+46.9\,mJ/m^2$. This, of course, signifies a strong repulsion, which is wholly due to polar (AB) repulsive forces: $\Delta G_{132}^{TOT} = \Delta G_{132}^{LW} + \Delta G_{132}^{AB}$, where $\Delta G_{132}^{LW} = -6.8\,mJ/m^2$ and $\Delta G_{132}^{AB} = +53.7\,mJ/m^2$. However, the repulsion energy expressed by $\Delta G_{132}^{TOT} = +46.9\,mJ/m^2$ applies only to the interaction between negligibly hydrated DEX and PEO molecules; these high repulsion energies therefore are operative only at very high DEX and PEO concentration. And DEX and PEO are surrounded by layers of water of hydration, which beyond the first layer are subject to an exponential decrease in the degree of orientation, which results in a decay in the actual value of ΔG_{132}^{TOT} with distance (ℓ). Thus, greater dilution leads to a decrease in the effective value of $|\Delta G_{132}^{TOT}|$. For all practical purposes, one may assume that at great dilution the value of ΔG_{132}^{TOT} (for $\gamma^{\ominus}$-monopolar polymers dissolved in water) approaches zero. The precise form of the function describing the decay of ΔG_{132}^{TOT}, going from very high to very low solute concentrations, is difficult to assess, because this depends on the different modes of decay of ΔG^{LW}, ΔG^{AB}, and, if present,

ΔG^{EL} with ℓ; see Appendix A. However, the assumption of a decay that is linearly proportional to the decrease in polymer concentration [10] serves rather well to predict the polymer concentration level at which phase separation commences. A minimum concentration of the dissolved polymers must be attained before phase separation can occur, because as long as the thermal motion energy of the dissolved (macro-) molecules is large enough to match the energy of interaction between dissolved molecules 1 and 2, any separation that might occur would be undone by thermal remixing. Separation therefore can only prevail over remixing when $\Delta G^{TOT}_{1W2} \geq +0.5\,kT$. It is therefore reasonable to estimate that phase separation sets in when $\Delta G^{TOT}_{1W2} \geq +0.5\,kT$ [29], where ΔG is converted from energy per unit surface area to energy per minimum contactable surface area (Sc) per molecule pair (see earlier) and expressed in units of kT. Thus the onset of aqueous phase separation (corresponding to the "critical point" in the phase diagram, cf. Albertsson [17]) is not only a function of the concentration of each polymer species, but also of their minimum contactable surface area Sc. And as the dimensions of polymer molecules in aqueous solution are a function of their molecular weight, the onset of aqueous phase separation also depends on the respective molecular weight of each polymer species; the higher the molecular weight of a polymer, the lower its minimum concentration required for the onset of phase separation [17,61].

Whilst the molecules of a given monosized polymer A dissolved in water, repel each other with a free energy $\Delta G^{TOT}_{AWA} = +a$, these dissolved polymer molecules of one species, when alone, will not give rise to a phase separation, because the repulsive energy is everywhere the same. The same holds for any other single monosized species of polymer molecules, e.g., B, dissolved in water, which repel one another with a free energy $\Delta G^{TOT}_{BWB} = +b$. However, when aqueous–aqueous solutions of A and B are mixed together in sufficiently high concentrations, a sorting out of the two solutions will occur into two compartments (phases), one of which contains mainly molecules of repulsive energy a, and the other comprising mainly molecules of repulsive energy b. It is thus possible to obtain a separation into as many phases in aqueous systems as there are different dissolved polymer species (each with a different repulsive energy, +a, +b, +c, etc.). Albertsson [17] gives an illustration of an aqueous 18-phase system. Polydispersity can give rise to multiple phase formation, because polymer molecules of the same species but of different molecular dimensions also have different Sc values and thus different values for ΔG^{TOT}_{iwi} (when expressed in kT).

It is important to note that whereas it is always necessary, in aqueous phase separation of the coacervation type, that at least one of the solutes be of macromolecular size, in certain cases a second solute can be a microsolute. Bungenberg de Jong [58] mentions a phase-separation system comprised of agar and ethanol, as well as a system of isoelectric gelatin and Na_2SO_4. Albertsson [17] describes an aqueous phase-separation system consisting of DEX and propanol; see also van Oss [18].

(For other modes of phase separation in liquids, see Appendix B.)

VII. Repulsive Interfacial Free Energies in Microemulsion Formation

A. Influence of EL and AB Forces in Stabilization with Ionic and Nonionic Surfactants

Repulsive interfacial free energies ΔG_{iji} of materials i immersed in a liquid j signify that the interfacial tension between condensed-phase material i and liquid j *has a negative value* (cf. Eq. 4). It was shown in Sec. V that, e.g., in aqueous media, a positive value of ΔG_{iwi} is essential for the stability of suspended particles or droplets and for a significant solubility of large molecules and micelles. It can also be argued that a positive value of ΔG_{iwi} is also the driving force for the spontaneous formation of microemulsions [6]. (For general discussions on microemulsions see, e.g., Prince [15], Friberg and Venable [62], Martellucci and Chester [63].) A possible role of positive ΔG_{iwi} values in the guise of initially negative interfacial tensions was proposed by Stoeckenius et al. [12] of Schulman's group, by Prince [14], and by Overbeek [64]. Nonetheless, the possible existence of negative interfacial tensions was treated with great prudence, if not trepidation, and it was thought that they could only exist locally and temporarily as a result of fluctuations; see, e.g., Prince [14] and Overbeek [64]. Miller and Scriven [65] discounted the possibility of the occurrence of negative interfacial tensions by stressing the importance of repulsive electrical double-layer interactions. Still, microemulsions can occur in the absence of ionic surfactants—e.g., through the use of nonionic surfactants [62]—and even through the sole use of what is normally designated as a cosurfactant—e.g., 2-propanol [66], which can give rise to a water-in-oil (w/o) microemulsion.

Thus the influence of a net Lewis acid–base repulsion (implying a negative interfacial tension) appears to be crucial in the formation of many, and possibly all, microemulsion systems. It should also be noted that the high positive ΔG^{EL} values, which can be obtained by the action of ionic surfactants, become very much smaller at the high ionic strengths ("brine") that are frequently employed in microemulsion studies (usually with a view to emulate the conditions that often prevail in petroleum recovery). At these high ionic strengths, the positive ΔG^{EL} values caused by the adsorption of ionic surfactants tend to become negligible for droplets with a small radius (e.g., $R \approx 10$ to 20 nm), in comparison with the positive ΔG^{AB} values engendered by the high $\gamma^{\ominus}$ values of the surfactant. The influence of salts with *monovalent* ions on ΔG^{AB} is small in comparison with their influence on ΔG^{EL}; see, e.g., Chheda et al. [67]. However, the influence of *plurivalent* counterions on ΔG^{AB} is considerable [68]. The use of plurivalent cations is generally avoided in studies of microemulsions involving anionic surfactants.

B. Stabilizing Influence of Alcohols as Cosurfactants

It can be seen in Table 2 that an alcohol such as ethanol has an exceedingly hydrophilic polar moiety, which manifests itself in an extremely high $\gamma^{\ominus}$ value, giving rise to a very strongly negative interfacial tension γ_{iw} with water (even more strongly negative than the γ_{iw} of PEO). The $\gamma^{\ominus}$ value of methanol is only

Table 2. Surface and Interfacial Tension Values, Surface Potentials, and Surface Free Energies of Interactions of the Hydrophilic Moieties of Surfactant- or Ethanol-coated Particles or Droplets of R = 10 nm

Hydrophilic moiety of surfactant or cosurfactant	γ_i^{LW} (in mJ/m²)	$\gamma_i^{\oplus}$ (in mJ/m²)	$\gamma_i^{\ominus}$ (in mJ/m²)	γ_{iw}^{d} (in mJ/m²)	ψ_0 (in mV)	ΔG_{iwi}^{LW}[e] (in kT)	ΔG_{iwi}^{AB}[h] (in kT)	ΔG_{iwi}^{EL}[i] (in kT)	ΔG_{iwi}^{TOT} (in kT)
Aerosol-oT (docusate sodium)	≈ 42	0	47.7	−16.0[a]	≈ −100	−8.0 (−0.63)[f]	+175 (+8.1)[f]	≈ +12 to +110 (+0.015 to 14.0)[f,g]	+167 to +277 (+7.5 to +21.5)
Sodium dodecylsulfate (SDS)	34.6[a]	0	46.0[a]	−16.0	− 129[a]	−3.6 (−0.28)[f]	+163 (+7.6)[f]	≈ +27 to +183 (+0.034 to +23.2)[f,g]	+168 to +342 (+7.3 to +30.5)
Poly(ethylene oxide)	43.0[b]	0	64.0[b]	−26.2	0	−8.7 (−0.68)[f]	+278 (+12.9)[f]	0	+269 (+12.2)
Ethanol	≈ 18.8[c]	0.019[c]	68.0[c]	−31.3	0	−0.27 (−0.02)	+293 (+13.6)	0	+293 (+13.6)

[a]From Ref. 39.
[b]From Ref. 6.
[c]From Ref. 50, Table XIII-4.
[d]Eq. (10).
[e]Eqs. (2) and (4); ΔG_{iwi} values at closest range; see Appendix A-1.
[f]Figures in paretheses indicate ΔG_{iwi} values at $\ell = 2$ nm.
[g]First figures at high ionic strength (≥ 1 M NaCl), second figures at lower ionic strength (0.1 M NaCl).
[h]Eq. (16); see Appendix A-2.
[i]See Appendix A-3.

slightly higher than that of ethanol [50, Table XIII-4] and the $\gamma^{\ominus}$ value of propanol, or even of higher alcohols, is unlikely to differ greatly from that of ethanol. Thus, aliphatic alcohols would appear to be among the best conveyors of monopolar ($\gamma^{\ominus}$) AB repulsion to small droplets or particles when directionally attached to them at the water interface. It is therefore not surprising that alcohols are excellent microemulsion-forming agents as cosurfactants and even as sole surfactants. The only reason why one hesitates to designate them as full-fledged surfactants is that alcohols do not form micelles, as a consequence of the small surface area of their hydrophilic moiety.

VIII. Phase Separation Phenomena in Microemulsions

A. Phase Separation Between Brine and the Middle Phase

In a typical microemulsion system composed of octane (as the oil phase), aerosol-OT (as the surfactant), ethanol (as the cosurfactant), and NaCl [69], it is striking to note—when comparing the compositions of the brine (bottom) and the middle phase—that about 95% of the surfactant is in the middle phase, whereas about 77% of the ethanol is in the brine phase [69]. Also, about 87% of the NaCl remains in the (appropriately named) brine phase.

In analogy with the systems discussed at the end of Section VI, it appears plausible that in the case described by Good et al. [69], two overlapping phase separations of the coacervation type had occurred between the brine and the middle phase: (1) a phase separation between the ethanol and the surfactant, mainly in the guise of micelles (if one may use that term in this context) gorged with octane, and (2) a phase separation between the salt and the surfactant. One advantage of this arrangement is that the ethanol and the salt in the brine compartment, through their repulsive action on the surfactant-coated microemulsion droplets or particles, keep the middle phase compartment relatively small in volume and concentrated in surfactant, while at the same time providing a reserve supply of cosurfactant, which is probably interspersed among the surfactant molecules. The salt in the brine compartment also exerts a dehydrating action on the ethanol molecules, which enhances their repulsive action on the surfactant molecules [10].

B. Multiple Phase Separations in the Middle Phase

As shown in Sec. VI [17], multiple phases can occur in aqueous multicomponent systems. Now, microemulsions can be considered to be multiple-component systems when droplets of different sizes exist. This is because even if the interfacial tensions between, e.g., oil droplets and water are the same, when droplets of different sizes exist, then the free energies of repulsion (ΔG_{iwi}) between such droplets of different sizes (expressed as energy per droplet pair) will differ (cf. Appendix A). Even though only one, or at most two, phase boundaries are normally visible with the naked eye, it was expected that with an optical system that visualizes differences in refractive index, such as schlieren optics, more phase boundaries would become apparent, if multiple phases indeed occur. This was first verified in an analytical ultracentrifuge (Beckman–Spinco,

Model E), and multiple phases were visible early on in each run (still at low g forces), and persisted for at least 1 hour at about 100,000 g [19–21,69]; see Figures 1 and 2. However, it could not be entirely ruled out that the high pressures occurring during centrifugation and ultracentrifugation, or other high g-force effects, might have contributed to the observed multiple phase separations. It was therefore also attempted to use schlieren optics to study

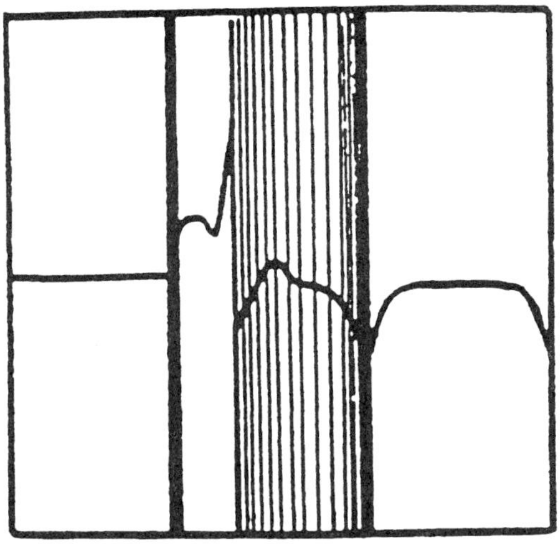

Figure 1. Drawn reproduction of schlieren picture of analytical ultracentrifugation (Beckman Model E instrument) of microemulsion (*n*-octane, brine, and docusate sodium); 100,000 × g; 14 min from start; 70° schlieren angle. (From Refs. 20 and 69, and from Ref. 50.)

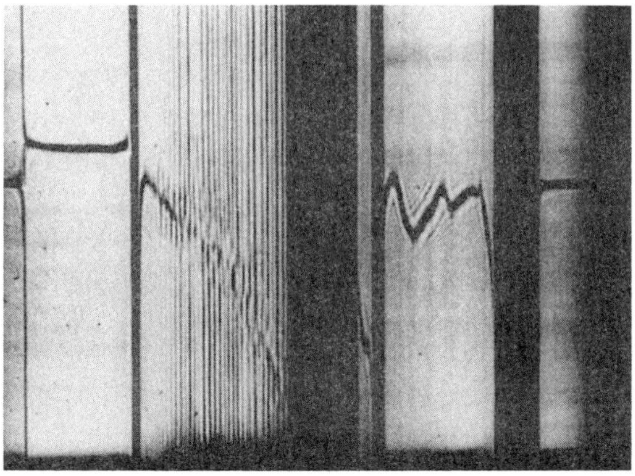

Figure 2. Schlieren picture of analytical ultracentrifugation of a microemulsion as for Figure 1; 100,000 × g; speed just attained. As in Figure 1; the multiple phases in the middle (microemulsion) phase are clearly visible. From left to right: air phase, hydrocarbon phase, microemulsion phase, and brine phase. (From Ref. 21, and from Ref. 50.)

microemulsions at ambient gravity. To that effect a moving boundary ("Tiselius") electrophoresis apparatus (Beckman–Spinco, Model H) was used, with or without an electric field. Without an electric field, and at ambient gravity, multiple phases can also be observed; they usually start within about 24 hours, and they persist for several days; see Figure 3.

In the presence of an electric field, most peaks (phases) were observed traveling toward the anode (Figure 4) [21]. However, a few peaks traveled in the opposite direction. The existence of positively charged droplets in the presence of negatively charged droplets, all freely suspended in a liquid medium, is not as improbable as might be thought. For particles of opposite charges to coexist without interacting and neutralizing each other, it suffices for them to have a higher positive value of ΔG^{AB}_{1W2} than the negative value of their ΔG^{EL}_{1W2} (assuming ΔG^{LW}_{1W2} to be negligible). Taking the highest observed negative ζ-potential of $-60\,mV$ and the highest estimated positive ζ-potential of $+7\,mV$ would yield an attractive potential $\psi_0 \approx 30.6\,mV$, which for $R = 100\,nm$ yields $\Delta G^{EL}_{1W2} \approx -1\,kT$; whereas for $\gamma_{12} = -16\,mJ/m^2$, $\Delta G^{AB}_{132} \approx +1500\,kT$ (at close range) so that ΔG^{TOT}_{132} remains strongly repulsive in spite of opposing electric charges. In brine the decay with distance of ΔG^{EL} is steeper than the decay of ΔG^{AB}.

Based on the $(\gamma^{\ominus})$ monopolarity of the surfactants used in microemulsions, and on the observations made on microemulsions with schlieren optics, it may be concluded that in addition to a moderate EL repulsion and to a relatively weak LW attraction, microemulsion formation and maintenance is due predominantly to a strong AB repulsion.

Figure 3. Schlieren picture of sedimentation at ambient gravity of microemulsion (as for Figure 1), taken after 120 hr (Beckman Model H instrument). The multiple phases persisting in the microemulsion (middle) phase are clearly visible. (From Ref. 21, and from Ref. 50.)

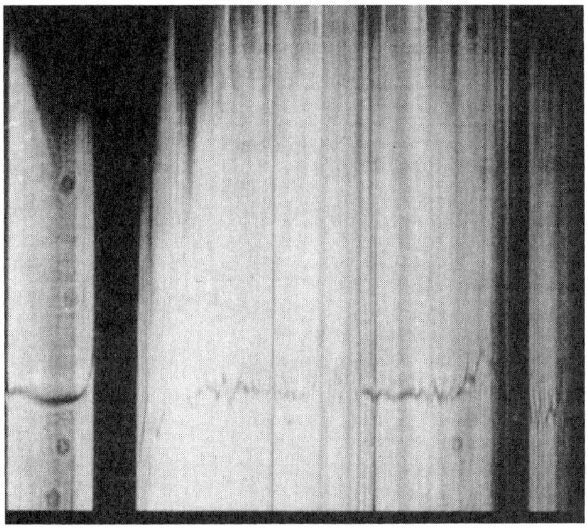

Figure 4. Schlieren picture of electrophoresis of microemulsion (as for Figure 1), at about 10 V/cm; as in Figure 3, taken after 3 min (Beckman Model H Tiselius electrophoresis instrument). Shortly after connecting the DC field to the microemulsion phase, drastic structure changes occurred in the microemulsion phase. Small peaks can be seen to cover the whole microemulsion phase. These small peaks quickly sorted into three major peaks and moved to the positive electrode. Observations made during the run did reveal some peaks moving in the opposite direction. However, this opposite movement cannot be seen clearly in the photographs, mainly because the peaks were constantly changing size and shape and could not be easily followed. The average zeta-potential of particles calculated based on the mobility of the largest peak was −59.3 mV. (From Ref. 21, and from Ref. 50.)

Appendices

A. Equations for Decay with Distance (ℓ) of LW, AB, and EL Interactions

1. LW Interactions

Configuration	ΔG_{ℓ}^{LW}
Flat parallel plates:	$\frac{-A^*}{12\pi\ell^2}$
Two spheres of radius R:	$\frac{-AR}{12\ell}$

*$A = 24\pi\ell_0^2\gamma^{LW}$ (ℓ_0 is the minimum equilibrium distance $\ell = \ell_0$; $\ell_0 \approx 0.157$ nm)

2. *AB Interactions*

Configuration	ΔG_{ℓ}^{AB}
Flat parallel plates:	$\Delta G_{\ell_0}^{AB''} \exp\left(\frac{\ell_0 - \ell}{\lambda}\right)^*$
Two spheres of radius R:	$\pi R \lambda \Delta G_{\ell_0}^{AB''} \exp\left(\frac{\ell_0 - \ell}{\lambda}\right)^*$

$^*\Delta G_{\ell}^{AB''}$ indicates the polar (AB) free energy of interaction between two parallel flat plates measured at the minimum equilibrium distance $\ell = \ell_0$; $\ell_0 \approx 0.157\,\text{nm}$.

3. *EL Interactions (for $\psi_0 \leq 30\,mV$)*

Configuration	ΔG_{ℓ}^{EL}
Flat parallel plates:	$\frac{1}{\kappa} \cdot 64nkT\gamma_0^2 \exp(-\kappa\ell)$
Two spheres of radius R:	$0.5\epsilon R\psi_0^2 \ln[1 + \exp(-\kappa\ell)]$

where $1/\kappa$ is the Debye length, i.e., the thickness of the diffuse ionic double layer:

$$\frac{1}{\kappa} = \left(\frac{\epsilon KT}{4\pi e^2 \Sigma v_i^2 n_i}\right)^{1/2}$$

where ϵ is the dielectric constant of the liquid (for water, $\epsilon = 80$ at 20°C), k is Boltzmann's constant ($k = 1.38 \times 10^{-23}$ J per °K), T is the absolute temperature in degrees Kelvin, e is the charge of the electron ($e = 4.8 \times 10^{-10}$ e.s.u.), v_i is the valence of each ionic species, and n_i is the number of ions of each species per cubic centimeter of bulk liquid;

$$\gamma_0 = \frac{\exp(ve\psi_0/2kT) - 1}{\exp(ve\psi_0/2kT) + 1}$$

where v is the valence of the counterions per cubic centimeter of bulk liquid and ψ_0 is the surface potential at the exact particle/liquid interface;

$$\psi_0 = \zeta\left(1 + \frac{z}{R}\right) \exp(\kappa z)$$

where ζ is the potential at the slipping plane (ζ usually is measured by electrophoresis [60]) and z is the distance from the particle's surface to the slipping plane; z is generally of the order of $\approx$0.3 to 0.5 nm [50].

B. Different Modes of Phase Separation in Liquids

1. *Phase Separations Caused by Attractive Forces*

Phase separation between two immiscible liquids. When two largely immiscible liquids (1 and 2) are brought together, or mixed together, they separate out into two separate phases (each phase comprised of one of the liquids), in spite of the fact that the two liquids exert a considerable attraction on each other. Liquid immiscibility prevails when the work of *adhesion* between the two liquids (W_{12}) is smaller than the average work of *cohesion* of the two liquids (W_{11} and W_{22}). Thus, immiscibility occurs when $W_{12} < (W_{11} + W_{22})/2$, i.e., when $W_{12} < (\gamma_1 + \gamma_2)$. Conversely, miscibility obtains when $W_{12} > (W_{11} + W_{22})/2$, i.e., when $W_{12} > (\gamma_1 + \gamma_2)$. This follows from the Dupré equation (cf. Eq. 6) for interactions *in vacuo* (recall that $W = -\Delta G$):

$$W_{12} = \gamma_1 + \gamma_2 - \gamma_{12}$$

This also indicates (tautologically) that miscibility prevails when $\gamma_{12} < 0$, and immiscibility occurs when $\gamma_{12} > 0$.

Complex coacervation. In complex coacervation (usually studied in aqueous media), the phase separation is caused by the attraction between two solutes dissolved in one liquid medium. It gives rise to the formation of complexes, which are either insoluble (causing flocculation, or precipitation) or soluble (causing complex coacervation) [18]. The attraction between the solutes is usually electrostatic in nature, but can also be a consequence of Lewis acid–base interactions [70,71] and in rare instances even of a specific antigen–antibody reaction [18].

Although LW attractions may play a minor subsidiary role in complex coacervation, no cases are known in which LW attractions represent the sole (or even a major part of the) driving force of complex coacervation. In aqueous media, LW attractive forces are much weaker than the EL or AB interactions that are the main attractive forces between solutes giving rise to complex coacervation (cf. Sec. IV.A.1). In apolar media, where LW forces are the only ones operative, complex coacervation is not known to occur. This is, *inter alia*, owing to the fact that the solutes participating in complex coacervation are, by definition, soluble in the solvent. Thus the molecules of any species of solute attract each other with a free energy $\Delta G_{iwi}^{LW} > -1\,kT$ [72] and are therefore unlikely to exert a stronger attraction on other solutes that are also present in the same solution.

In complex coacervation, the phase separation usually occurs as a consequence of the migration of the formed complexes to the lower side of the container under the influence of gravity.

2. Phase Separations Caused by Repulsive Forces

Apolar systems. In completely apolar systems (e.g., polyisobutylene in carbon tetrachloride), or in systems containing a monopolar polymer in an apolar solvent (e.g., polystyrene in carbon tetrachloride), or in systems comprising monopolar polymers *and* a monopolar solvent, all of the same sign of polarity (e.g., polymethylmethacrylate and polystyrene in methylethyl ketone or tetrahydrofuran) miscibility or separation depends solely on LW interactions [16]. In such apolar systems, composed of two polymers (1 and 2) dissolved in one solvent (3), phase separation will occur when $\Delta G_{132} > 0$, which happens when $\gamma_1^{LW} > \gamma_3^{LW} > \gamma_2^{LW}$ or when $\gamma_1^{LW} < \gamma_3^{LW} < \gamma_2^{LW}$ [16]. When neither of these conditions is fulfilled, i.e., when $\Delta G_{132} < 0$, no phase separation will ensue. When more than two polymers occur in an apolar system and when conditions are favorable for phase separation (as outlined earlier), polymer X will be found either in the phase of polymer 1 or in the phase of polymer 2. Thus, irrespective of the number of polymer species involved, in apolar systems no more than two phases can occur [16,18,26,50].

Polar systems. Polar systems that are monopolar, with all components of the same sign of polarity (e.g., polymethylmethacrylate and polystyrene dissolved in methylethyl ketone), behave like apolar systems (see App. B.2, Apolar systems, above). However, with monopolar polymers dissolved in *bipolar* solvents (e.g., polymethylmethacrylate and polystyrene dissolved in chlorobenzene) *or* with monopolar polymers dissolved in a monopolar solvent of *opposite polarity* (e.g., polymethylmethacrylate and polystyrene dissolved in chloroform), *or* with at least one *bipolar* polymer (e.g., cellulose acetate) and one monopolar polymer (e.g., polystyrene) or even an apolar polymer (e.g., polyisobutylene) dissolved in a monopolar solvent (e.g., tetrahydrofuran), *AB repulsions are the main driving forces for phase separation, often overriding a* (smaller) *LW attraction* [16,50].

Coacervation, or aqueous phase separation. Given the γ_w^{LW} (of water) of $21.8\,\mathrm{mJ/m^2}$ (at 20°C), which is lower than the γ_i^{LW} of all, or most, water-soluble polymers or other relevant solutes, the LW interaction between such polymers and/or solutes dissolved in water is practically always *attractive*, i.e., $\Delta G_{1w2}^{LW} < 0$. Thus any repulsive energy causing phase separation between two (or more) polymers dissolved in water is due to AB and/or EL interactions. In practice, AB interactions are quantitatively the most important and qualititatively often the *only* driving force, e.g., in the typical and important instance of the dextran-polyethylene oxide (DEX-PEO) system. (DEX has a ζ-potential very close to zero [60]), and the ζ-potential of PEO (although irrelevant to the DEX-PEO system) is also quite low. Contrary to apolar systems, aqueous phase separation can give rise to as many different phases as there are different polymers dissolved in the system (see Sec. VI).

References

1. H. C. Hamaker, *Rec. Trav. Chim. Pays-Bas 55*:1015 (1936); *56*:3 (1937); *56*:727 (1937).
2. E. J. W. Verwey and J. Th. G. Overbeek, *Trans. Faraday Soc. 42B*:117 (1946).
3. E. J. W. Verwey and J. Th. G. Overbeek, *Theory of the Stability of Lyophobic Colloids*, Elsevier, Amsterdam, 1948.
4. B. V. Derjaguin and L. D. Landau, *Acta Physicochimica USSR 14*:633 (1941).
5. C. J. van Oss, M. K. Chaudhury, and R. J. Good, *Advan. Colloid Interface Sci. 28*:35 (1987).
6. C. J. van Oss, M. K. Chaudhury, and R. J. Good, *Chem. Rev. 88*:927 (1988).
7. C. J. van Oss, R. J. Good, and M. K. Chaudhury, *Langmuir 4*:884 (1988)
8. C. J. van Oss and R. J. Good, *Langmuir 8*:2877 (1992).
9. C. J. van Oss and R. J. Good, *J. Dispersion Sci. Technol. 12*:95 (1991).
10. C. J. van Oss, M. K. Chaudhury, and R. J. Good, *Separation Sci. Tech. 22*:1515 (1987).
11. C. J. van Oss, R. F. Giese, and P. M. Costanzo, *Clays Clay Miner. 38*:151 (1990).
12. W. Stoeckenius, J. H. Schulman, and L. M. Prince, *Kolloid-Z. 169*:170 (1960).

13. J. T. Davies and E. K. Rideal, *Interfacial Phenomena*, Academic Press, New York, 1963, p. 362.
14. L. M. Prince, *J. Colloid Interface Sci. 23*:165 (1967).
15. L. M. Prince, in *Microemulsion, Theory and Practice* (L. Prince, ed.), Academic Press, New York, 1977, pp. 45, 104.
16. C. J. van Oss, M. K. Chaudhury, and R. J. Good, *Separ. Sci. Technol. 24*:15 (1989).
17. P.-Å. Albertsson, *Partition of Cell Particles and Macromolecules*, Wiley-Interscience, New York, 1986.
18. C. J. van Oss, *J. Dispersion Sci. Technol. 9*:561 (1988).
19. R. J. Good, C. J. van Oss, G. Broers, and X. Yang, *J. Colloid Interface Sci. 110*:604 (1986).
20. R. J. Good, C. J. van Oss, J. Ho, X. Yang, G. Broers, and M. Cheng, *Colloids Surfaces 20*:187 (1986).
21. X. Yang, Middle Phase Emulsion Study, Ph.D diss., SUNY, Buffalo, 1990.
22. L. A. Girifalco and R. J. Good, *J. Phys. Chem. 61*:904 (1957).
23. R. J. Good and L. A. Girifalco, *J. Phys. Chem. 64*:561 (1960).
24. C. J. van Oss, R. J. Good, and M. K. Chaudhury, *J. Colloid Interface Sci. 111*:378 (1986).
25. C. J. van Oss, *J. Dispersion Sci. Technol. 11*:491 (1990).
25a. C. J. van Oss, R. J. Good, and H. J. Busscher, *J. Dispersion Sci. Technol. 11*:75 (1990).
26. C. J. van Oss, *Colloids Surfaces 78*:1 (1993).
27. C. J. van Oss, *Separ. Purif. Meth. 4*:167 (1975).
28. C. J. van Oss and R. J. Good, *J. Dispersion Sci. Techn. 9*:355 (1988).
29. C. J. van Oss, K. Arnold, R. J. Good, K. Gawrisch, and S. Ohki, *J. Macromol. Sci. Chem. A27*:563 (1990).
30. H. K. Christenson, in *Modern Approaches to Wettability* (M. E. Schrader and G. I. Loeb, eds.), Plenum, New York, 1992, p. 29.
31. C. J. van Oss, in *Biophysics of the Cell Surface* (R. Glaser and D. Gingell, eds.), Springer Verlag, New York, 1990, p. 131.
32. J. L. Parker and P. M. Claesson, *Langmuir 10*:635 (1994).
33. P. Becher, in *Nonionic Surfactants* (M. J. Schick, ed.), Dekker, New York, 1967, p. 604.
34. C. J. van Oss, in *Protein Interactions* (H. Visser, ed.), VCH Verlag, Weinheim, Germany, 1992, p. 25.
35. C. J. van Oss and R. F. Giese, Tech. Program Proc. Clay Minerals Soc. Meeting San Diego, 1993; *Clays Clay Minerals, 43*: 474 (1995).
36. S. J. Gregg and K. S. W. Sing, *Adsorption, Surface Area and Porosity*, Academic Press, New York, p. 79.
37. R. S. Snyder, M. Bier, R. N. Griffin, A. J. Johnson, H. Leidheiser, F. J. Micale, S. Ross, and C. J. van Oss, *Separ. Purif. Methods 2*:259 (1973).
38. C. J. van Oss and J. M. Singer, *J. Reticuloendothelial Soc. 3*:29 (1966).
39. C. J. van Oss and P. M. Costanzo, *J. Adhesion Sci. Tech. 6*:477 (1992).
40. C. J. van Oss and J. M. Singer, *J. Colloid Interface Sci. 21*:117 (1966).
41. V. A. Parsegian, N. Fuller, and R. P. Rand, *Proc. Natl. Acad. Sci. USA 76*:2750 (1979).
42. V. A. Parsegian, R. P. Rand, and D. C. Rau, *Chem. Scripta 25*:28 (1985).
43. V. A. Parsegian, R. P. Rand, N. L. Fuller, and D. C. Rau, *Meth. Enzymology 127*:400 (1986).
44. D. H. Napper, *Polymeric Stabilization of Colloidal Dispersions*, Academic Press, New York, 1983.
45. C. J. van Oss, *J. Dispersion Sci. Techn. 12*:201 (1991).

46. H. A. Abramson, L. S. Moyer, and M. H. Gorin, *Electrophoresis of Proteins*, Reinhold, New York; Reprinted: Hafner, New York, 1964.
47. P. M. Claesson, Forces Between Surfaces Immersed in Aqueous Solutions, Ph.D diss., Royal Institute of Technology, Stockholm, 1986.
48. J. N. Israelachvili, *Intermolecular and Surface Forces*, Academic Press, London, 1991.
49. R. H. Ottewill, in *Nonionic Surfactants* (M. J. Schick, ed.), Dekker, New York, 1967, p. 627.
50. C. J. van Oss, *Interfacial Forces in Aqueous Media*, Dekker, New York, 1994.
51. H. Stephen and T. Stephen, *Solubilities of Inorganic and Organic Compounds*, Vol. 1, Part 1, Pergamon Press, Oxford, 1963.
52. C. J. van Oss and R. J. Good, *J. Dispersion Sci. Techn. 17* (1996) (in press).
53. A. W. Adamson, *Physical Chemical of Surfaces*, 5th ed., Wiley-Interscience, New York, 1990, p. 509.
54. P. Becher, in *Nonionic Surfactants* (M. J. Schick, ed.), Dekker, New York, 1967, p. 478.
55. H. F. Huisman, *Proc. Meth. Acad. Sci. Ser. B 67*:388 (1964).
56. M. W. Beijerinck, *Zentralbl. Bakt. 2*:627 (1896).
57. M. W. Beijerinck, *Kolloid-Z. 7*:16 (1910).
58. H. G. Bungenberg de Jong, in *Colloid Science*, Vol. II (H. R. Kruyt, ed.), Elsevier, Amsterdam, 1949, pp. 232, 335, 433.
59. H. G. Bungenberg de Jong and H. R. Kruyt, *Kolloid-Z. 50*:39 (1930).
60. C. J. van Oss, R. M. Fike, R. J. Good, and J. M. Reinig, *Analyt. Biochem. 60*:242 (1974).
61. H. Walter, D. E. Brooks, and D. Fisher (eds.), *Partitioning in Aqueous Two-Phase Systems*, Academic Press, New York, 1985.
62. S. E. Friberg and R. L. Venable, in *Encyclopedia of Emulsion Technology*, Vol. 1 (P. Becher, ed.), Dekker, New York, 1983, p. 287.
63. S. Martellucci and A. N. Chester, *Progress in Microemulsions*, Plenum Press, New York, 1989.
64. J. Th. G. Overbeek, *J. Chem. Soc.—Faraday Disc. 65*:7 (1978).
65. C. A. Miller and L. E. Scriven, *J. Colloid Interface Sci. 33*:360, 371 (1970).
66. G. D. Smith, C. E. Donelan, and R. E. Barden, *J. Colloid Interface Sci. 60*:488 (1977).
67. P. Chheda, D. Grasso, and C. J. van Oss, *J. Colloid Interface Sci. 153*:226 (1992).
68. W. Wu, C. J. van Oss, and R. F. Giese, *Colloids Surfaces A, 89*:241 (1994).
69. R. J. Good, J. T. Ho, G. E. Broers, and X. Yang, *J. Colloid Interface Sci. 107*:290 (1985).
70. A. Dobry, *Bull. Soc. Chim. Belg. 57*:280 (1948).
71. K. L. Smith, A. E. Ninslow, and D. E. Peterson, *Ind. Eng. Chem. 51*:13161 (1959).
72. C. J. van Oss and R. J. Good, *J. Macromolec. Sci. Chem. A26*:1183 (1989).
73. J. Wood and R. Sharma, *Langmuir 11*:4797 (1995).

5
A Model of Micro- and Fine-Emulsion Formation

CHRISTIAN B. OLEKSIAK* and HENRI L. ROSANO / Department of Chemistry, The City College of the City University of New York, New York, New York

I. Introduction

A. Characteristics Common to Micro- and Fine Emulsions

The emulsion systems considered here all involve three or more components: a hydrophobic component ("oil"), a hydrophilic component ("water"), and one or more surfactants. The mathematical model of emulsion formation developed in this chapter takes into consideration the following characteristics common to micro- and fine emulsions.

1. The surfactant or surfactants are required to reduce the interfacial free energy and facilitate the formation of the interface, and to provide an interfacial film capable of ensuring dispersion stability.
2. Lowering of the interfacial free energy is beneficial during emulsion formation, but, at equilibrium, the interfacial free energy cannot be lower than a certain critical value if one is to obtain the curling of the interface that gives rise to the droplets.
3. The decrease in droplet size that accompanies the increase in the number of surfactant molecules required is limited to a value R_0, the natural radius of curvature of the interface.
4. During the preparation of the dispersion, as the volume of the dispersed phase is increased, one passes from a monodispersed system (with low dispersed phase volume) to a polydispersed system (with a dispersed phase volume larger than a limiting volume beyond which, in

**Current affiliation*: Kraft Foods, Glenview, Illinois.

the case of microemulsion systems, percolation occurs [1–3]) and finally, at even higher dispersed-phase volumes, to phase separation, phase inversion, or equilibrium with the excess dispersed phase.

B. Importance of the Rigidity of the Interface

Rosano [4] developed a formula for calculating the amount of surfactant needed to emulsify a certain volume V of dispersed phase. Assuming that all the surfactant molecules end up at the interface, we have

$$R = \frac{3V}{n\sigma} \tag{1}$$

where R is the radius of a droplet (nm), V is the volume of the dispersed phase (nm^3), n is the number of surfactant molecules, and σ is the cross-sectional area of a surfactant molecule (nm^2/molecule).

However, experimental results [5] show that, above a certain concentration of surfactant molecules, rather than a smooth transition to smaller particle sizes, we observe a phenomenon of creaming due to reversible flocculation by free surfactant. The maximum value of n before creaming occurs, inserted in Eq. (1), yields the value R_0, the natural radius of curvature of the interface. This value, R_0, represents the particular radius of curvature at which (in the absence of other forces acting on the system) the structure of a given surfactant molecule and the three-dimensional structure of the interface together result in steric constraints that prevent the adsorption of additional surfactant molecules at the interface [6]. The excess surfactant molecules then form micelles in the solvent phase that may foster the instability of the emulsion[5].

It is thus important to develop a model taking into account not only the interfacial free energy but also a term relative to the three-dimensional structure of the interface, namely, K_R, the rigidity of the interface.

II. Previous Model and Its Limitations

A. Chan–Rosano Model

In a recent paper, Chan and Rosano [7] developed a mathematical model for the formation of oil-in-water microemulsions, based on the various components of the difference in total free energy of the system G, before and after emulsification. We have

$$G = NG_T$$

where G_T is the total free energy of formation of a droplet and N is the total number of droplets.

Thus,

$$G_T = G_{SH} + G_A + G_B + G_I \tag{2}$$

where G_M is the total free energy of formation of a droplet (J); G_{SH} is the free energy of formation of the interfacial sheath structure (J); G_A is the work required to expand the interface (J); G_B is the interfacial bending energy (J); and G_I is the free energy of interaction between droplets (J).

On the basis of these different terms of free energy, Rosano and co-workers found

$$G_T = -\epsilon \frac{4}{3}\pi R^3 \Delta G_M + 4\pi R^2 \left[\gamma_I + \frac{K_R}{2}\left(\frac{1}{R} - \frac{1}{R_0}\right)^2 \right] + 16\pi^2 K_R \left(\frac{R - R_{min}}{R_{min}}\right)^2 \left(\frac{R - R_{max}}{R_{max}}\right)^2 \tag{3}$$

where ΔG_M is the free energy of formation of the interfacial sheath structure, in joules per cubic meter of interface. The interface is an ideal mixture of octane and dodecane (50:50 mole:mole). The free energy of mixing is then calculated to be $9 \times 10^6\,J/m^3$; ϵ is a correction factor representing the fraction volume of the interface relative to the total volume of the droplet. If we assume that, for microemulsions ($R_0 = 9\,nm$), the thickness of the interface represents 60% of the radius of the droplet, then the volume of the interface will be the total volume of the droplet, corrected by $\epsilon = 0.94$; R is the radius of the microemulsion droplets (m); γ_I is the interfacial free energy (N/m); R_0 is the natural radius of curvature of the interface (m); K_R is the rigidity constant of the interfacial sheath (J); R_{min} is the radius of the smallest droplet that can be found in a stable emulsion, for a given value of the rigidity constant K_R (m); and R_{max} is the radius of the largest droplet that can be found in a stable emulsion, for a given value of the rigidity constant K_R (m).

Determination of K_R, R_{min}, and R_{max} is based on a separate consideration of G_{TS}, the total free energy per droplet, assuming no droplet interaction—in other words, omitting G_I from G_T. If we plot G_{TS} versus R at various K_R, for a given set of R_0, ΔG_M, and γ_I (see Fig. 1), we can make the following observations.

1. K_R can vary only between two limits, $K_{R\,min}$, under which the curves exhibit no minimum (the energy decreases asymptotically and the system is not stable), and $K_{R\,max}$, above which the maximum of the curve is so highly positive that only a narrow window of possible radii remains, in contrast to the commonly observed polydispersity.
2. For a satisfactory value of K_R we are very unlikely to observe droplets of radii smaller than the radius corresponding to the first optimum of the curve (minimum) or larger than the radius corresponding to the second optimum of the curve (maximum). This last observation permits the identification of these two optima as R_{min} and R_{max}, respectively.

The use of G_{TS} was a satisfactory approach to the free energy of formation of a droplet until the phemomenon of percolation was observed [2]. In that case, a term relative to the energy cost of the fusion between two droplets (namely, G_I) has to be considered. As just noted, the radii of the smallest and the largest droplets that can interact are, respectively, R_{min} and R_{max}. The interaction between two droplets of radius R_1 and R_2 involves only work against the deformation of the interface occurring when its surface is increased from R_1 and R_2 to R. The quantity G_I is then approximated by the product of the rigidity constant

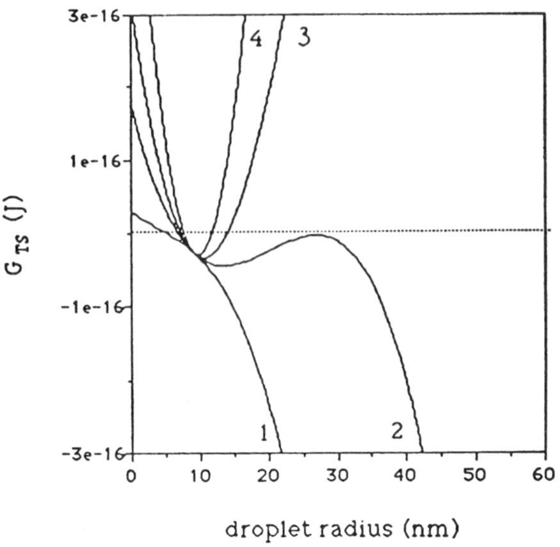

Figure 1. G_{TS} vs. droplet radius for various values of the rigidity constant ($G_{TS} = G_{SH} + G_A + G_B$, $R_0 = 9\,nm$, $\Delta G_M = 9 \times 1^6\,J/m^3$, $\epsilon = 0.94$, $\gamma_I = 0.1 \times 10^{-3}\,N/m$). Curve 1: $K_R = 5 \times 10^{-18}\,J$. Curve 2: $K_R = 2.75 \times 10^{-17}\,J$. Curve 3: $K_R = 5 \times 10^{-17}\,J$. Curve 4: $K_R = 10^{-16}\,J$.

of the interface K_R and the product of the relative increases in surface area when the droplets fuse. An exact if tedious way to compute this energy contribution would be to consider all the possible two-droplet encounters resulting in a droplet of radius R, and to average the energy contributions over all such encounters. Instead, we approximate this value by considering that this average can be expressed as the energy contribution of the encounter between the most probable droplet (droplet of radius R_{min}, heavily represented because it corresponds to the minimum in total free energy) and the least probable one (droplet of radius R_{max}, poorly represented because it corresponds to the maximum in total free energy). Comparison of Figs. 2 and 3 reveals that G_I leads to a model that accounts for the experimental results related to polydispersity [8]. For given values of G_{SH}, γ_I, R_0, and K_R, the total energy of the system exhibits two minima, having comparable magnitudes and radii R_a and R_b, where R_b ranges in value from $1.8R_a$ to $2.2R_a$. Thus, for a particular set of parameters, the system will have two possible stable states, one with droplets of radius R_a and one with droplets of radius R_b, coexisting in the emulsion.

B. Invalidity of the Model at Low Interfacial Free Energy

If we now consider that, for given values of G_{SH}, γ_I, and R_0, K_R varies slightly from one droplet to another owing to different molecular configurations (the rigidity of the interface is related to its structure), it is found that the system exhibits not two, but a whole set of energy minima covering a range of radii, accounting for the experimentally observed polydispersity [8] (Fig. 2).

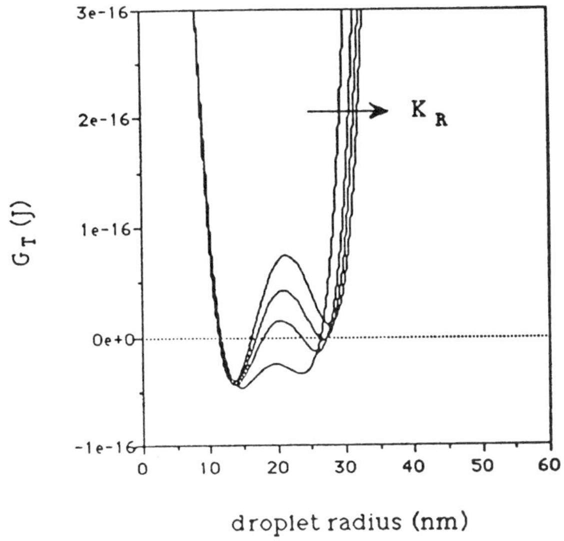

Figure 2. G_T vs. droplet radius for various values of the rigidity constant. K_R increases from 2.6×10^{-17} to 2.8×10^{-17} J ($G_T = G_{SH} + G_A + G_B + G_I$, $R_0 = 9$ nm, $\Delta G_M = 9 \times 10^6$ J/m^3, $\epsilon = 0.94$, $\gamma_I = 0.1 \times 10^{-3}$ N/m).

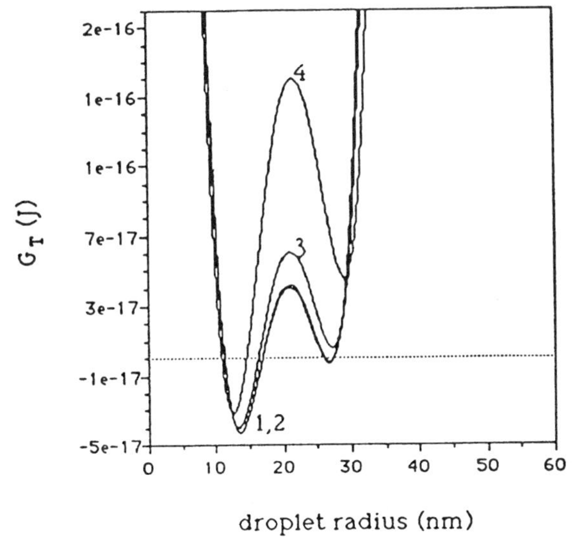

Figure 3. G_T vs. droplet radius for various values of the interfacial free energy γ_I. ($G_T = G_{SH} + G_A + G_B + G_I$, $R_0 = 9$ nm, $\Delta G_M = 9 \times 10^6$, J/m^3, $\epsilon = 0.94$, $K_R = 2.75 \times 10^{-17}$ J). Curves 1 and 2: $\gamma_I = 0.01$ and 0.1 mN/m. Curve 3: $\gamma_I = 1$ mN/m. Curve 4: $\gamma_I = 5$ mN/m.

Keeping the rigidity K_R constant, but varying the interfacial free energy γ_I between values as low as 0.01 mN/m and as high as 5 mN/m (Fig. 3), it may be observed (according to the Chan–Rosano model) that whatever the interfacial free energy, the system exhibits two energy minima (comparable in magnitude and radius), which account for polydispersity and stability. However, very low interfacial free energies have been proven experimentally to be detrimental to the stability of the microemulsion (type I to II or II to III transition [9,10]). As mentioned by Friberg and Solans [11], a low interfacial free energy results in a low Laplace pressure ($P = 2\gamma/R$), and it is reasonable to conclude that the droplets, unable to maintain a regular spherical geometry, will have a tendency to flatten and thus to increase their surface of contact (Fig. 4). The resulting coalescence of pairs of droplets leads to droplets whose Laplace pressure is lower, since the radius is increased during the coalescence. It is easy to imagine that all of the droplets will coalesce rapidly, with a resultant phase separation, or by formation of a bicontinuous phase: The emulsion is thus unstable.

The same reasoning applies at high interfacial free energies, when the Laplace pressure is sufficiently high so that the surface of contact between two droplets is too small to allow fusion between them, resulting in a system that, if stable, does not experience percolation (Fig. 4). Intermediate Laplace pressure values will characterize a system that is still stable above the percolation threshold, where coalescence is possible, owing to a high probability of encounters between droplets, but limited, owing to a moderate area of surface contact. Such a system breaks down on further increase of the emulsified phase because of the greater probability of encounters of "doubled droplets" (resulting from the percolation of two droplets). The resultant low Laplace pressure rapidly leads to phase separation, as explained earlier (Fig. 4).

III. Improvement of the Model

Since the probability of interaction between two droplets, which depends on the Laplace pressure, is responsible for the stability of the emsulsion, it seems reasonable to weight the fourth term of Eq. (2), i.e., G_I, by a dimensionless factor reflecting the effect of the surfactant layer on the Laplace pressure. This factor

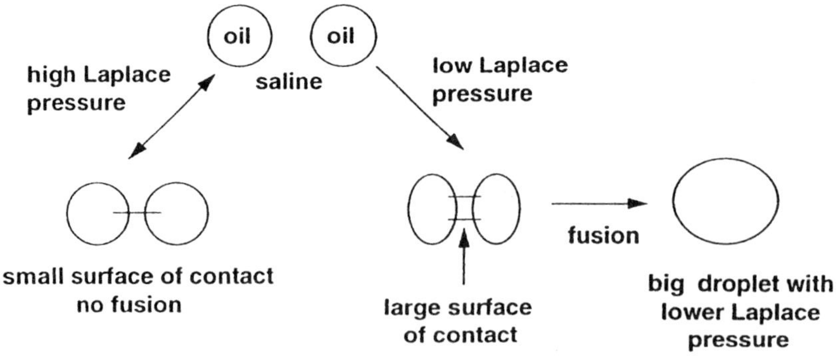

Figure 4. Effect of the Laplace pressure on the probability of fusion of two droplets.

is the ratio of the Laplace pressure (arising from the difference between pressures inside and outside the droplet) applied to the droplet in the presence of the surfactant film to that applied on it in the absence of surfactant. We call this factor L_P.

$$L_P = \frac{2\gamma_I/R}{\gamma_{I,0}/R} = \frac{\gamma_I}{\gamma_{I,0}} \tag{4}$$

where γ_I equals the interfacial free energy of the interface with surfactant layer present (N/m) and $\gamma_{I,0}$ is the interfacial free energy of the bare interface (N/m).

Equation (2) can now be written as

$$G_T = G_{SH} + G_A + G_B + L_P G_I \tag{5}$$

and Eq. (3) becomes

$$G_T = -\epsilon\frac{4}{3}\pi R^3\,\Delta G_M + 4\pi R^2\left[\gamma_I + \frac{K_R}{2}\left(\frac{1}{R} - \frac{1}{R_0}\right)^2\right] + 16\pi^2 K_R \frac{\gamma_I}{\gamma_{I0}}\left(\frac{R - R_{min}}{R_{min}}\right)^2\left(\frac{R - R_{max}}{R_{max}}\right)^2 \tag{6}$$

Analyzing Fig. 5, which corresponds to the plot of Eq. (6), the following may be concluded.

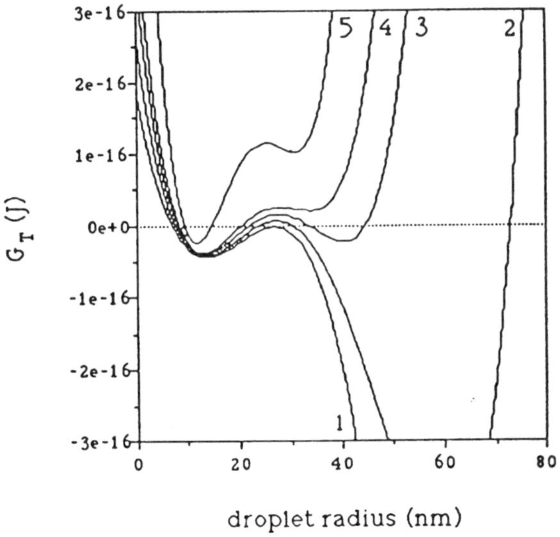

Figure 5. G_T vs. droplet radius for various values of the interfacial free energy γ_I ($G_T = G_{SH} + G_A + G_B + L_P G_I$, $R_0 = 9\,\text{nm}$, $\Delta G_M = 9 \times 10^6\,\text{J/m}^3$, $\epsilon = 0.94$, $K_R = 2.75 \times 10^{-17}\,\text{J}$). Curve 1: $\gamma_I = 0.1\,\text{mN/m}$. Curve 2: $\gamma_I = 1\,\text{mN/m}$. Curve 3: $\gamma_I = 1.5\,\text{mN/m}$. Curve 4: $\gamma_I = 2\,\text{mN/m}$. Curve 5: $\gamma_I = 10\,\text{mN/m}$.

1. At low interfacial free energy (0.01 mN/m), the total free energy exhibits two minima, but the second minimum occurs at a radius far greater than the radius corresponding to the first minimum (4000 nm as against 13.6 nm) and at a free energy far smaller than the free energy of the first minimum (-6×10^{-7} J as against -0.4×10^{-17} J). The only physical significance of such a curve is that the system will be entirely in a state corresponding to the second minimum, and the radius value is too great to be realized physically in a droplet. The end result is thus a phase separation or a bicontinuous phase.
2. At high interfacial free energy (10 mN/m), the radii are of physically realizable magnitude (13.6 nm and 32 nm), but the first minimum is much lower than the second, and the system will be entirely in a state corresponding to the first minimum. Note that when $R_0 = 9$ nm, the second minimum rapidly becomes positive for relatively low values of the interfacial free energy (>2.5 mN/m) and the system is not stable for the corresponding value of the radius.
3. At intermediate interfacial free energy (1.5–2 mN/m), both the radii and the magnitude of the total free energies at the minima are comparable; the system is stable and displays percolation, which accounts for the experimentally observed polydispersity.

The most recent model was developed using experimental data relative to microemulsions ($R_0 = 9$ nm). When a fine emulsion is considered ($R_0 = 90$ nm), the volume of a 4-nm-thick interface represents 16% of the volume of a droplet. Provided the cohesion of the interface is high enough, the surfactant sheath will resolve into 90-nm droplets at a larger value of the rigidity constant than in the case of 9-nm droplets. It is interesting to note that if both the rigidity of the interface and the free energy of formation of the interface (which accounts for the cohesion of the resulting interface) are increased, the three conclusions just listed remain valid for a fine emulsion, as shown Fig. 6.

IV. Conclusions

This model has the merit of being based on physical chemical properties common to micro- and fine emulsions. The crucial system-determining roles played by two experimentally measurable properties [12], namely, interfacial free energy and rigidity of the interface, suggests that further work, taking into account the effect of interparticle forces (van der Waals, hydration, and electrostatic forces) along these lines, will facilitate prediction of emulsion stability based on these properties.

Acknowledgments

C. O. would like to thank Alliance Pharmaceutical Corp., San Diego, Calif., and Kraft General Foods, Tarrytown, N.Y., for financial assistance, and Prof. John Arents for computer programming and helpful suggestions.

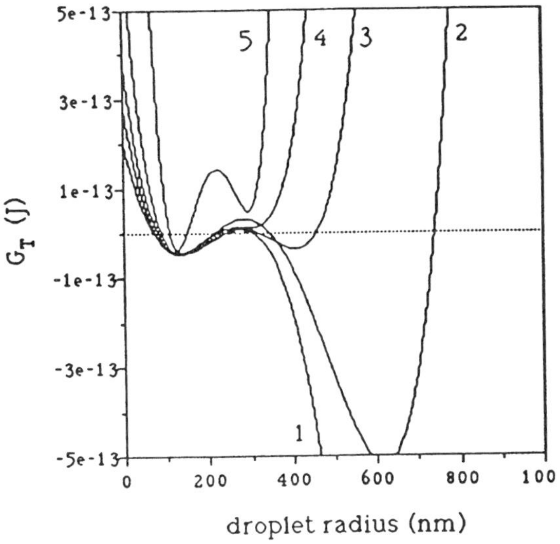

Figure 6. G_T vs. droplet radius for various values of the interfacial free energy γ_I ($G_T = G_{SH} + G_A + G_B + L_P G_I$, $R_0 = 9\,\text{nm}$, $\Delta G_M = 9 \times 10^6\,\text{J/m}^3$, $\epsilon = 0.16$, $K_R = 3 \times 10^{-14}\,\text{J}$). Curve 1: $\gamma_I = 0.1\,\text{mN/m}$. Curve 2: $\gamma_I = 1\,\text{mN/m}$. Curve 3: $\gamma_I = 2\,\text{mN/m}$. Curve 4: $\gamma_I = 5\,\text{mN/m}$. Curve 5: $\gamma_I = 40\,\text{mN/m}$.

All graphs appearing in this chapter were generated by Wolfrarn Research, Inc.'s *Mathematica 2.0.*

References

1. A. M. Cazabat, D. Langevin, J. Meunier, and A. Pouchelon, *Phys. Rev. Lett. 43*:L89 (1982).
2. K. E. Bennett, J. C. Hatfield, C. W. Macosko, and L. E. Scriven, in *Microemulsions* (I. D. Robb, ed.), Plenum Press, New York, 1982, pp. 65–75.
3. J. L. Cavallo and H. L. Rosano, *J. Phys. Chem. 90*:6817 (1986).
4. H. L. Rosano, *J. Soc. Cosmetic Chemists 25*:609 (1974).
5. M. P. Aronson, *Colloids Surf. 58*:195 (1991).
6. P. G. De Gennes and C. Taupin, *J. Phys. Chem. 86*:2294 (1982).
7. S. Y. Chan and H. L. Rosano, *J. Dispersion Sci. Technol. 9*(5,6):523(1988–89).
8. S. A. Safran, G. S. Grest, A. L. R. Bug, and I. Webman, in *Microemulsion Systems* (H. L. Rosano and M. Clausse, eds.), Surfactant Science Series 24, Dekker, New York, 1987, Chap. 14, pp. 235–243.
9. H. Saito and K. Shinoda, *J. Colloid Interface Sci. 32*:647 (1970).
10. R. L. Reed and R. N. Healy, in *Improved Oil Recovery by Surfactant and Polymer Flooding* (D. O. Shah and R. S. Schechter, eds.), Academic Press, New York, 1977, pp.
11. S. E. Friberg and C. Solans, *Langmuir 2*:121 (1986).
12. J. B. F. A. Smeulders, C. Blom, and J. Mellema, *Physical Review A. 42*:3483 (1990).

6
Monodisperse Emulsions: Properties and Uses

T. G. MASON / Université de Bordeaux I/CRPP, Bordeaux, France

A. H. KRALL / Department of Physics and Astronomy, University of Pennsylvania, Philadelphia, Pennsylvania

HU GANG / Exxon Research and Engineering Company, Annandale, New Jersey

J. BIBETTE / Université de Bordeaux I/CRPP, Bordeaux, France

DAVID A. WEITZ / Department of Physics and Astronomy, University of Pennsylvania, Philadelphia, Pennsylvania

---

I. Introduction

Emulsions are formed by mixing two immiscible fluids in the presence of a surfactant that coats the surface of the droplets, stabilizing them against coalescence. While emulsions can be made kinetically stable through the choice of appropriate fluids and surfactants, they are nevertheless inherently nonequilibrium systems. As a result, energy of some form must be added to produce the emulsion. This is typically in the form of shear energy that is added through the mixing process used to create the droplets. An inevitable consequence of this is that the droplet sizes of an emulsion are polydisperse, complicating the study of its physical properties. For example, the consistency of the emulsion, and thus its rheological properties, is strongly dependent on the volume fraction of the

---

Research conducted while the authors were at Exxon Research and Engineering Company, Annandale, New Jersey.

dispersed phase.* As the volume fraction is increased, the dispersed droplets must pack together; this packing is directly controlled by the droplet sizes and the distribution of these sizes.

Perhaps even more important, many of the properties of an emulsion depend on the pressure difference between the fluid inside the droplet and the continuous phase. This pressure difference, or the Laplace pressure, results from the interfacial tension between the two fluids, and is inversely related to the radius of curvature of the droplets and hence to their size. The Laplace pressure represents, among other things, the pressure required to distort the shape of the droplet significantly, and thus it plays a key role in determining the rheological properties of the emulsion. In a polydisperse emulsion, each droplet has a different Laplace pressure, and therefore its distortion is resisted by a different internal pressure. As a consequence, any detailed understanding of the properties of an emulsion is greatly complicated by this polydispersity of the droplet sizes. A monodisperse emulsion, comprised of droplets of the same size, would offer considerable opportunity for elucidating the properties of emulsions, avoiding the complications inherent in typical, polydisperse emulsions.

In this chapter, we discuss the production of such a monodisperse emulsion and illustrate how it can help to elucidate critical properties of emulsions. The monodisperse emulsion is made by a fractionation technique: Beginning with a polydisperse emulsion, droplets of a very narrow size range are separated from the remaining droplets to form the monodisperse emulsion. By separating different fractions of the initial distribution, monodisperse emulsions can be made with varying droplet sizes, enabling a detailed study of the dependence of the properties of an emulsion on the droplet size. These monodisperse samples can be exploited to study a wide range of fundamental properties of emulsions. Here, we focus on the dependence of emulsion properties on both droplet size and droplet volume fraction.

Of primary concern are the physical properties of emulsions that are most strongly influenced by the droplet size and volume fraction. In particular, we focus on the rheology of emulsions. As the volume fraction of the dispersed phase is increased, an emulsion changes from a fluid to a paste and ultimately to an elastic solid. The origin of this behavior is the packing of the droplets and their deformation in shape as the volume fraction is increased. Thus, by using a monodisperse emulsion, the packing geometry, and its consequences, can be more easily determined. Furthermore, the effects of change in the droplet size can be studied directly.

In addition to the study of their physical properties, monodisperse emulsions can also serve as useful systems for measuring other important properties of emulsions. For example, the stability of droplets can be monitored very simply, since changes in the droplet size are easily discernible. Furthermore, the consequences of droplets' interactions on their structure and the resultant properties of the emulsion are most easily studied by having monodisperse droplets. Thus, in this chapter we also illustrate how the monodisperse emulsions can aid in the study of other general properties of emulsions, including their stability and the consequences of varying interactions between emulsion droplets.

---

*See Chap. 3 of this volume.

This chapter is organized as follows. In Sec. II we discuss the fractionation technique for purifying the emulsions to obtain monodisperse droplets. The equation of state of these emulsions, or the dependence of the volume fraction of the droplets on the externally imposed osmotic pressure, is discussed in Sec. III. Section IV discusses the droplet structure as their volume fraction is varied. Then, in Sec. V, we exploit the monodisperse droplets to study the elasticity or rheological behavior. Finally, in Sec. VI, we illustrate other uses of monodisperse droplets to study properties of emulsions, including their stability against coalescence upon application of large osmotic pressures, the colloidal nature of the droplets, and the resultant structures that are induced when the interaction between the droplets becomes attractive. In addition, we review how the monodisperse emulsions can be made magnetic and used to determine the magnitude of colloidal forces as a function of the droplet separation. A brief conclusion closes the chapter.

II. Purification of Monodisperse Emulsions

Virtually all methods for making emulsions require the addition of some form of shear energy to produce the droplets. As a result, there is no precise control over the distribution of the droplet sizes produced. This complicates the study of the properties of emulsions, making it more difficult to distinguish between physical effects that depend on the size or packing of the droplets and other variables. Thus, to enable more precise control of the droplet size and to allow for the study of the properties of truly monodisperse emulsions, a method has been developed to purify polydisperse emulsions to produce emulsions with highly monodisperse droplet sizes [1–3]. This is accomplished with a fractionation technique, which exploits both the colloidal nature of relatively small emulsion droplets and their resultant phase behavior. This fractionation technique, applied repeatedly to a crude emulsion, separates the droplets of a single size and results in an emulsion with highly monodisperse droplets.

The key to the purification technique is an attractive interaction between droplets that depends on the droplet size. This interaction is the depletion attraction [4,5], which can occur in mixtures of colloidal particles of two very different sizes; the presence of the smaller particles can induce an attractive interaction between the larger particles, even though the direct interaction between any two particles is purely repulsive. This interaction is purely entropic in nature and can be understood physically by considering the entropic contribution to the free energy of the system. This will be dominated by the much larger number of small particles and will decrease as the total volume accessible to the centers of these small particles increases. The centers are excluded from the volume occupied by the larger particles, as well as by a thin shell surrounding the surfaces. The width of this shell is the distance between the surface of the large particle and the center of the small particle when it approaches as close to the large one as possible. If the repulsive interaction is very short-ranged, then this width is simply the radius of the smaller particles. However, if two large particles approach one another closer than this width, the shells of the two particles overlap; although the small particles are excluded from this overlap region, the total volume accessible to them is actually increased, thereby decreasing the entropic energy of the system. Therefore, it is energetically favorable for the large and small particles to phase separate, with the small particles inducing an attractive interaction between the large ones.†

The magnitude of the depletion attraction can be determined by considering the osmotic pressure exerted by the small particles on the larger ones [4,5]. Since the concentration of smaller particles is depleted in the volume between the two larger ones, there is an uncompensated pressure on the larger droplets forcing them together. Physically, we expect this attractive interaction to increase as the volume fraction of smaller particles increases, since this increases the osmotic pressure exerted on the larger droplets. In addition, we expect the magnitude of the attraction to increase with the volume of the excluded region, and therefore to increase with the radius of the larger droplets. Moreover, since the attraction is purely entropic in nature, we expect the energy scale of the attraction between any pair of large particles to be set by the thermal energy k_BT, where k_B is Boltzmann's constant. The attractive interaction can only extend over a distance equal to twice the width of the shell, and must reach a maximum in energy when the two large particles approach one another as closely as possible. Thus, when the two large particles touch, the magnitude of the attractive interaction is [2,5]

$$U_d = 3k_BT\phi_m\left(\frac{a}{a_m}\right) \tag{1}$$

where a is the radius of the larger particle, a_m is the width of the excluded shell of smaller particles around the larger ones, and ϕ_m is the volume fraction of the smaller particles. The attractive energy depends on the ratio of the radii of the two particles and the volume fraction of the smaller ones. Thus, by suitably adjusting the volume fraction of the smaller particles, the magnitude of the attraction can be controlled.

We exploit this attractive interaction to fractionate a crude emulsion [2]. We use an emulsion comprised of silicone oil droplets in water, stabilized with the surfactant sodium dodecylsulfate (SDS). These emulsions are highly stable against coarsening, because of the extremely low solubility of the silicone oil in water; they are also highly stable against coalescence, provided the concentration of SDS is sufficiently large. The SDS is soluble only in the continuous water phase; above a critical micelle concentration (CMC) C^*, it forms micelles in the water. These micelles have a radius of the order of 20 Å and are the smaller particles for the depletion attraction; the oil droplets are the larger particles. By adjusting the concentration of SDS above C^*, the volume fraction of the micelles can be controlled. For any given volume fraction ϕ_m, the magnitude of the depletion attraction will be greater than k_BT for oil droplets larger than some radius, and will cause these larger droplets to flocculate; the density mismatch between the oil and the water will cause these flocs to cream. The creamed droplets can be skimmed from the emulsion and redispersed in a separate suspension. By this means, the droplet distribution can be divided into two parts; droplets larger than some size will be in one part, while smaller droplets

†See, also, Chaps. 1 and 4 of this volume.

will be in the other. By repeating this fractionation several times, a suspension of highly monodisperse droplets can be obtained.

The critical micelle concentration for SDS in water is $C^* = 8\,\text{mM}$; thus, the micelle concentration is $C_m = (C - C^*)/\nu_m$ for $C > C^*$, where ν_m is the aggregation number of monomers in each micelle. Observations of the flocculation and subsequent creaming of emulsions indicate that the droplets remain dispersed by Brownian motion unless the depth of the potential well exceeds about $5k_BT$, assuming $a_m \approx 20\,\text{Å}$. This criterion, when combined with Eq. (1), determines the minimum SDS concentration to produce flocculation of droplets larger than radius a,

$$C \approx \frac{5\nu_m}{4\pi N_0 a a_m^2} + C^* \qquad (2)$$

where N_0 is Avogadro's number. Thus, for example, the SDS concentration to cause droplets having $a = 0.5\,\mu\text{m}$ to cream is $C \approx 30\,\text{mM}$.

To produce a monodisperse emulsion we begin with a crude polydisperse silicone-oil-in-water emulsion with an average droplet size of about one micron, and a droplet volume fraction of $\phi \approx 0.1$. For the first purification step, we add SDS to set $C \approx 30\,\text{mM}$ and allow the emulsion to cream for about 12 hr. We skim the cream containing the large droplets and redisperse it at $\phi \approx 0.1$. For the second purification step, we set different SDS concentrations in the two daughter emulsions. With the emulsion containing the large droplets, we reduce the SDS concentration by 5 mM to flocculate only the largest droplets. With the daughter emulsion containing the smaller droplets, we raise the SDS concentration by 5 mM to flocculate still-smaller droplets. The 5-mM concentration difference assures high size selectivity and a useful volume of cream. By repeating the separation five or six times, several monodisperse emulsions having different droplet radii can be fractionated from a single polydisperse emulsion.

The uniformity of the droplet size distribution can be seen by comparing the initial polydisperse emulsion at $\phi \approx 0.65$, shown in the top optical microscope image in Fig. 1, with a monodisperse emulsion at $\phi \approx 0.55$ obtained after six purification steps, shown in the bottom image in Fig. 1. The polydisperse emulsion is highly disordered, with a wide range of droplet sizes. By contrast, the droplet size of the purified emulsion is highly uniform, permitting the development of highly ordered packing, induced by the shear when the microscope slide and cover slip are pressed together. The degree of this order suggests that the residual polydispersity, defined as the ratio of the size distribution's width to its average, is about $\sigma_d \approx 0.1$.

The purification technique can produce monodisperse emulsions with droplet radii that range from $a \approx 0.1\,\mu\text{m}$ to $a \approx 2\,\mu\text{m}$. The lower limit is set by the relative ratio of the micelle to droplet radii; this ratio becomes too large when $a < 0.1$, μm and no amount of additional surfactant can cause creaming. The upper limit is set by the precision in controlling the surfactant concentration to select a particular droplet size. The size selectivity of the depletion attraction requires a much higher degree of accuracy in setting C for larger droplet sizes, so extra purification steps are typically required to make a monodisperse emulsion with $a > 0.6\,\mu\text{m}$. For $a > 2\,\mu\text{m}$, the size selectivity is so poor that other purification methods are needed. A necessary consequence of the repeated separations is

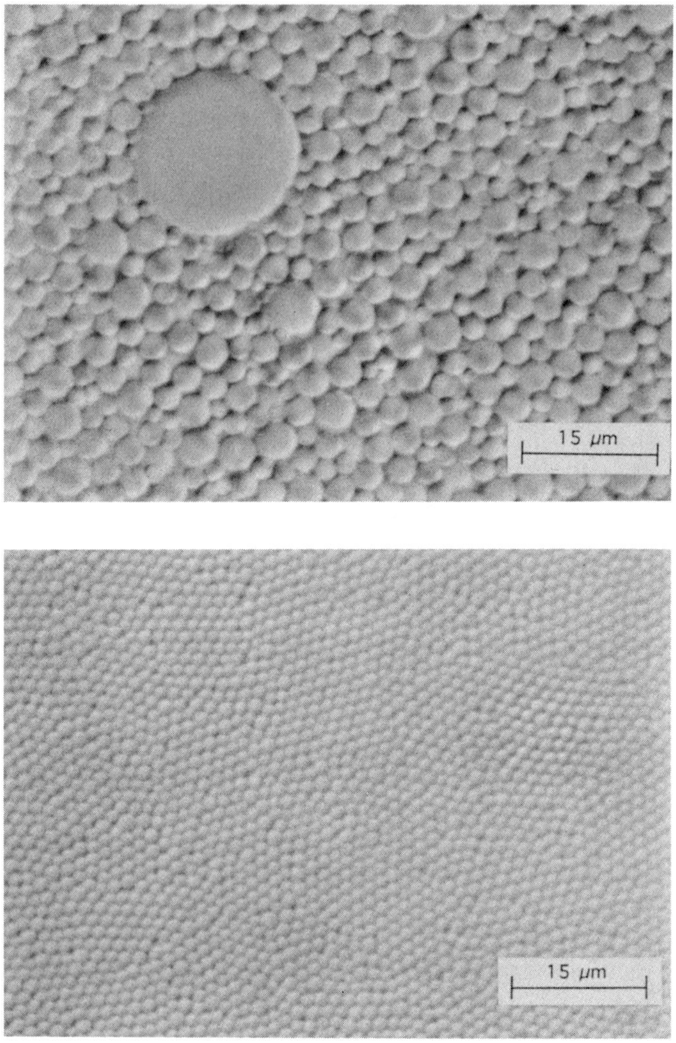

Figure 1. *Top*: Optical micrograph of a polydisperse emulsion at $\phi \approx 0.65$. The average droplet radius is about 3 μm. *Bottom*: Optical micrograph of a monodisperse emulsion at $\phi \approx 0.55$ with a ≈ 0.8 μm, obtained from a polydisperse emulsion after six separation steps. The polycrystalline ordering is caused by the shear exerted on the droplets when preparing the microscope slide.

the relatively small amount of monodisperse emulsion at a given droplet size. About 2% of a typical, initially polydisperse emulsion can be fractionated into a monodisperse sample.

This techique is not, in principle, limited to silicone-oil-in-water emulsions. Virtually any oil can be used, provided the emulsion is stable against both coarsening and coalescence. For example, we have made emulsions from many linear chain alkanes; however, for chain lengths shorter than about 10 (decane), we find the emulsion to be so unstable against coarsening as to be unable to purify it. In addition, although other surfactants can, in principle, serve in the purification, we typically find it simpler to exploit the well-known phase behavior of SDS for the fractionation and to replace the surfactant once the emulsion has been purified. This can be accomplished by repeated washing of the emulsion in solutions of the new surfactant. The emulsion droplets are dispersed at $\phi \approx 0.1$ in water containing a higher concentration of the desired surfactant. The suspension is then centrifuged to concentrate the emulsion droplets, and the water is replaced with fresh solution containing the new surfactant. After several washes, all the SDS is replaced by the new surfactant. By this means, we have employed a wide variety of surfactants, including cationic, anionic, and nonionic surfactants.

III. Equation of State

The equation of state of any material is the relationship between the pressure and the volume. For a colloidal system, the relevant equation of state is the relationship between the volume fraction of the colloidal particles and the osmotic pressure, $\Pi(\phi)$ [6]. The osmotic pressure of an emulsion is the pressure exerted by its droplets on the walls of a semipermeable container that allows the free exchange of water with a reservoir [7]; the osmotic pressure allows the droplets to be deformed without also requiring hydrostatic compression. When compressed, the droplets become nonspherical and develop locally flattened facets where they press against their neighbors; this increases their total surface area while conserving their volume. Therefore, the osmotic pressure must work against the oil–water interfacial tension, σ, to raise the interfacial curvature of the droplets. Thus, the osmotic pressure represents the energy per unit volume stored in the deformed interfaces of the droplets as their volume fraction increases [7]. It is a macroscopic reflection of the microscopic droplet structure and interaction forces and determines how the droplets deform as they pack to higher volume fractions. The scale for the osmotic pressure is set by the Laplace pressure of the undeformed droplet, which represents the pressure required to deform the droplets significantly. Thus, all energy densities can be scaled by σ/a.

The osmotic equation of state has been measured for emulsions, but using a polydisperse distribution of droplets [8]. Polydispersity introduces different microscopic-length scales, creating differences in the Laplace pressure required to deform each droplet. Furthermore, in polydisperse emulsions, the connection between the droplet deformation and the volume fraction is obscured because small droplets can fit in the interstices between larger droplets. Thus, the use of an emulsion with monodisperse droplets and a uniform Laplace pressure can facilitate our ability to interpret the osmotic pressure data and lead to new insight about how the microscopic structure and forces between the droplets give rise to their equation of state. Furthermore, the packing of the droplets can be more conveniently related to the droplet volume fraction than in polydisperse emulsions.

To measure $\Pi(\phi)$, we use a monodisperse emulsion having radius $a = 0.48\,\mu\text{m}$ and surface tension $\sigma = 9.8$ dynes/cm. The Laplace pressure is $\sigma/a = 2 \times 10^5$ dynes/cm^2, which is a substantial fraction of an atmosphere. To measure the equation of state, we set the osmotic pressure to a known value and measure the resultant volume fraction of the emulsion. The volume fraction is determined by weighing the emulsion before and after the water is evaporated. The osmotic pressure is set by two different techniques, allowing an extended range of pressures to be reliably accessed. Low pressures are set by allowing the emulsion to cream, either in the earth's gravitational field or in a centrifuge. For higher pressures, we use a dialysis technique with a hydrophilic polymer to remove the water and increase the volume fraction. In both techniques, the SDS concentration is fixed at $C = 10\,\mathrm{mM}$, slightly above the critical micelle concentration; this inhibits coalescence but keeps the depletion attraction negligible [9].

For osmotic pressures much less than the Laplace pressure, we set Π by creaming, using a column of emulsion that initially has a uniform volume fraction. Since the total volume of droplets is known, the osmotic pressure at the top of the cream is determined by their total density mismatch, independent of the spatial variation of ϕ within the column,

$$\Pi = \Delta\rho\, \mathrm{g} z \phi_{\mathrm{i}} \tag{3}$$

where $\Delta\rho \approx 0.06\,\mathrm{g/cm^3}$ is the density difference between the oil and the water, g is the acceleration due to gravity, z is the height of the column, and ϕ_{i} is the initial volume fraction. To increase the accessible range of osmotic pressures, we use a centrifuge to vary g. By fixing $\phi_{\mathrm{i}} \approx 0.1$ and varying g between 10^4 and $10^6\,\mathrm{cm/sec^2}$, we can set osmotic pressures between 5×10^2 and $5 \times 10^4\,\mathrm{dynes/cm^2}$. Equilibration occurs after several days at low speeds and after about half a day at the highest speeds. After centrifugation, a thin layer of emulsion is skimmed from the top of the cream and ϕ is determined. For the lowest osmotic pressures, near $\Pi \approx 10^2\,\mathrm{dynes/cm^2}$, the earth's gravitational field is used, and equilibration takes about half a year.

For higher osmotic pressures, we set Π via a polymer dialysis technique [9]. An initially dilute emulsion is enclosed in a semipermeable cellulose dialysis bag and immersed in a reservoir containing a solution of dextran with a molecular weight of 500,000 and an average radius of gyration of 300 Å. The dialysis bag has a cutoff of 12,000, making its average pore diameter approximately 70 Å; therefore, neither the dextran molecules nor the emulsion droplets can flow through the membrane, whereas the water and SDS do. Dextran is extremely hydrophilic, driving water from the emulsion into the reservoir, thereby concentrating the emulsion as its osmotic pressure is equilibrated with that of the dextran solution; this occurs within about 20 days. The osmotic pressure depends on the dextran concentration [10] and can be varied between 10^4 and 10^6 dynes/cm^2. The polymer dialysis technique is limited by coalescence at the highest osmotic pressures and by the extremely long equilibration times at the lowest osmotic pressures.

The data for the osmotic equation of state are shown in Fig. 2. The data obtained by centrifugation are shown by the solid symbols, while those obtained by dialysis are shown by the open symbols; good agreement is found where the data overlap. There is a dramatic rise of Π over several orders of magnitude near $\phi_{\mathrm{c}} \approx 0.60$, although the exact nature of the rise is obscured by the scatter in the data, which arises primarily from the measurement of the volume fraction. This

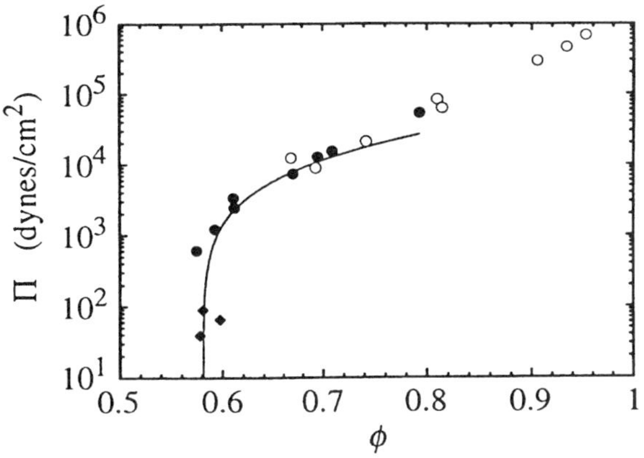

Figure 2. Osmotic equation of state $\Pi(\phi)$ of a monodisperse emulsion having $a = 0.48\,\mu m$ measured via dextran dialysis (open circles), centrifugation (solid circles), and creaming in the earth's gravitational field (solid diamonds). The solid line is a fit to $\Pi(\phi)$ using the linear model of Eq. (4) for $\phi \leq 0.75$.

rise corresponds to the volume fraction where the droplets can no longer remain spherical in shape and begin to deform as they are packed; an osmotic pressure is required to cause this deformation. After this sharp rise, Π rises more gradually between $\phi \approx 0.63$ and $\phi \approx 0.75$ and then begins to diverge again as ϕ approaches 1.

To describe the observed behavior when the droplets are weakly deformed near ϕ_c, we consider the compression of a single droplet in a cube. If the length L of each side of the cube is decreased so that it is slightly less than the diameter of the droplet, $L = 2a - \Delta l$, six flat facets, each having area πd^2, will result where the droplet touches the cube. The difference in the cube's volume from that needed to contain an undeformed droplet is $\Delta V \approx 3\,\Delta l(2a)^2$ to the first order in Δl, while the area of the facet is $d^2 = a\,\Delta l$. For small distortions, the radius of the droplet, and hence its Laplace pressure, will not change significantly. the force applied by the walls of the cube to create the flat facets must balance the force applied by the Laplace pressure on the walls, $(2\sigma/a)\pi d^2 = 2\sigma\pi\,\Delta l$. Thus, the osmotic pressure is given by

$$\Pi = \left(\frac{\pi}{3}\right)\left(\frac{\sigma}{a}\right)\left(\frac{\Delta V}{2a^3}\right)$$

The droplets begin to touch at a critical volume fraction ϕ_c, so the fractional volume change $\Delta V/(2a)^3$ is $\phi - \phi_c$. Thus, the equation of state near ϕ_c is

$$\Pi(\phi) = B_1\left(\frac{\sigma}{a}\right)(\phi - \phi_c) \qquad \phi \geq \phi_c \tag{4}$$

where the constant $B_1 = \pi/3 \approx 1.05$ depends on the cubic packing; in a random emulsion, the neighboring droplets form the walls and B_1 can vary and can even depend weakly on ϕ as the average number of facets increases. However, the essential feature of this model will remain unchanged; this is the linear dependence of the force on the distance that the droplet has been deformed, and thus on the increase in the volume fraction from ϕ_c. In this sense, the facet behaves as a harmonic spring whose force increases linearly with displacement. A more detailed calculation of the precise geometry of the distorted droplet yields an additional logarithmic correction to the harmonic interaction between droplets [11]; however, the essential nature of the linear variation above ϕ_c is not changed. A fit using Eq. (4) for $\phi < 0.75$ is shown by the solid line in Fig. 2. The fit is in good agreement with the data over the lower volume fractions, with $\phi_c = 0.58$ and $B_1 = 0.57$. The prefactor B_1 is within a factor of 2 of our prediction for a simple cubic lattice, reflecting the difference in the coordination number for the droplets in a random emulsion.

In the strong compression limit, the droplets are highly compressed and become nearly polyhedral in shape, so the emulsion resembles a biliquid foam, and Π diverges. Here, the Laplace pressure of a droplet is no longer simply σ/a but becomes $\sigma(r_1^{-1} + r_2^{-1})$, where r_1 and r_2 represent the local radii of curvature, which become infinite on the flat facets and are finite only at the droplet edges and corners. Assuming that most of the water is contained within the droplet edges, and not in the films or in the corners, results in [7]

$$\Pi(\phi) = B_2\left(\frac{\sigma}{a}\right)(1-\phi)^{-1/2} \tag{5}$$

near $\phi \approx 1$, where $B_2 = 0.58$ for an ordered FCC lattice of droplets. This equation predicts $\Pi = 6 \times 10^5$ dynes/cm^2 at $\phi = 0.96$, in good agreement with our measurement of $\Pi = 7 \times 10^5$ dynes/cm^2. An extensive test of the ϕ dependence predicted in Eq. (5) is precluded by coalescence of the droplets.

A comparison of our results with those for a polydisperse emulsion illustrates the advantage of using a monodisperse emulsion. The osmotic pressure for a polydisperse emulsion vanishes at a volume fraction of 0.715 [8], whereas the osmotic pressure for a monodisperse emulsion vanishes at a significantly lower volume fraction, 0.58. This reflects the importance of the droplet packing in determining the equation of state; only with droplets of uniform size can the packing conditions be well defined, even in the absence of any long-range order of the droplets. In addition, monodisperse droplets, with a uniform Laplace pressure, allow a critical test of simple theoretical hypotheses for the equation of state. The essential concepts can then be extended to the more commonly encountered polydisperse emulsions; these are that the compression due to the osmotic pressure causes the droplets to form flat facets, and that these facets behave as harmonic springs whose force is linear in the amount of the deformation. However, the effective spring constant scales as σ/a, and it is the same for all droplets in a monodisperse emulsion but varies from droplet to droplet in a polydisperse one.

IV. Droplet Structure

Many properties of an emulsion depend critically on the packing of the droplets. In a polydisperse emulsion, the packing is highly disordered and difficult to characterize. By contrast, in a monodisperse emulsion, the structure of the droplets can attain some degree of order; moreover, this can be well characterized. This is most conveniently done with light scattering, provided the strong multiple scattering, which gives the emulsion its white appearance, is eliminated. This can be accomplished by adding glycerol to the water to match the index of refraction of the continuous phase with that of the droplets. This makes even highly concentrated emulsions perfectly clear. Both static and dynamic light scattering prove useful. Static light scattering is the measurement of the average scattered intensity as a function of angle and provides information about the structure of both the droplets and their packing. Dynamic light scattering (DLS) is a measure of the fluctuations in the scattered intensity and provides information about the dynamics of the droplets.

Dynamic light scattering is typically used to determine both the average size and the degree of polydispersity of colloidal systems. Since these emulsion are so highly monodisperse, we exploit the strong angular dependence of the intensity of light scattered from particles whose size is comparable to optical wavelengths to determine the remaining polydispersity [12]. Spherical scatterers of radius a produce a deep scattering intensity minimum at the angle θ such that $qa \cong 4.5$. Here q is the scattering wavevector defined by $q = (4\pi n/\lambda) \sin(\theta/2)$, where λ is the wavelength and n is the index of refraction of the emulsion's continuous phase. Thus, in the vicinity of the minimum, the detected light is primarily due to scattering from those particles whose sizes deviate significantly from the average. As a result, the apparent particle radius $R_a(q)$ obtained with DLS is anomalously angle-dependent near $q\bar{a} = 4.5$, where $\bar{a}$ is the unweighted average radius of the droplets. Moreover, the apparent radius is sensitive to the details of the particle size distribution, allowing a precise determination of the width of the distribution. Thus, by measuring $R_a(q)$ around $q\bar{a} = 4.5$, it is possible to obtain a very sensitive measure of both the average size of the droplets and their polydispersity.

A very dilute sample ensures that there is no multiple scattering whatsoever; even small amounts of multiple scattering can result in an overestimation of the polydispersity. The values of $R_a(q)$ are found via DLS by determining the initial decay rate Γ of the measured correlation function through a cumulants analysis [13] and with $\Gamma = k_B T q^2 / 6\pi\eta R_a(q)$; here, k_B is Boltzmann's constant, T is the temperature, and η is the continuous phase viscosity. Typical data for $R_a(q)$ are shown in Fig. 3 for a volume fraction of 7×10^{-6} for an emulsion that had undergone nine steps of purification. The solid curve is calculated from [12]

$$R_a(q) = K_1 \frac{\int dR'\, n(R'; \bar{a}, \sigma_d) I(R', q)}{\int dR'\, n(R'; \bar{a}, \sigma_d) I(R', q)/R'} \tag{6}$$

This expression reflects a ratio of average radii weighted by the distribution of the droplets, $n(R'; \bar{a}, \sigma_d)$; the numerator is the average radius determined from the scattered intensity as measured by static light scattering, while the denominator is the average radius of the particles as measured by DLS. This ratio is very sensitive to the strong angular dependence of the scattered intensity $I(R', q)$ in the vicinity of $q\bar{a} = 4.5$. To apply Eq. (6), we calculate $I(R', q)$ with the Mie theory of scattering [14]. For $n(R'; \bar{a}, \sigma_d)$ we assume a normalized Gaussian centered at $\bar{a}$. We treat the average radius $\bar{a}$ and the polydispersity σ_d as fitting parameters, where we define

$$\sigma_d \equiv \frac{[\langle (a - \bar{a})^2 \rangle]^{0.5}}{\bar{a}} \tag{7}$$

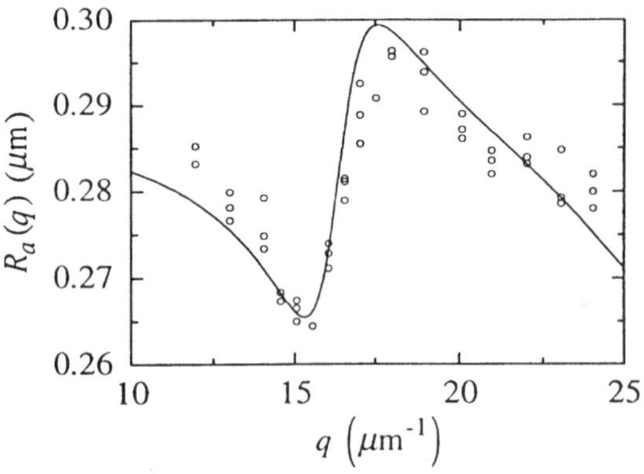

Figure 3. The apparent radius of the emulsion droplet measured with dynamic light scattering as a function of wave vector in the vicinity of the minimum of the scattered intensity, where the results are highly sensitive to the small, remaining polydispersity in the droplet sizes. The solid curve is a fit to Eq. (6), and yields an average radius of 0.265 μm and a polydispersity of 7% of the average radius.

The prefactor K_1 is also adjustable and accounts for the effect of a layer of the continuous phase around the droplet, whose thickness is of the order of the Coulombic screening length, which contributes to the drag, and hence to $R_a(q)$, without contributing to the scattering intensity [15]. From the fit to the data shown in Fig. 3 we find $\bar{a} = 0.265\,\mu m$, $\sigma = 0.07$, and $K_1 = 1.06$. This provides a very sensitive measure of the polydispersity, implying that the width of the distribution of droplets varies by only about 7% by radius. This is a clear demonstration of how highly monodisperse these emulsions are.

To investigate the structure and packing of the droplets further, we eliminate multiple scattering and make this emulsion transparent by choosing for the continuous phase an aqueous mixture that has the same index of refraction as the oil. This technique allows us to use conventional light scattering to study concentrated emulsions. The refractive index of a typical silicone oil is about 1.4; this is matched by a 52% mass fraction mixture of glycerol in water. However, a perfectly transparent system is not obtained. There are two reasons for this. First, since the indices of the oil and the continuous phase mixture have different temperature coefficients, the condition of exact index matching holds at only one temperature for each fixed composition chosen for the continuous phase. In order to get stable results, the measurements must be made at a temperature differing from the exact-match temperature by an amount that is large compared with the temperature fluctuations and drifts allowed by the thermostat. The second reason is that the near match of the oil and continuous phase indices allows us to observe the scattering due to the thin surfactant layer adsorbed onto the droplets, a contribution that is completely negligible when the contrast between the oil and continuous phase is high. These two effects result in samples that scatter strongly enough to prevent us from maintaining rigorously singly scattering conditions at the droplet volume fractions of interest. However, the only effect of the multiple scattering is to cause some smearing of the minima in the intensity; this has the same effect on the angular dependence of the scattering as a small amount of polydispersity.

In Fig. 4, we plot the static scattering I(q) as a function of q, measured after index-matching the same emulsion used to obtain the dynamic scattering data discussed earlier. The volume fraction of this index-matched sample is 0.2. These data can be analyzed with the Rayleigh–Gans theory of scattering from systems with low optical contrast [14]. The solid curve is given by

$$I(q) = K_2 \int dR'\, n(R';\, \bar{a},\, \sigma_d) \left\{ \frac{4\pi}{3} R'^3 P_c(qR') + 4\pi R'^2 \beta P_s(qR') \right\}^2 \tag{8}$$

where K_2 is an arbitrary factor proportional to the incident intensity. The Rayleigh–Gans form factors for a sphere and for a thin spherical shell are $P_c(x) = 3j_1(x)/x$ and $P_s(x) = j_0(x)$; here, j_1 and j_0 are spherical Bessel functions. The first term within the braces is proportional to the strength of the scattered electric field contributed by the droplet's oil core. The second term is the contribution of the surfactant shell relative to that of the core. The parameter β is the optical thickness of the shell, which is the true thickness enhanced by a factor of the optical contrast between the surfactant material and the continuous phase, relative to the contrast between the oil and the continuous phase. This enhancement is large because the oil and the continuous phase are nearly index-matched; its sign changes as one moves through the exact-match temperature. The solid curve shown in Fig. 4 has been calculated with $\beta = -0.050\ \mu m$, $\bar{a} = 0.256\ \mu m$, and $\sigma_d = 0.12$. We attribute to multiple scattering the small differences between these values for $\bar{a}$ and σ_d and those we find from DLS, and believe that the DLS data provide a more accurate measure. We find the same DLS values upon rediluting the index-matched emulsion in water, indicating that the emulsion does not degrade upon addition of glycerol.

As the volume fraction ϕ is raised further, the intensity of the scattering does not rise in simple proportion to ϕ. Instead, it is modified by an additional

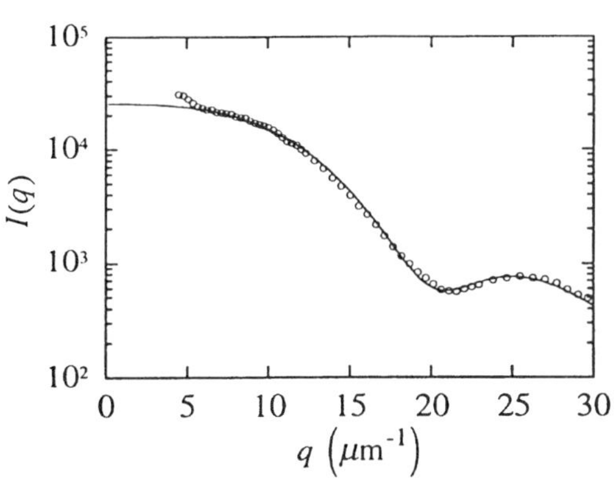

Figure 4. The scattered light intensity as a function of wave vector for an index-matched emulsion at a volume fraction of 0.2. The solid line is a fit to the Rayleigh–Gans form for a coated sphere, given by Eq. (8). These data provide the form factor for the droplets of this emulsion.

factor of S(q), the static structure factor. This provides information about correlations in the positions occupied by neighboring droplets; it is the Fourier transform of the droplet radial distribution function [16]. In Fig. 5 we plot the intensity of the light scattered from a sample of the index-matched stock discussed earlier after its volume fraction was increased to $\phi = 0.7$ by centrifugation. At this volume fraction, the droplets in the emulsion are in contact with each other but the deviation from spherical shape is small. Moreover, the measurement was made at the same incident intensity and temperature that we used in collecting the data from the more dilute index-matched sample (Fig. 4), so we can calculate S(q) as the ratio of the intensities times the inverse ratio of the concentrations of the concentrated and dilute samples. The structure factor so obtained is plotted in Fig. 6. It is similar to the structure factor of a typical

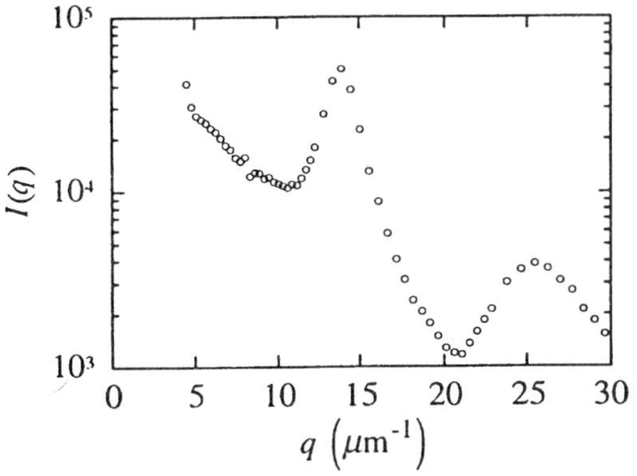

Figure 5. The scattered intensity as a function of wave vector for an index-matched emulsion at a volume fraction of 0.7.

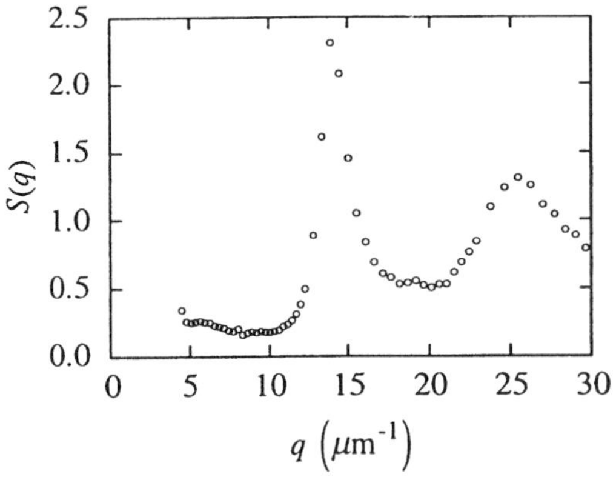

Figure 6. The structure factor for an index-matched emulsion of volume fraction 0.7, calculated from the ratio of the measured scattering intensity, shown in Fig. 5, to the form factor of the droplets, determined from the scattering from a low-volume-fraction emulsion, shown in Fig. 4.

atomic liquid. It shows the strong tendency for a given droplet to lie within a cage of neighbors whose centers are at distance $2\bar{a}$ from the center of the given droplet, and shows the rapid loss of order at distances beyond the cage. The evidence for this is the peak in the intensity around $qa \approx \pi$. A peak at a similar position is always found in the structure factor of a dense glass [16].

These and similar data obtained from different volume fractions confirm that concentrated emulsions with volume fractions above 0.6 may be considered as colloidal glasses because they combine liquidlike structure as exhibited by S(q) with solidlike elastic properties. The sample just discussed, for example, is stiff enough that we can invert the cell containing it without causing appreciable flow of the sample over times of less than an hour. Emulsions with ϕ less than 0.4 are both structurally and rheologically liquidlike. Disordered structure is therefore typical for these monodisperse emulsions. However, under certain conditions, monodisperse emulsions can develop long-range order, forming colloidal crystals. These crystals exhibit Bragg scattering of visible light in much the same fashion as atomic crystals exhibit Bragg scattering of x-rays. In these instances we have the interesting example of a solid with long-range order comprised entirely of liquid components. Figure 7 shows a photograph of an emulsion that has formed colloidal crystals. In a homogeneous state, this sample has a droplet volume fraction of about 0.25. If it is left undisturbed for several months, however, the droplets cream to the top half of the sample volume. The crystals visible in the photograph have formed on the interface between the concentrated and dilute parts of the sample. The opalescent appearance and the small regions of different colors represent Bragg diffraction of the light, indicative of the colloidal crystals. Crystals can form throughout the bulk in homogeneous samples of emulsions with ϕ near 0.5. This behavior is similar to that observed for colloidal suspensions of hard spheres [17], suggesting that the crystallization is entropically driven. Though these emulsion crystals are quite beautiful in appearance, they occur only in rare instances. However, the fact that they do appear is further evidence of the high degree of monodispersity of these emulsions.

V. Shear Rheology

One of the most important properties of emulsions is their viscoelasticity; emulsions can vary from low-viscosity liquids to highly elastic solids (see also Chap. 3 and Ref. 17a). At low volume fractions, an emulsion typically flows freely, and its behavior is dominated by that of the continuous phase. At high volume fractions, by contrast, an emulsion can become an elastic solid. However, its behavior is not dominated by that of the dispersed phase. Instead, its behavior is determined by the large interfacial area of the droplets and the deformation of their shapes; the further deformation of the shapes of the droplets upon application of a strain stores energy in the same way that a spring does, making the emulsion an elastic solid. Thus the shape and packing of the droplets plays a critical role in determining their elastic behavior. The use of monodisperse droplets, with their uniform Laplace pressure and their relatively well-controlled droplet packing, can greatly facilitate the development of a detailed understanding of the origin of the elasticity of emulsions.

Figure 7. Picture of the colloidal crystals formed by a monodisperse emulsion. The top plate shows the index-matched emulsion, creamed at the top of the cell. At the interface between the creamed droplets and the pure fluid, the monodisperse droplets have formed small colloidal crystals, as seen in the increasingly closer views in the two lower plates. The bottom plate shows the crystals most clearly; the bright regions of different colors result from Bragg scattering from the colloidal crystals.

Since emulsions can behave both as fluids and as solids, the rheological description relating stress to strain can change dramatically with volume fraction. Most emulsions are viscoelastic, exhibiting simultaneously characteristics of both a viscous fluid and an elastic solid. At low volume fractions their rheology is typically dominated by a single Newtonian viscosity, whereas at higher volume fractions a richer variety of viscoelastic and plastic yield properties are needed to describe their rheology. These include the linear viscoelastic modulus determining the degree of elastic energy storage and viscous energy dissipation due to a small shear [18], the yield stress required to initiate irreversible flow

[19], and the additional viscous stress needed to maintain flow at a steady strain rate [20]. By using monodisperse emulsions, we can make meaningful measurements of these quantities on well-defined emulsions, thereby determining their dependence on volume fraction and droplet size. In addition, since the droplet size, and thus the internal Laplace pressure, is well defined, we can also compare our results to theoretical predictions. The understanding developed for the monodisperse droplets can then be generalized to account for the behavior of less ideal polydisperse emulsions.

The most important way of characterizing the rheological behavior of an emulsion is through its elastic shear modulus. This is the response of an emulsion to a small shear strain; it sets the scale for all elastic properties of the emulsion. This is measured experimentally by applying a small, sinusoidal strain at a fixed frequency and measuring the resultant stress response of the emulsion. The portion of the response that is in phase with the applied strain yields the elastic, or storage, modulus $G'(\omega)$ and reflects the energy storage in the emulsion. The portion of the response that is out of phase with the applied strain yields the viscous, or loss, modulus $G''(\omega)$ and reflects the energy loss in the emulsion. It is convenient to represent their complex sum, $G^*(\omega) = G'(\omega) + iG''(\omega)$, as the complex viscoelastic modulus. This reflects the microscopic mechanisms of energy storage and dissipation inherent in the emulsion without the structural rearrangements of the droplets caused by large-scale flow. This modulus also determines the propagation and damping of stress waves, while its frequency dependence establishes the time scales for stress relaxation.

When the deformation is increased out of the linear stress–strain regime, concentrated emulsions exhibit irreversible flow. This transition is marked by a yield stress τ_y and a yield strain γ_y at which the structural rearrangements of the droplet positions become important. The emulsion will flow if an applied stress exceeds the yield stress, producing a deformation that will exceed the yield strain. The flow of concentrated emulsions subjected to steady shear at a fixed strain rate $\dot{\gamma}$ can be characterized by the additional viscous stress τ_ν in excess of the yield stress. However, inhomogeneities in the spatial variations in the strain rate within the emulsion can sometimes preclude a simple rheological description using a single strain rate.

To determine the dependence of the rheological properties on droplet size, we use four monodisperse emulsions with different droplet radii; their characteristics are listed in Table 1. For all, we set the SDS concentration at

Table 1. Properties of the Monodisperse Emulsions Used for the Shear Rheology Measurements. (Silicone Oils Used: Polydimethylsiloxane (PDMS) and Polyphenylmethylsiloxane (PPMS)

Radius a (μm)	Oil	Density ρ_0 (g/cm^3)	Viscosity η_0 (cP)
0.25	PDMS	0.937	12
0.37	PDMS	0.937	12
0.53	PPMS	1.047	235
0.74	PDMS	0.937	12

C = 10 mM, which is low enough to eliminate any strong depletion attraction but high enough to ensure the stability of the droplets against coalescence. All rheological measurements are made at room temperature using a controlled-strain rheometer [21]. At high volume fractions we employ a cone and plate geometry, but for $\phi \leq 0.60$ we use a double wall Couette geometry with a large surface area to increase the rheometer's stress sensitivity. Standing waves in the gap are negligible over the frequency range we probe. Creaming is also negligible due to the small droplet sizes and relatively short measurement times. A vapor trap prevents water evaporation to maintain the volume fraction at the initial value. In some previous investigations of the viscoelasticity of emulsions [22,23], slip at the walls of the rheometer was pronounced, preventing a direct measurement of emulsion elasticity and instead requiring substantial analysis of the data to extract the desired moduli. By contrast, we use emulsions with much smaller droplets, which have a much larger modulus; we find that we are able to eliminate slip entirely by roughening the walls of the rheometer. Thus we obtain a direct measure of the stress–strain relationship of the emulsions.

To ensure that we measure the true viscoelastic moduli, we first establish the strain regime where the stress response is linear. Using a frequency of $\omega = 1$ rad/sec, we measure the response as the maximum value of the applied strain is varied. Typical examples of the resultant strain dependence of the storage and loss moduli are shown in Fig. 8 for an emulsion with a = 0.53 μm and for volume fractions of $\phi = 0.57$, 0.60, and 0.77. The moduli are independent of strain for $\gamma \leq 0.02$ for the two lowest volume fractions, so the linear regime exists only at very small strains. However, the linear regime extends beyond $\gamma \approx 0.1$ for the most concentrated emulsion. In the linear regime, G′ is greater than G″, reflecting the emulsion's dominantly elastic nature. However, at larger strains, the loss modulus rises noticeably, indicating the transition to plastic

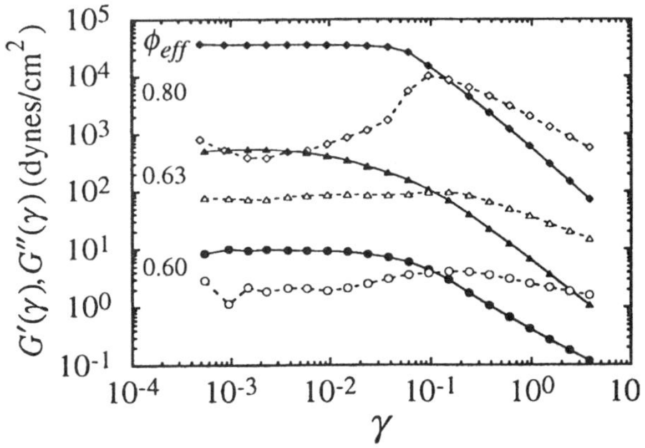

Figure 8. The dependence on the peak strain of the storage modulus G′ (solid symbols) and loss modulus G″ (open symbols) for a monodisperse emulsion with a radius of 0.53 μm at $\phi = 0.77$ (diamonds), 0.60 (triangles), and 0.57 (circles). The data were obtained at a frequency of 1 rad/sec. The data for the highest volume fraction are obtained using a cone-and-plate geometry, while the lower volume fraction data are obtained using a double-wall-Couette geometry.

flow. Beyond the yield strain, marked by the dramatic downturn in G′, we observe that the temporal stress waveform is no longer sinusoidal but becomes flattened at the peaks [24]. There, the apparent G″ dominates the apparent G′, indicating the dominance of energy loss introduced by the flow.

Using the data from the strain sweeps, such as those shown in Fig. 8, we can limit the maximum value of the strain, to ensure linear response, and probe the frequency dependence of the shear moduli. Typically, we set the maximum strain to 0.01 for the higher volume fraction emulsions, but restrict it to 0.005 for volume fractions around 0.6, where the linear regime of strain is a minimum. Results for the frequency dependence of the storage modulus $G'(\omega)$ and the loss modulus $G''(\omega)$ for several different volume fractions of an emulsion with $a = 0.53\,\mu m$ are shown in Figs. 9a and 9b, respectively. Both moduli increase significantly with volume fraction; however, the storage modulus is always considerably greater than the loss modulus, for all volume fractions and for all frequencies measured, consistent with the elastic nature of these emulsions. The storage modulus exhibits a frequency-independent plateau at all volume

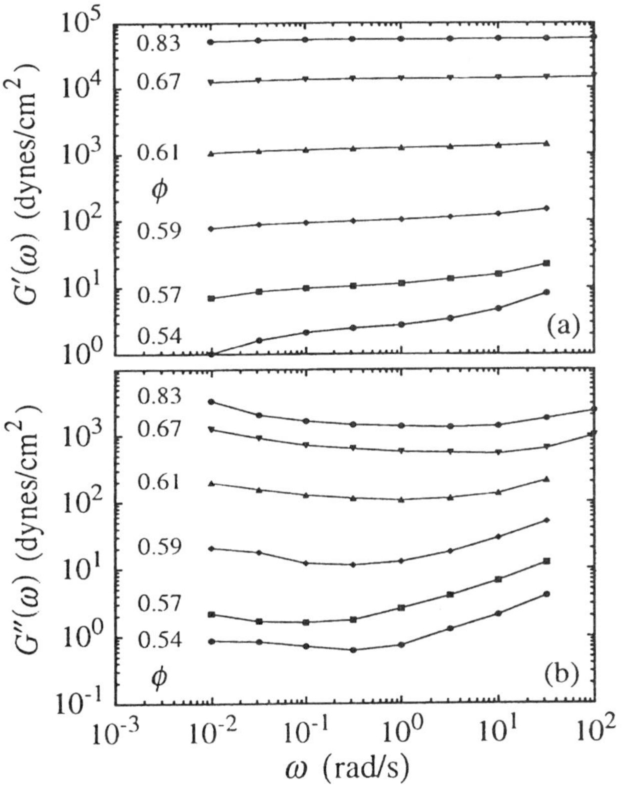

Figure 9. The frequency dependence of (a) the storage modulus G′ and (b) the loss modulus G″ for a monodisperse emulsion having a radius of $0.53\,\mu m$ for several volume fractions. The peak strains used were: $\phi = 0.83$, $\gamma = 0.03$; $\phi = 0.67$, $\gamma = 0.01$; $\phi = 0.61$, $\gamma = 0.005$; $\phi = 0.59$, $\gamma = 0.005$; $\phi = 0.57$, $\gamma = 0.015$; $\phi = 0.54$, $\gamma = 0.025$; these values ensured that measured moduli were independent of γ.

fractions; at the highest ϕ this plateau extends over the full four decades of frequency probed, while at the lowest ϕ its range is very narrow, with $G'(\omega)$ rising at higher frequencies and falling at lower frequencies. We can characterize the volume fraction dependence of $G'(\omega)$ by its value at the midpoint of the plateau, G_p'; this is a true plateau value at higher ϕ, while it is more of an inflection point at the lowest values of ϕ. The loss modulus exhibits a shallow, but nevertheless distinct, minimum for all volume fractions, and we can characterize the ϕ dependence of $G''(\omega)$ by means of its value at this minimum, G_m''. The minimum is shallow at the highest ϕ but becomes more pronounced at lower ϕ, and the frequency of the minimum increases with volume fraction. Although concentrated emulsions are primarily elastic, there are clearly internal dynamics that lead to the observed frequency dependence of the storage and loss moduli; these must reflect the dynamics of the droplets themselves. In particular, the existence of a well-defined minimum in $G''(\omega)$ at high ϕ contrasts with the monotonic rise found for a foam [25], illustrating the importance of using monodisperse emulsions for unraveling the controlling features of the behavior. This minimum must be due to the combination of relaxations at both higher and lower frequencies than are probed by our measurements. The high-frequency rise in $G''(\omega)$ reflects the increasing importance of the solvent viscosity; the low-frequency rise must reflect the viscous dissipation caused by configurational rearrangements of the droplets. If the droplets are highly concentrated, they cannot rearrange easily, so this relaxation is pushed to very low frequencies, leading to the lower frequency of the minimum in $G''(\omega)$.

To investigate how the droplet size controls the rheology, we repeat these measurements for three other emulsions, having radii $a = 0.25$, 0.37, and $0.74\,\mu m$, and plot the values of G_p' as a function of ϕ in Fig. 10. Each emulsion shows a dramatic increase in elasticity around $\phi \approx 0.58$, with G_p' increasing by

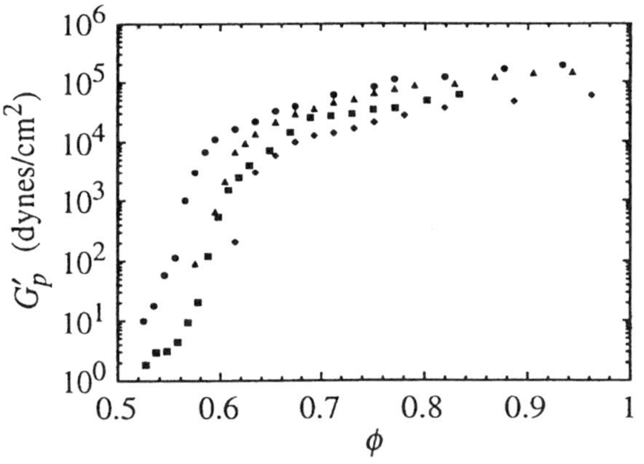

Figure 10. The plateau storage modulus G_p' as a function of volume fraction for monodisperse emulsions having radii $a = 0.25\,\mu m$ (circles), $0.37\,\mu m$ (triangles), $0.53\,\mu m$ (squares), and $0.74\,\mu m$ (diamonds). The data for the emulsions having $a = 0.37\,\mu m$ and $a = 0.74\,\mu m$ were taken entirely with the cone-and-plate geometry; the data for the other two emulsions were taken using a cone-and-plate geometry for $G_p' > 10^2$ dynes/cm^2 and a double-wall-Couette geometry for $G_p' < 10^2$ dynes/cm^2.

over three decades as ϕ increases by less than 0.05; there is an additional increase by about another decade as ϕ increases further. In addition, there is clear trend with droplet radius: At any given ϕ, the value of G_p' is larger as a decreases. In addition, the rapid rise in G_p' occurs at smaller volume fractions for the smaller droplets. To investigate the size dependence, we scale the data by the characteristic energy density of each droplet, σ/a, and plot the results in Fig. 11. At high ϕ, this scaling collapses the data for the different droplet sizes, suggesting that the characteristic value of the asymptotic modulus at high volume fractions is set by the Laplace pressure. At low ϕ, by contrast, systematic deviations persist, suggesting that a different mechanism is responsible for the rise in G_p'.

We hypothesize that this difference is a result of the nature of the repulsive interaction between the droplets caused by our use of an ionic surfactant to stabilize the interfaces. The resultant screened Coulomb interaction at the interfaces leaves a thin layer of water between the droplets, even when they are highly compressed [26]. The thickness h of this shell of water decreases with volume fraction as the osmotic pressure increases the force between the droplets. To account properly for the effects of droplet packing on the rheology, the volume of water in this irreducible film must be included with the droplet volume fraction. Moreover, since the SDS concentration in each emulsion is the same, the thickness of the film should not depend on droplet size; however, it *relative* importance will depend on droplet size. Thus, we correct the droplet volume fraction to include this irreducible shell of water, by defining an effective volume fraction,

$$\phi_{\text{eff}} = \phi\left(1 + \frac{h(\phi)}{2a}\right)^3 \tag{9}$$

This approximation is strictly valid only for spherical droplets, but it is accurate to better than 10% even for the highly deformed droplets at high osmotic pressures. Moreover, this volume fraction adjustment is larger for smaller droplets.

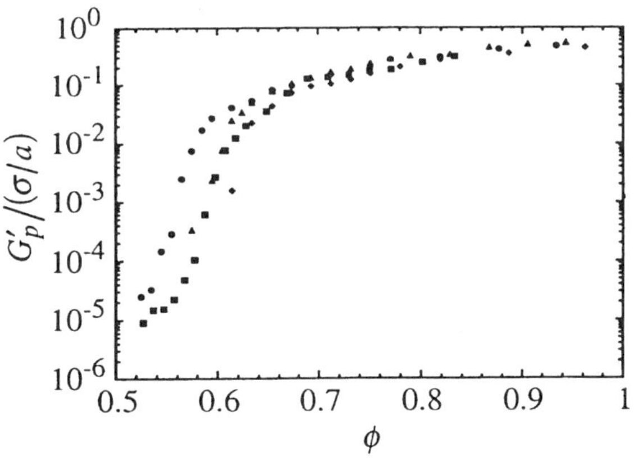

Figure 11. The volume fraction dependence of the plateau storage modulus G_p', scaled by σ/a, for four emulsions having radii $a = 0.25\,\mu m$ (circles), $0.37\,\mu m$ (triangles), $0.53\,\mu m$ (squares), and $0.74\,\mu m$ (diamonds).

The film thickness should depend on the forces between the droplets and therefore on the volume fraction. At the lowest volume fractions, where the packing constraints do not force the droplets to deform, h should have its largest value; thus, we assume that the film thickness is fixed at some maximum value, h_{max}, below the volume fraction of random close packing of spheres, $\phi_c = 0.63$. The film thickness should decrease as the forces between the droplets increase with increasing volume fractions. Ultimately, at the highest volume fractions, where the droplets are nearly polyhedral in shape, the thickness of the film presumably resembles that of a Newton black film for SDS at a similar electrolyte concentration, defining the minimum film thickness, $h_{min} \approx 50\,\text{Å}$ [27]. Since the dependence of these forces on volume fraction is not known, we interpolate linearly between h_{max} at ϕ_c and h_{min} at $\phi = 1$:

$$h(\phi) = \begin{cases} h_{max} & \phi \le \phi_c \\ h_{max} - (h_{max} - h_{min})\dfrac{\phi - \phi_c}{1 - \phi_c} & \phi > \phi_c \end{cases} \tag{10}$$

By adjusting the value of h_{max}, we can make all the data for G'_p collapse onto a single master curve; this is shown by the solid symbols in Fig. 12. The maximum film thickness that gives the best data collapse is $h_{max} = 175\,\text{Å}$; this is consistent with maximum separation observed in force–distance measurements for monodisperse emulsions at the same SDS concentration [28]. The uniform film thickness raises the volume fraction more for smaller droplets, producing the universal dependence of the scaled elasticity with ϕ_{eff}. For the scaled data, the dramatic rise in G'_p occurs near $\phi_c \approx 0.63$, which corresponds with random close packing of disordered but undeformed monodisperse spheres. This value of ϕ_c is not a result of our assumption of $h(\phi)$, but is determined by the measured elastic onset of the largest emulsion. For $a = 0.74\,\mu m$, the adjustment to ϕ is a mere 1%

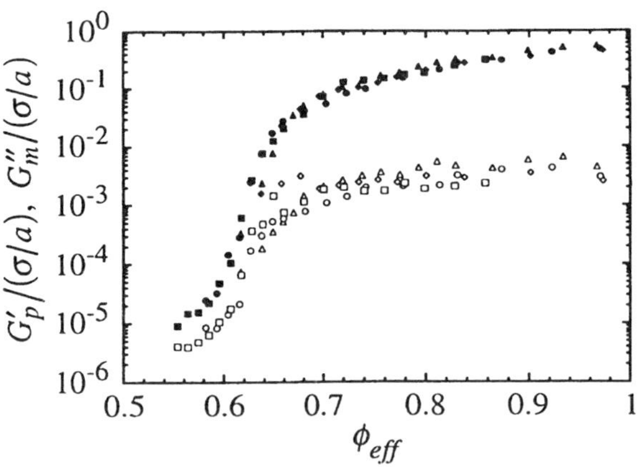

Figure 12. The plateau modulus G'_p (solid symbols) and the minimum in the loss modulus G''_m (open symbols), each scaled σ/a, as functions of ϕ_{eff} for monodisperse emulsions having radii $a = 0.25\,\mu m$ (circles), $0.37\,\mu m$ (triangles), $0.53\,\mu m$ (squares), and $0.74\,\mu m$ (diamonds).

because the volume of the water shell is almost negligible compared to the droplet volume.

We also show the ϕ_{eff} dependence of G''_m for the different emulsions by the open symbols in Fig. 12. They also exhibit scaling behavior, albeit not as precise as that for G'_p. In the dry limit at $\phi_{eff} \approx 1$, G''_m approaches $4 \times 10^{-3}\sigma/a$, about two orders of magnitude lower than G'_p. This difference is less as ϕ_{eff} decreases, with the scaled storage and loss moduli becoming much closer in value below ϕ_c. However, even there the storage modulus is greater than the loss modulus. Finally, to test if the oil viscosity influences energy storage or loss in concentrated emulsions, we also varied the droplet viscosity from $10\,\text{cP} < \eta_0 < 10^3\,\text{cP}$. The dependencies of $G'(\omega)$ and $G''(\omega)$ on ϕ_{eff} are not affected.

The scaling of the data for the different droplet sizes clearly shows that the elasticity of an emulsion is controlled by the Laplace pressure and hence by the deformation of the interfaces. Although this fact has been well appreciated previously, the clear demonstration indicated by the scaling behavior is made possible only by our use of monodisperse emulsion droplets. At $\phi_{eff} \approx 1$, the elasticity of the highly compressed emulsion is similar to that of a dry foam and is completely determined by σ/a. For a disordered monodisperse foam, the elastic modulus is predicted to be $0.55\sigma/a$ [29–31]. We find that G'_p approaches $0.6\sigma/a$, in excellent agreement with this prediction; this is the first measure of this asymptotic limit using well-defined, monodisperse droplets.

The origin of the volume fraction dependence of the elasticity of compressed emulsions is not well understood [23,32]. Some insight can be obtained from the behavior at higher volume fractions; we show this behavior by replotting the scaled data for G'_p logarithmically as a function of $\phi - \phi_c$ in Fig. 13. The storage modulus exhibits a nearly linear rise that persists up to $\phi_{eff} \approx 1$. We speculate that the droplet disorder inherent in our monodisperse emulsion creates locally

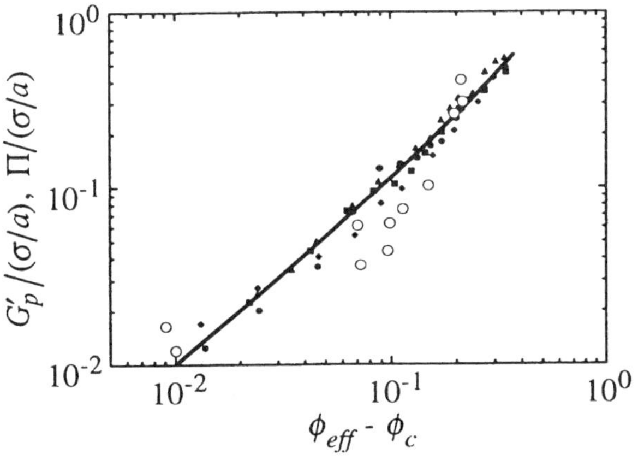

Figure 13. The scaled plateau modulus $G'_p/(\sigma/a)$ and the scaled osmotic pressure $\Pi/(\sigma/a)$ as functions of $\phi_{eff} - \phi_c$. The plateau modulus is measured for monodisperse emulsions having radii $a = 0.25\,\mu m$ (circles), $0.37\,\mu m$ (triangles), $0.53\,\mu m$ (squares), and $0.74\,\mu m$ (diamonds), while the osmotic pressure is measured for a monodisperse emulsion with $a = 0.48\,\mu m$ (large open circles).

weak regions where droplet interfaces do not make contact and cannot support shear. The number of these weak regions increases when the droplets become more weakly packed as $\phi_{\mathrm{eff}} \rightarrow \phi_c$, similar to lattices of interconnected springs that lose their rigidity gradually as springs are removed [33]. If we assume that the number of these weak regions increases in inverse proportion with volume fraction, we would expect to see a linear increase in the modulus with ϕ above ϕ_c, by analogy to the network spring model [33]. However, the number of springs should depend on volume fraction, so we would expect

$$G_p' \sim \frac{\sigma}{a}\,\phi(\phi - \phi_c) \tag{11}$$

This behavior is indeed observed, as shown by the solid line in Fig. 13. This behavior is similar in spirit to a hypothesis that has been invoked to account for the behavior observed in computer simulations of the elasticity of a two-dimensional soap froth [34]. The essentially linear rise in the elasticity that we measure is similar to that observed for polydisperse emulsions [23]; in that case, however, $\phi_c \approx 0.71$. This is considerably larger than $\phi_c \approx 0.63$, which is expected for random close packing of spheres. This difference can be attributed to the greater packing efficiency in polydisperse emulsions. Smaller droplets can fit between the larger droplets without deforming, thus raising the overall volume fraction at which deformation begins to occur. Thus, we can clearly distinguish the essential role played by the packing of the droplets through the use of monodisperse emulsions.

While this picture of missing springs appears to describe the data well, it may not capture adequately the origin of the behavior as ϕ_{eff} approaches very close to ϕ_c. The percolation effects inherent in this model require a decreasing number of facets to cause the loss of elasticity. However, even at ϕ_c, each sphere should have, on average, several nearest neighbors, restricting the total number of additional facets that can be formed with increasing ϕ_{eff} and thus the total change in the modulus that can be achieved. It is possible the droplets become increasingly susceptible to rearrangements of their positions as ϕ approaches ϕ_c; right near ϕ_c, some droplets must possess very weak facets, which can become unbalanced upon application of a very small shear. Restoring the balance to ensure that the net force is zero may cause rearrangement of the droplet position. This would account for the vanishingly small γ-independent G_p' near ϕ_c, shown in Fig. 8.

Perhaps the most surprising aspect of the behavior of the elasticity of emulsions is shown when we compare the volume fraction dependence of the storage modulus with the osmotic pressure. This is done in Fig. 13, where we plot the osmotic pressure σ/a as a function of ϕ_{eff}. Remarkably, both the scaled Π and G_p' exhibit nearly the same behavior. This is surprising, since the osmotic pressure and shear elasticity are fundamentally different properties. Each represents an energy density associated with droplet deformation in a fixed configuration; however, the osmotic pressure represents the energy density required to maintain the deformed droplets at a fixed volume fraction, while the plateau elasticity represents the energy density stored through additional droplet deformation due to a imposed shear of unity at fixed volume fraction. Thus, the origin of this agreement is not clear; it suggests that the magnitude of the osmotic pressure may be intrinsically related to that of the shear modulus.

Finally, even below ϕ_c, the storage modulus still dominates the loss modulus, at least over the range of frequencies probed by our rheometer; it is not until $\phi_{eff} \approx 0.55$ that the minimum disappears entirely and G_p' and G_m'' are no longer well defined. In the region between $\phi_{eff} \approx 0.55$ and ϕ_c, the packing constraints do not force the droplets to be deformed; nevertheless, they still exhibit a predominant elasticity. The origin of this behavior can be understood by analogy to hard spheres, which cannot deform. They also exhibit a predominant elastic modulus at these volume fractions; this is a result of the colloidal glass transition [35]. At these volume fractions, the elastic modulus is purely entropic in origin, and its energy scale should be set by k_BT/a^3 [36]. While this is much smaller than the scale of the interfacial energy σ/a, this contribution to the modulus should diverge as the volume fraction approaches ϕ_c. The limited data for the lowest volume fractions precludes testing for this different scaling with size, which is required to confirm the entropic origin of the elastic modulus in this regime.

The linear viscoelastic modulus of an emulsion sets the essential energy scale for the elasticity, while its frequency dependence reflects the relaxation processes that determine the equilibrium dynamics. However, equally important for the practical rheology of emulsions is their behavior when they are sheared so much that they flow. This behavior is characterized by a yield stress τ_y and yield strain γ_y where the stress and strain amplitudes cease to be linearly proportional. Examples of the departure from linearity for a = 0.53 μm at several volume fractions are shown in Fig. 14, where we plot the peak stress measured as a function of the peak of the applied sinusoidal strain. At the lowest γ the slope of the stress–strain relationship is 1 at all ϕ_{eff}, but at the highest γ the

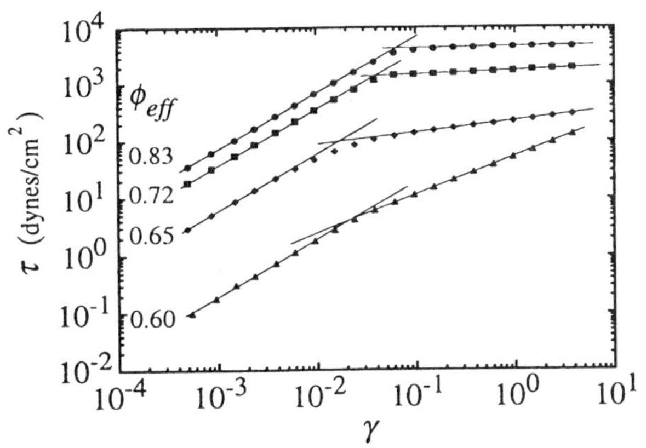

Figure 14. The peak stress τ of the periodic response plotted as a function of the peak strain γ of an applied sinusoidal shear excitation at $\omega = 1$ rad/sec for a monodisperse emulsion having a radius of 0.53 μm for several different ϕ_{eff}. At low γ the stress response is sinusoidal, while at high γ above the knee, it is nonsinusoidal. At each ϕ_{eff}, a line having slope of 1 on the log–log plot is extrapolated through the data at low γ until it intersects a line of lower slope drawn through the data at high γ. The intersection defines the yield stress τ_y and yield strain γ_y for that ϕ_{eff}.

slope decreases significantly, since the resultant stress is nonsinusoidal, with flattened peaks due to plastic flow. A good approximation of both the yield stress and the yield strain can be obtained by the intersection of power law extrapolations from the high and low γ regimes. The deviation from linearity is much more abrupt at high ϕ_{eff} than at low ϕ_{eff}, making the yield strain more difficult to determine precisely there.

The variation of the yield strain with ϕ_{eff} for all droplet radii is shown in Fig. 15. The yield strain is independent of the droplet radius, indicating that it is controlled by the droplet packing. It increases from about $\gamma_y \approx 0.03$ at $\phi_{\text{eff}} \approx 0.62$ to more than $\gamma_y \approx 0.1$ near $\phi_{\text{eff}} \approx 1$. Near ϕ_c, relative displacements of the droplets that are nearly two orders of magnitude less than their diameters are sufficient to cause the emulsion to yield. In the high volume fraction limit, the measured yield strains for monodisperse emulsions are consistent with measurements for polydisperse emulsions [22]. However, even near $\phi_{\text{eff}} \approx 1$, they are more than a factor of 5 lower than predictions of $\gamma_y \approx 0.6$ based on computer calculations of ordered lattices of monodisperse droplets [37]. This suggests that the droplet disorder plays a major role in determining the yield properties. A single global topological rearrangement is the only mechanism by which an ordered emulsion can yield; by contrast, a disordered emulsion can yield through local droplet rearrangements. As the volume fraction decreases, there is an increase in the number of weak regions most susceptible to yield, resulting in a lower value of the yield strain.

Unlike the yield strain, the yield stress increases dramatically with ϕ_{eff}; this is shown in Fig. 16, where we plot the ϕ_{eff} dependence of the yield stress, normalized by σ/a, for all the droplets. The yield stresses are determined from the product of the yield strain with the geometric mean of G'_p and G''_m; this is equivalent to the previous measure obtained from the extrapolation of the power-law dependencies because $|G^*| \approx G'_p$ for all ϕ_{eff}, so $\tau_y \approx G'_p \gamma_y$. Since the yield strain is independent of droplet size, the yield stress exhibits the same scaling with σ/a as does G'_p. The yield stress increases dramatically near

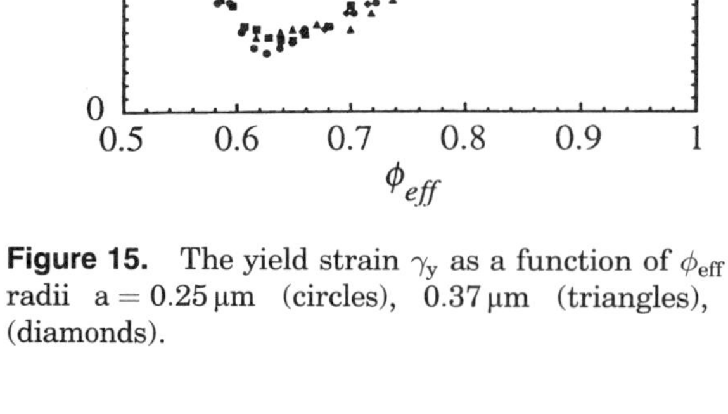

Figure 15. The yield strain γ_y as a function of ϕ_{eff} for monodisperse emulsions having radii $a = 0.25\,\mu m$ (circles), $0.37\,\mu m$ (triangles), $00.53\,\mu m$ (squares), and $0.74\,\mu m$ (diamonds).

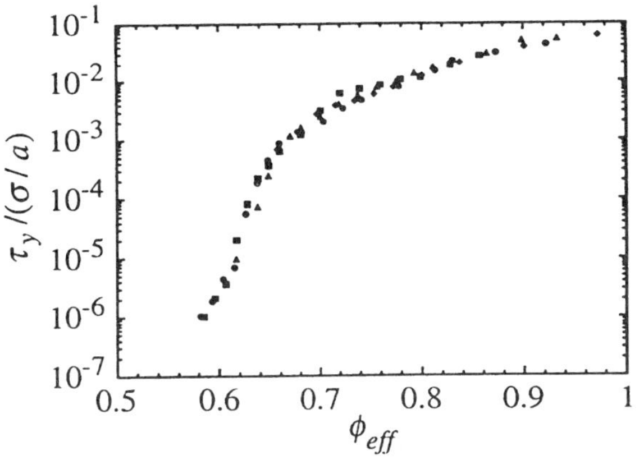

Figure 16. The scaled yield stress $\tau_y/(\sigma/a)$ as a function of ϕ_{eff} for monodisperse emulsions having radii $a = 0.25\,\mu m$ (circles), $0.37\,\mu m$ (triangles), $0.53\,\mu m$ (squares), and $0.74\,\mu m$ (diamonds).

$\phi_c \approx 0.63$, continuing to rise for higher ϕ_{eff}. However, the rise is not quite as sharp as the rise in G'_p; this reflects the increase in the yield strain at larger ϕ_{eff}. In the highly compressed limit as $\phi_{eff} \to 1$, the yield stress approaches $0.08\sigma/a$, about 10% of G'_p. The measured value of τ_y can be compared with theoretical predictions near $\phi_{eff} \approx 1$. Computer calculations of lattices of monodisperse droplets [37] predict $\gamma_y \approx 0.6$, while calculations of the elasticity of an ordered network of films [30,31] predict $G'_p = 0.03\sigma/a$; this gives $\tau_y \approx 0.5\sigma/a$, significantly larger than that observed. This supports the concept that the flow of emulsions begins by the local yielding of droplets in weaker regions.

After yielding, concentrated emulsions exhibit plastic flow; this can be investigated using steady-shear techniques. Weakly concentrated emulsions may flow homogeneously with a uniform strain rate; by contrast, strongly concentrated emulsions may fracture and flow inhomogeneously, so there is no uniform strain rate that can characterize the behavior. To study the flow of the emulsions, we measure the total stress resisting a steady shear. The homogeneity of the strain is determined by painting a stripe on the exposed surface of the emulsion before shearing and watching its evolution during shear [22]. For $\phi_{eff} < 0.70$, the painted stripe deformed continuously below and above the yield strain. However, for higher ϕ_{eff}, an abrupt discontinuity in the stripe developed when the strain exceeded the yield strain, indicating inhomogeneous flow and the existence of a fracture surface. The yield region occurred at random locations in the sample. The thickness of the yield region was only a small fraction of the gap distance, and the strain rate was nearly zero everywhere in the emulsion except in this small region. Consequently, the actual strain rate was much higher than the quoted average strain rate.

We measure the total stress resisting the steady flow for an emulsion having $a = 0.25\ \mu m$ by sweeping the strain rate from $\dot{\gamma} = 10^{-2}\,\sec^{-1}$ to $\dot{\gamma} = 10^{2}\,\sec^{-1}$. The results for a series of volume fractions between $0.55 \le \phi_{eff} \le 0.87$ are shown in Fig. 17. To ensure steady plastic flow, the yield strain is exceeded before the stress is measured. Since $\gamma_y \le 0.2$ for all ϕ_{eff}, we apply the shear at the lowest

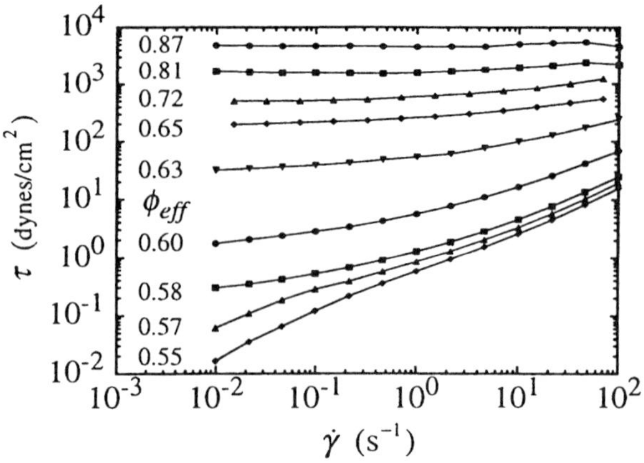

Figure 17. The variation of the shear stress with strain rate for steady flow for a monodisperse emulsion with $a = 0.25\,\mu m$ from several different ϕ_{eff}. For $\phi_{eff} < 0.72$, the strain field is homogeneous and the strain rate is well defined; for $\phi_{eff} \geq 0.72$, the strain field is inhomogeneous and the strain rates are not well defined.

strain rate and wait 1 min, producing an initial strain of 0.6, which exceeds γ_y. The shear stress is then averaged over 1 min during the plastic flow. At any strain rate, the stress varies over many orders of magnitude from the lowest to the highest ϕ_{eff}. For $\phi_{eff} > 0.70$, the flow is inhomogeneous, and the stress is nearly independent of the strain rate; however, we emphasize again that the true strain rate in the thin yield region is much larger than the average value shown in Fig. 17. Moreover, the stress does not always vary continuously with the strain rate. By contrast, for $0.58 < \phi_{eff} < 0.70$, the strain rate is homogeneous and the stress increases monotonically with the strain rate from a plateau at $\dot{\gamma} = 10^{-2}\,\text{sec}^{-1}$. Although we could not probe arbitrarily low strain rates, we define a yield stress for steady shear by the plateau stress at the lowest measurable strain rate. For $\phi_{eff} < 0.58$, the stress drops at low $\dot{\gamma}$, as indicated by the appearance of inflection points, and the plateau yield stress disappears completely.

These results provide a different measure of the yield stress, based on steady flow rather than oscillatory strain. Thus, we compare the different measures of τ_y for an emulsion with $a = 0.25\,\mu m$ in Fig. 18, where the yield stress measured with the oscillatory strain is shown by the open symbols and that measured with the steady strain is shown by the solid symbols. For $\phi_{eff} \leq 0.70$, there is excellent agreement between the two methods; at higher volume fractions, by contrast, the steady-flow measurement of the yield stress is typically about a factor of 3 lower than the dynamic oscillatory measurement, although there is some variation for different ϕ_{eff} due to the inhomogeneous flow. The reason why the steady-shear yield stress is consistently lower may be due to the past shear history. The oscillatory shear measurements were performed first, and entailed maximum strains of $\gamma = 3$, which may have aligned the emulsion structure along the direction of flow.

When the flow is homogeneous, we can calculate the viscous stress by subtracting the yield stress from the total measured stress. The results are shown

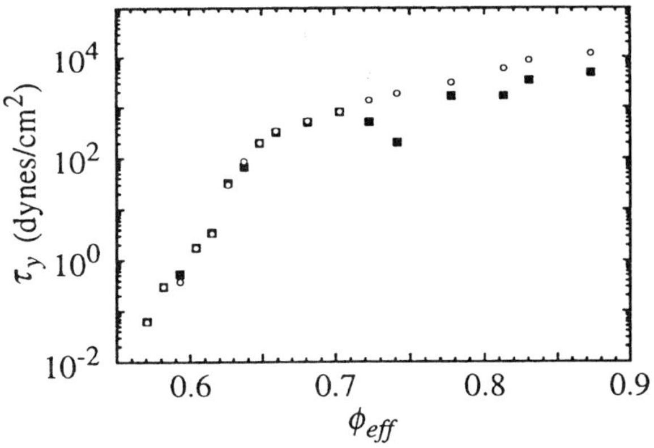

Figure 18. A comparison of the yield stress τ_y determined by measurements using a steady shear (solid squares) and an oscillatory shear (open circles) as a function of $\phi_{\rm eff}$ for a monodisperse emulsion having radius a $= 0.25\,\mu$m. The steady-shear yield stresses are determined from the plateaus at the lowest measured strain rate of $\dot{\gamma} = 10^{-2}\,\text{sec}^{-1}$ as shown in Fig. 17, while the oscillatory yield stresses are determined from stress–strain plots similar to those shown in Fig. 14.

for the four lowest volume fractions in Fig. 19. The viscous stress increases as a power law with the strain rate, with the exponent depending on $\phi_{\rm eff}$. The exponent decreases from about 2/3, shown by the solid line at $\phi_{\rm eff} = 0.58$, to about 1/2, shown by the solid line at $\phi_{\rm eff} = 0.63$. These exponents are insensitive to variations in the yield stress because of the large extent of the data. For $\phi_{\rm eff} \geq 0.65$, the exponents of τ_ν are less than 1/2 at low $\dot{\gamma}$, but the dependence

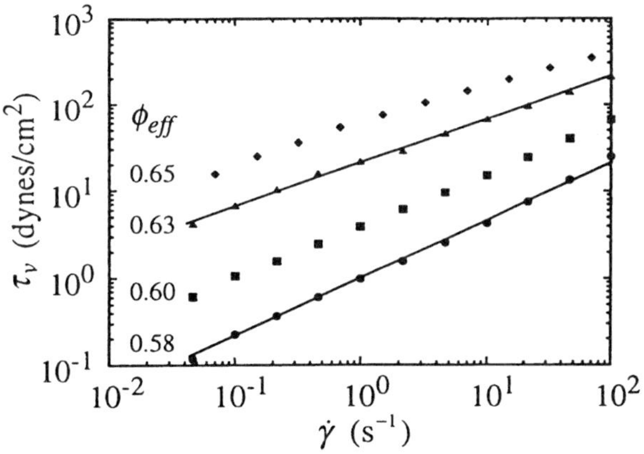

Figure 19. The viscous stress τ_y as a function of strain rate $\dot{\gamma}$ for a monodisperse emulsion having radius a $= 0.25\,\mu$m from $\phi_{\rm eff} = 0.5$ to $\phi_{\rm eff} = 0.65$. The viscous stress is obtained by subtracting the yield stress from the total stress measured at the lowest strain rate. The line drawn through the data for $\phi_{\rm eff} = 0.58$ has slope 2/3, while the line for $\phi_{\rm eff} = 0.63$ has slope 1/2.

deviates from a power law at high $\dot{\gamma}$. Such power-law behavior of the viscous stress has been predicted for concentrated emulsions [20,38,39], using a model for the behaviour of the viscous stress during plastic creeping flow of a two-dimensional, ordered emulsion. The flow of the water around the highly deformed droplets results in a viscous stress that varies as $\dot{\gamma}^{2/3}$. A single power law cannot adequately describe the behavior observed at all volume fractions; moreover, the agreement of the predicted power law for $\phi_{\text{eff}} = 0.58$ is surprising, since the droplets are not deformed and thus the dominant dissipation does not occur between flattened facets, as is assumed in the theory. The decrease in the exponent with volume fraction suggests that the increased packing of the droplets makes the flow even less like that of a viscous fluid. These results can also be compared with measurements of the viscous stress for polydisperse emulsions [40], where the viscous stress was found to vary as $\dot{\gamma}^{1/2}$ for $0.84 < \phi < 0.98$, although only an average $\dot{\gamma}$ was quoted despite a spatially inhomogeneous flow. Our measurements for monodisperse emulsions at $\phi_{\text{eff}} = 0.65$ and $\phi_{\text{eff}} = 0.63$ agree with these other measurements. It is conceivable that the differences in the packing between monodisperse and polydisperse droplets lead to the same $\dot{\gamma}^{1/2}$ behavior being observed at such markedly different volume fractions. However, at lower ϕ_{eff}, the exponent increases to about 2/3 for the monodisperse emulsions, suggesting an evolution toward purely viscous behavior with an exponent of 1.

The measurement of the steady-shear stress at low strain rates provides useful information about the time scales for stress relaxation in the emulsion. For $\phi_{\text{eff}} \leq 0.57$, the appearance of inflection points and the disappearance of the plateau in the shear stress at low $\dot{\gamma}$ are evidence that the droplets can partially relax to an unsheared configuration in less than 100 sec. Since these effective volume fractions are much less than random close packing, the droplets are essentially undeformed. Therefore, it is likely that this relaxation is driven by thermal motion, leading to rearrangements of the droplets. Since the droplets rearrange faster than the shear can perturb them, the yield stress does not exist at $\dot{\gamma} = 10^{-2}\,\text{sec}^{-1}$. This relaxation corresponds with the disappearance of the plateau storage modulus and the rise of the loss modulus at similarly low frequencies for $\phi_{\text{eff}} \approx 0.55$, as determined by oscillatory measurement.

The droplet packing is an essential feature in determining an emulsion's rheological behavior. This packing will be profoundly affected if the interaction between the droplets is changed from repulsive to attractive. This can be accomplished by adding SDS to induce a stronger depletion attraction between the droplets. We investigate the effects of an attractive interaction on the rheological behavior using a monodisperse emulsion having $a = 0.25\,\mu\text{m}$. We set $C = 200\,\text{mM}$, far above the concentration necessary to produce flocculation; we estimate that the attractive potential between droplets is greater than $10k_BT$. We measure both $G'(\omega)$ and $G''(\omega)$ for different volume fractions, and again are able to identify a plateau in the storage modulus and a minimum in the loss modulus; in this case, however, these quantities persist down to $\phi = 0.44$. The results for G'_p and G''_m are plotted by solid and open symbols, respectively, in Fig. 20. At high volume fractions, $\phi \geq 0.6$, the value of G'_p for the attractive emulsion is essentially unchanged from that for the identical emulsion with a much lower SDS concentration, where the interdroplet interactions are repul-

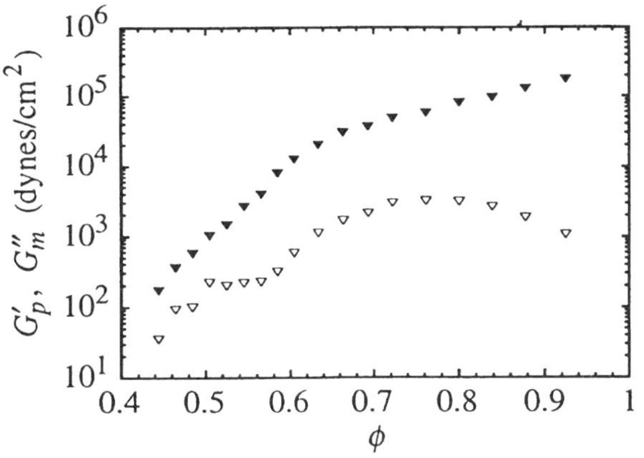

Figure 20. The plateau storage modulus G'_p (solid symbols) and minimum in the loss modulus G''_m (open symbols) as functions of volume fraction ϕ for an emulsion having a radius of 0.25 μm and a surfactant concentration of $C = 200$ mM, sufficient to lead to strong depletion attraction between the droplets.

sive. However, as the volume fraction decreases from 0.6, G'_p drops much more gradually for the attractive emulsion than for the repulsive one and persists to much lower volume fractions. This precludes defining a critical volume fraction for the attractive emulsion. Instead, this behavior suggests that the rigidity at low ϕ results from the formation of extended networks of droplets, similar to a colloidal gel structure, as a result of the attractive interactions. These networks can support stress, leading to the measured G'_p; however, since they do not completely fill all of the space, the magnitude of the modulus is decreased, reflecting the ϕ dependence of these networks. As the volume fraction increases to $\phi \approx 0.58$, volume constraints force the droplets to be essentially close packed, and the effects of the interdroplet interaction no longer distinguish between the attractive and repulsive cases.

In contrast with G'_p, the loss modulus, as characterized by G''_m, is higher for the attractive emulsion than for the identical repulsive emulsion, at all values of ϕ. It also exhibits a more gradual rise with increasing volume fraction but seems to have a peak around $\phi \approx 0.78$, decreasing at still higher volume fractions. The behavior of G''_m may reflect the formation of a different phase of the SDS in the water at these high concentrations. While viscoelastic itself, this phase will not influence the much larger storage modulus but may affect the weaker loss modulus, resulting in the observed behavior of G''_m.

We conclude this section by emphasizing again the utility of using these monodisperse emulsions to measure the rheological behavior. The use of monodisperse droplets ensures that the Laplace pressure in each droplet is identical; it also ensures that the packing of the droplets, although disordered, is nevertheless independent of droplet size. Moreover, the packing of uniform spheres is well understood, allowing the effects of the droplet packing to be rationalized in a simple manner. The scaling of the elastic modulus of emulsions with different droplet sizes confirms the dependence on Laplace pressure, and identifies the crucial role played by the droplet packing. The similarity between the osmotic pressure and the elastic modulus sets a challenge that must be addressed by any microscopic model for the rheological behavior of emulsions.

VI. Other Uses of Monodisperse Emulsions

A. Stability of Emulsions

Emulsions are not thermodynamically stable systems; nevertheless they can be kinetically stable almost indefinitely. Thus, the origin of the stability of an emulsion, and its control through the choice of its constituents are critical for the effective use of the emulsions. Monodisperse emulsions can very helpful in controlling the stability and in studying the fundamental mechanisms that lead to their instability.

Very generally, there are two fundamental mechanisms that lead to instability in emulsions. The first is a coarsening mechanism (Ostwald ripening); this entails the molecular diffusion of the fluid in the dispersed phase through the continuous phase, allowing transfer from one droplet to another. Since the fluid in the droplets is typically under pressure due to the Laplace pressure, there is a driving force that causes the fluid in the droplets to diffuse into the continuous phase. Furthermore, the internal pressure varies inversely with the size of the droplets, so it is larger the smaller the droplets are. This provides a driving force that causes fluid to diffuse from the smaller droplets to the larger droplets, causing the smaller droplets to shrink and the larger ones to grow. The net result is a growth in the average size of the droplets. Monodisperse droplets can, in principle, greatly reduce this coarsening by reducing the effective driving force because the droplets are identical in size. However, even with the most monodisperse droplets, there is still some remnant polydispersity, and thus some coarsening would occur; thus the monodispersity will only slow the rate of coarsening, not eliminate it entirely. However, the rate of coarsening can also be significantly decreased by reducing the solubility of the droplet fluid in the continuous phase.

The second fundamental mechanism that causes instability is coalescence, by which the films separating two droplets rupture, creating a single larger droplet. Coalescence is prevented by the stability imparted to the droplets by the surfactant coating; this is the critical role of the surfactant. As a result, knowledge about the effectiveness of a surfactant in preventing coalescence is crucial for producing any emulsion. Monodisperse droplets can be very effective in determining the resistance to coarsening of an emulsion. A key to their utility is the ease with which coalescence can be observed; if all the droplets begin with the same size, it is very easy to identify the doubling in the droplet size that results from even very small amounts of coalescence, and thereby precisely determine the conditions that lead to coalescence.

An example of this use of monodisperse emulsions is the determination of the criteria for stability of repulsive emulsions [9]. The emulsions were comprised of silicone oil droplets in water, stabilized by SDS. The silicone oil is completely insoluble in water, eliminating the possibility of coarsening; thus, the only mechanism leading to instability was coalescence. The stability was studied as a function of the SDS concentration, droplet size, and externally applied osmotic pressure. Because the droplets were initially the same size, the internal Laplace pressure was the same in each. The stability was monitored by examining a small sample of the emulsion in a microscope; coarsening was easily observed by the presence of some droplets that were (at least) twice as large as their neighbors. Since stability is, in a sense, a kinetic phenomenon, a criterion was set that an emulsion would be considered stable if no coalescence was observed for 2 weeks; while somewhat arbitrary, none of the essential conclusions are affected by the exact criterion chosen.

The results indicated that the emulsions required a minimum concentration of SDS in the water to be stable at any volume fraction. This concentration corresponded to slightly less than the CMC of SDS; this suggests that the droplet interfaces must be fully covered by surfactant in order to ensure stability and that this is achieved near the CMC for SDS and this oil–water combination. We expect that this minimum concentration will be somewhat dependent on both the surfactant and the oil. Once this minimum concentration was exceeded, the emulsions remained stable until rather large osmotic pressures were applied, corresponding to large volume fractions. However, at sufficiently high volume fractions, the emulsions again became unstable to coalescence. This volume fraction varied with droplet size, increasing signficantly as the droplet size grew larger. Since the droplets are monodisperse and each has the same internal Laplace pressure, a simple physical picture can be developed to account for this behavior. As the volume fraction of the larger droplets is increased, they develop flat facets where they are squeezed against their neighbors. At the same time, the radius of curvature, which now exists only at the edges or vertices of the facets, decreases, leading to an increase in the internal Laplace pressure. However, this pressure is then essentially identical for each droplet because the droplets are monodisperse. Moreover, it is this Laplace pressure that balances the force of the neighbors flattening the facets. The emulsion becomes unstable when this pressure exceeds some critical value; this value depends on the surfactant and the oil, but not on the droplet size. However, as the droplet size increases, the emulsion must be compressed to a higher volume fraction in order that this critical pressure be exceeded.

This example illustrates the utility of monodisperse emulsions for studying stability. While the use of repulsive emulsions is perhaps the simplest case, the techniques can easily be extended to other conditions as well. The key is the ease with which any instability can be monitored because of the uniform size of the droplets.

B. Adhesive Emulsions

Many emulsions have attractive, rather than repulsive, interactions between the droplets, which leads to flocculation. The use of a monodisperse emulsion greatly simplifies the identification of the nature and origin of the structure of the flocs. In particular, it allows a direct analogy to other colloidal aggregation phenomena to be made, providing a framework for understanding the interactions and structures in emulsions.

The silicone-oil-in-water emulsion can be made attractive by adding additional salt to screen the Coulombic repulsion of the droplets [41,42]. The droplet interactions become highly sensitive to temperature; below some critical value, T_c, the interaction changes from strongly repulsive to strongly attractive, with adhesive energies much larger than 50 k_BT. This attractive interaction causes the droplets to stick together and flocculate. However, it is completely reversible; when the temperature is raised above T_c, the droplets again become repulsive and redisperse.

An example of the structures obtained are shown in the microscope images in Fig. 21. These images were obtained using an emulsion with droplets of radius $a = 0.6\,\mu m$ and with an SDS concentration of 10 mM and NaCl concentration of 0.5 M. This resulted in a critical temperature of $T_c \approx 25°C$. The images were obtained after the emulsion was quenched from well above T_c to 20°C. The result is gelation; the droplets form a random but well-connected network. As shown in the top picture in Fig. 21 for $\phi = 0.01$, at lower volume fractions the structure of the gel consists of a highly random, tenuous network at short length scales, with branches that are, at most, only a few droplets thick. By contrast, at higher volume fractions, the interconnected branches at short length scales are thicker and less tenuous, although the network still remains disordered, as shown in the bottom picture in Fig. 21 for $\phi = 0.3$. These networks are analogous to colloidal gels, where similar structures are observed, albeit at lower volume fractions [43,44]. Interestingly, these emulsion gels have a dominant characteristic length scale. This is seen qualitatively in the microscope images in Fig. 21; it is seen more quantitatively in light-scattered studies of the gels. The intensity of the light scattering at small angles from these gels exhibits a pronounced ring of brighter scattering at very low angles [41]; the angular position of the peak provides a direct measure of this characteristic length scale. This length decreases with the volume fraction of the emulsion [42]; the dependence on ϕ provides direct evidence for the nature of the gelation process and the structure of the gel.

The gel that forms is weak, but it is nevertheless still rigid. This rigidity is somewhat surprising, since the droplets are fluid; and even though they are attractive, the surfaces should still be lubricated and hence fluidlike. Thus, it is unlikely that chains of single droplets are themselves rigid. However, tetragonally packed structures of lubricated droplets can form the basis for a rigid network. Indeed, careful observation of the chains under a microscope suggest that even the thinnest and most tenuous network structures are actually assemblies of tetragonally packed droplets. Thus, the adhesion does not introduce shear rigidity in a single bond between two individual droplets.

The origin of the adhesive interaction is not well understood, but we speculate that it is related to the surface crystallization of surfactant in the films between the droplets. The Kraft temprature for SDS crystallization between two proximate interfaces is mediated by the concentration of NaCl. Below the Kraft temperature, crystallized bridges of SDS form between droplets only where they touch and not over their entire surfaces. These crystallized bridges may resemble mobile islands on the droplet surfaces and thus may be responsible for the strong adhesion. However, they do not introduce shear rigidity, because the SDS can still flow over the droplet surface.

While the discussion here has focused on the structure of the emulsions at lower volume fractions, monodisperse emulsions are useful in determining the

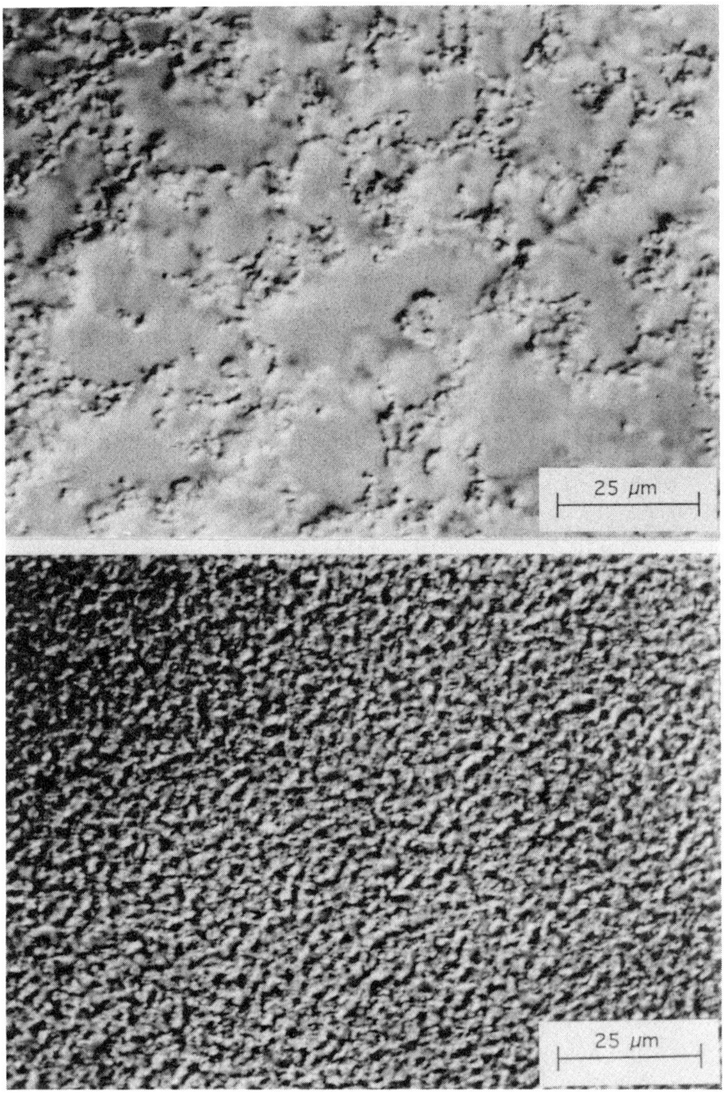

Figure 21. Microscope pictures of gel structures formed when the interaction between the droplets is made highly attractive. The volume fractions are 0.01 for the top picture and 0.3 for the bottom picture.

nature of the structures at higher volume fractions as well [42]. The use of uniformly sized droplets allows the origin of the structure—and its relationship to the droplet interactions—to be determined with greater ease.

C. Ferrofluid Emulsions

Another elegant use of monodisperse emulsions allows the direct measurement of interparticle colloidal forces [28]. This is achieved by replacing the oil phase of the droplets by a ferrofluid, an oil that contains small magnetite particles at a few weight percent. The magnetite particles are about 100 Å in diameter and behave as small permanent magnets. Normally, their Brownian motion is sufficient to ensure that they are randomly oriented. However, upon application of an external magnetic field, they are preferentially oriented along the direction of the field. This preferential orientation of all the small magnets induces a large magnetic moment in each droplet. Thus the system is superparamagnetic: there is no magnetic moment in the absence of an applied field, but a very large one in the presence of an applied magnetic field.

The large induced magnetic moments lead to an attractive, dipolar interaction between the droplets, causing them to form chains. If the concentration is sufficiently low, these chains are only a single droplet thick. If the droplets are monodisperse, the chains are comprised of very uniformly sized spheres; this results in Bragg scattering from the chains. In fact it causes the suspension to change colors as different wavelengths are Bragg scattered [28]. By measuring the spectrum of this Bragg scattering, it is possible to determine accurately the separation of the droplet centers. If the droplets are highly monodisperse, it is then also possible to determine the separation between the surfaces of the droplets.

This ability to determine the separations of the droplets accurately provides the basis for a direct measurement of the colloidal forces between the droplets. The applied magnetic field induces an attractive interaction between neighboring droplets. The strength of this attraction can be calculated accurately and can be varied by changing the strength of the applied field. This attractive force must balance exactly the repulsive force between the droplets at their separation distance. Thus, by calculating the field-induced attractive interaction, it is possible to determine the repulsive force between the droplets. Moreover, using the Bragg scattering, it is possible to determine the separation of the droplets. Finally, by varying the strength of the applied magnetic field, it is possible to measure the force–distance relationship for the interdroplet interaction. This provides a direct measure of the forces between colloidal particles. It has great potential and can be used to probe the forces directly in order to determine the origin and range of stability of an emulsion using different surfactants. More generally, it provides a measure of the forces between colloidal particles under real conditions and can be used to explore their dependence on the surface chemistry and on the additives in the suspending fluid.

VII. Conclusions

In this chapter, we have discussed the uses of an emulsion that is comprised of droplets that are perfectly monodisperse in size. These droplets are made from a crude emulsion by successive fractionation steps that exploit the small attraction between larger particles resulting from a large concentration of smaller particles. This attraction is strongly dependent on the size ratio of large particles to small. By using micelles as the smaller particles and the emulsion droplets as the larger particles, the emulsion can be fractionated to obtain an emulsion with very uniformly sized particles. These monodisperse particles have the additional advantage that the internal Laplace pressure is identical in each droplet. As a result, any emulsion property that depends on Laplace pressure becomes much easier to study with these emulsions.

Besides describing the fractionation technique, in this chapter we discussed several different uses of monodisperse emulsions. We showed that the osmotic pressure and the equation of state of these emulsions exhibits a well-defined behavior that can be understood through simple concepts, since the Laplace pressure is the same in each droplet. The most common packing of the emulsions is disordered, forming a colloidal glass. We determined the structure of this glass by performing light-scattering measurements; these also provided a very sensitive measure of the remaining polydispersity of the droplets, which was found to be about 7% of the average radius. We also discussed the use of these monodisperse droplets to measure the rheology of emulsions. We showed that the behavior is strongly dependent on the packing of the droplets. Finally, we presented a brief discussion of several other uses of these monodisperse droplets to study the properties of emulsions.

Acknowledgments

We have greatly benefited from discussions and interactions with many colleagues, including Paul Chaikin, Gary Grest, Martin Lacasse, Dov Levine, David Morse, and Tom Witten. This work was partially supported by NASA.

References

1. J. Bibette, D. Roux, and F. Nallet, *Phys. Rev. Lett. 65*:2470 (1990).
2. J. Bibette, *J. Colloid Interface Sci. 147*:474 (1991).
3. J. Bibette, D. Roux, and B. Pouligny, *J. Phys. II (France) 2*:401 (1992).
4. S. Asakura and F. Oosawa, *J. Phys. Chem. 22*:1255 (1954).
5. A. Vrij, *J. Pure Appl. Chem. 48*:471 (1976).
6. W. B. Russel, D. A. Saville, and W. R. Schowalter, *Colloidal Dispersions*, Cambridge University Press, Cambridge, 1989.
7. H. M. Princen, *Langmuir 2*:519 (1986).
8. H. M. Princen, *Langmuir 3*:36 (1987).
9. J. Bibette, D. C. Morse, T. A. Witten, and D. A. Weitz, *Phys. Rev. Lett. 69*:2439 (1992).
10. V. A. Parsegian, R. P. Rand, N. L. Fuller, and D. C. Rau, *Meth. Enzym. 127*:400 (1986).
11. D. C. Morse and T. A. Witten, *Europhys. Lett. 22*:549 (1993).
12. P. N. Pusey and W. van Megen, *J. Chem. Phys. 80*:3513 (1984).
13. B. J. Berne and R. Pecora, *Dynamic Light Scattering with Applications to Chemistry, Biology, and Physics*, Wiley, New York, 1976.
14. M. Kerker, *The Scattering of Light and Other Electromagnetic Radiation*, Academic Press, New York, 1969.
15. G. A. Schumacher and T. G. M. van de Ven, *Farad. Disc. Chem. Soc. 83*:75 (1987).
16. J.-P. Hansen and I. R. McDonald, *Theory of Simple Liquids*, Academic Press, London, 1986.
17. P. N. Pusey and W. van Megen, *Nature (London) 320*:340 (1986).
17a. P. Sherman, in *Encyclopedia of Emulsion Technology*, Vol. 1 (P. Becher, ed.), Dekker, New York, 1983, Chap. 7.
18. J. D. Ferry, *Viscoelastic Properties of Polymers*, Wiley, New York, 1980.
19. Q. D. Nguyen and D. V. Boger, *Annu. Rev. Fluid Mech. 24*:47 (1992).

20. L. W. Schwartz and H. M. Princen, *J. Colloid Interface Sci. 119*:201 (1987).
21. R. W. Whorlow, *Rheological Techniques*, Ellis Horwood Ltd., Chichester, England, 1980.
22. H. M. Princen, *J. Colloid Interface Sci. 105*:150 (1985).
23. H. M. Princen, *J. Colloid Interface Sci. 112*:427 (1986).
24. A. S. Yoshimura and R. K. Prud'homme, *Rheol. Acta 26*:428 (1987).
25. S. A. Khan, C. A. Schnepper, and R. C. Armstrong, *J. Rheol. 32*:69 (1988).
26. H. M. Princen, M. P. Aronson, and J. C. Moser, *J. Colloid Interface Sci. 75*:246 (1980).
27. D. Exerowa, D. Kashchiev, and D. Platikanov, *Adv. Colloid Interface Sci. 40*:201 (1992).
28. F. L. Calderon, T. Stora, O. M. Monval, P. Poulin, and J. Bibette, *Phys. Rev. Lett. 72*:2959 (1994).
29. B. Derjaguin, *Kolloid Z. 64*:1 (1933).
30. D. Stamenovic and T. A. Wilson, *J. Appl. Mech. 51*:229 (1984).
31. D. Stamenovic, *J. Colloid Interface Sci. 145*:255 (1991).
32. T. G. Mason, J. Bibette, and D. A. Weitz, submitted (1995).
33. S. Feng and M. F. Thorpe, *Phys. Rev. B31*:276 (1985).
34. F. Bolton and D. Weaire, *Phil. Mag. B65*:473 (1992).
35. T. G. Mason and D. A. Weitz, submitted (1995).
36. J. F. Brady, *J. Phys. Chem. 99*:567 (1993).
37. A. M. Kraynik and D. A Reinelt, *J. Proc. XIth Int. Congr. on Rheology* (P. Moldenaerss, ed.), Elsevier, Brussels, Belgium, 1992, p. 675.
38. D. A. Reinelt and A. M. Kraynik, *J. Colloid Interface Sci. 132*:491 (1989).
39. D. A. Reinelt and A. M. Kraynik, *J. Fluid Mech. 215*:431 (1990).
40. H. M. Princen, *J. Colloid Interface Sci. 128*:176 (1989).
41. J. Bibette, T. G. Mason, H. Gang, and D. A. Weitz, *Phys. Rev. Lett. 69*:981 (1992).
42. J. Bibette, T. G. Mason, H. Gang, D. A. Weitz, and P. Poulin, *Langmuir 9*:3352 (1993).
43. G. Dietler, C. Aubert, D. S. Cannell, and P. Wiltzius, *Phys. Rev. Lett. 57*:3117 (1986).
44. M. Carpineti and M. Giglio, *Phys. Rev. Lett. 68*:3327 (1992).

7
HLB: Update III

PAUL BECHER / Consultant, Wilmington, Delaware

This is the third update to the original bibliography of HLB compiled by Griffin and the writer. The fact that over 100 titles have been added since Update II in 1988 suggests that hydrophile–lipophile balance is still an area of substantial research interest. Much more remains to be done.

Alander, J.
Wärnheim, T.
Model Microemulsions Containing Vegetable Oils. Part I: Nonionic Surfactant Systems.
J. Amer. Oil Chemists' Soc., **1989**, *66(11)*, 1656–1660.

Albonico, P.
Blute, I.
Borgarello, E.
Parker, W. O., Jr.
Acidic Microemulsions of Dodecane/*p*-Dodecylbenzene Sulfonic Acid/Water/Mineral Acid. Part I. HLB Acidity and Phase Behavior.
Colloids Surfaces, **1990**, *48*, 13–20.

Albonico, P.
See Parker, W. O., Jr.

Anazawa, I.
See Saito, Y.

Arai, T.
See Shinoda, K. (1)

Araki, M.
See Shinoda, K. (9)

Arkad, O.
Arkad, T.
Garti, N.
Quantitative Determination of Creaming in O/W Emulsions by Use of Absorption Measurements of Oil-Soluble Dyes.
Lebensm.-Wiss. u.-Technologie, **1986**, *19, 164–166.*

Arkad, T.
See Arkad, O.

Aserin, A.
Garti, N.
Brominated Surfactants. Brominated Surfactants as Emulsifiers and Weighting Agents.
Tenside, Surfactants, Deterg., **1986**, *23(6)*, 305–308.

Azemar, N. Solans, C. Parra, J. L. Calbet, J. Camp, S. A.	Formation of Emulsions at the HLB Temperature and Detersive Efficiency. *Jorn. Com. Esp. Deterg.*, **1988**, *19,* 121–135 (Spanish). Also, *Invest. Inf. Text. Tensioactivos*, **1989**, *32(4),* 186–187 (Spanish).
Azemar, N.	See Solans, C. (1)(2)
Barakat, Y. Gendy, T. S. Basily, I. K. Mohamad, A. I.	Polymeric Surfactants for Enhanced Oil Recovery. Part II—The HLB–CMC Relationship of Ethoxylated Alkylphenol-Formaldehyde Polymeric Surfactants. *Brit. Polym. J.*, **1989**, *21*, 451–457.
Barakat, Y.	See Gendy, T. S.
Bares, M.	See Silha, J.
Bartholomei, G. B.	See De Kanterewicz, R. J.
Basily, I. K.	See Barakat, Y.
Basu, A. K. Ghosh, S.	HLB Values of Nonyl Phenol Ethylene Oxide Adducts. *J. Inst. Chemists (India)*, **1990**, *62*, 29.
Beger, J.	See Miesiac, I.
Beger, J.	See Szymanowksi, J. (1)(6)
Benson, H. L.	See Raney, K. H.
Berjano, M.	See Gallegos, C.
Blute, I.	See Albonico, P.
Blute, I.	See Parker, W. O., Jr.
Bock, J.	See Robbins, M. L. (1)(2)
Borgarello, E.	See Albonico, P.
Borgarello, E.	See Parker, W. O., Jr.
Brooks, B. W. Richmond, H. N.	(1) Dynamics of Liquid–Liquid Phase Inversion Using Nonionic Surfactants. *Colloids Surfaces*, **1991**, *58*, 131–148.

Brooks, B. W.
Richmond, H. N.
(2) Phase Inversion in Non-Ionic Surfactant-Oil-Water Systems. I. The Effects of Transitional Inversion on Emulsion Drop Sizes.
Chem. Eng. Sci., **1994**, *49(7)*, 1053–1064.

Brooks, B. W.
Richmond, H. N.
(3) The Application of a Mixed Nonionic Surfactant Theory to Transitional Emulsion Phase Inversion. 1. Derivation of a Mixed Surfactant Partitioning Model.
J. Colloid Interface Sci., **1994**, *162*, 59–66.

Brooks, B. W.
Richmond, H. N.
(4) The Application of a Mixed Nonionic Surfactant Theory to Transitional Emulsion Phase Inversion. 2. The Relationship Between Surfactant Partitioning and Transitional Inversion—A Thermodynamic Treatment.
J. Colloid Interface Sci., **1994**, *162*, 67–74.

Buchert, P.
Candau, F.
Polymerization in Microemulsions: I. Formulation and Structural Properties of Microemulsions Containing a Cationic Monomer.
J. Colloid Interface Sci., **1990**, *136(2)*, 527–540.

Buchert, P. — See Candau, F. (2)

Bueno, J. H. F. — See Gomes de Oliveira, A.

Buzier, M. — See Ravey, J. C. (2)

Calbet, J. — See Azemar, N.

Calbet, J. — See Solans, C.

Callaborro, C. — See Gallegos, C.

Camp, S. A. — See Azemar, N.

Camp, S. A. — See Solans, C.

Candau, F.
Collin, D.
Kern, F.
(1) Properties of Copolymer Latexes Prepared in Inverse Microemulsions.
Makromol. Chem., Macromol. Symp., **1990**, *35/36*, 105–119.

Candau, F.
Buchert, P.
(2) Polymerization of Trimethylammonioethyl Methacrylate Chloride in Microemulsions: Formulation, Characterization and Rheological Behavior of the Microlatexes.
Colloids Surfaces, **1990**, *48(1–3)*, 107–122.

Candau, F. — See Buchert, P.

Candau, F. — See Holtzscherer, C. (1)(2)

Carlsson, A. — See Shinoda, K. (11)

Cavé, G. — See Rabaron, A.

Collin, D. — See Candau, F. (1)

Dabrowski, J.
Stanislawska, A.
Hydrophilic–Lipophilic Balance of Surfactants.
Przgl. Papier, **1990**, *46(2)*, 52–55 (Polish)

Davis, H. T. — See Kunieda, H. (4)

De Kanterewicz, R. J.
Elizalde, B. E.
Pilosof, A. M. R.
Bartholmei, G. B.
Water–Oil Absorption Index (WOAI): A Simple Method for Predicting the Emulsifying Capacity of Food Proteins.
J. Food Sci., **1987**, *52(5)*, 1381–1383.

de Luca, M. — See Depraetère, P.

del Pozo, A. — See del Pozo Carrascosa, A. (1)(2)(3)

del Pozo Carrascosa, A.
del Pozo, A.
(1) Conductimetric Graph of the Formation Process of Liquid Paraffin O/W Emulsions Prepared with Nonionic Emulsifiers. I. Conductometric Parameters and Required HLB.
Ciencia Indust. Farm., **1985**, *4(5)*, 141–144 (Spanish).

del Pozo Carrascosa, A.
del Pozo, A.
(2) Conductimetric Graph of the Formation Process of Liquid Paraffin O/W Emulsions Prepared with Nonionic Emulsifiers. II. Interpretation of the Graph.
Ciencia Indust. Farm., **1985**, *4(6)*, 167–173 (Spanish).

del Pozo Carrascosa, A.
del Pozo, A.
(3) Conductimetric Graph of the Formation Process of Liquid Paraffin O/W Emulsions Prepared with Nonionic Emulsifiers. III. Influence of Various Technological Factors on the Conductometric Parameters.
Ciencia Indust. Farm., **1985**, *4(7/8)*, 212–217 (Spanish).

Depraetère, P.
Grossiord, J. L.
de Luca, M.
Rolland, J. P.
Vaution, C.
Seiller, M.
Study of the Influence of the Addition of Electrolytes on the Viscosity and Zeta-Potential of Emulsions at the Critical HLB.
Congr. Int. Technol. Pharm., 5th, **1989**, *3*, 408–416 (French).

Dias de Andrade, S.
See Gomes de Oliveira, A.

Dickinson, E.
Emulsions.
Ann. Rep. Royal Soc. Chem., **1986**, *C*, 31–58.

Dokic, P.
See Rasulic, G.

Elizalde, B. E.
See De Kanterewicz, R. J.

Evans, D. F.
See Kunieda, H. (3)(4)

Fessi, H.
See Treiner, C.

Förster, Th.
Schambil, F.
Tesmann, H.
Emulsification by the Phase Inversion Temperature Method: The Role of Self-Bodying Agents and the Influence of Oil Polarity.
Int. J. Cosmetic Sci., **1990**, *12*, 217–227.

Fukuda, M.
See Shinoda, K. (11)

Gallegos, C.
Galaborro, C.
Munoz, J.
Berjano, M.
Guerrero, A.
High-HLB Sucrose Esters: A Pecular Viscous Behavior.
Commun. Jorn. Com. Esp. Deterg., **1991**, *22*, 283–295 (Spanish).

Garcia–Dominguez, J. A.
See Voelkel, A.

Garcia Fadrique, J.
See Romero Sambrano, Y.

Garti, N.
See Arkad, O.

Garti, N. — See Aserin, A.

Gasik, S. M. — See Ljubic, R. M. (2)(3)(4)

Gendy, T. S.
Barakat Y.
Mohamad, A. I.
Youssef, A. M.
Polymeric Surfactants for Enhanced Oil Recovery. Part V: Prediction Models for the Effect of Temperature and Length of Polyoxyethylene Chain or HLB on Surface and Thermodynamic Properties of Some Ethoxylated Polymeric Nonionics.
Polym. Int., **1991**, *24*, 235–240

Gendy, T. S. — See Barakat, Y.

Ghosh, S. — See Basu, A. K.

Goebel, J.
Adsorption of Nonionic Surfactants on the Liquid–Liquid Interface.
Abh. Akad. Wiss, DDR, Abt. Math., Naturwiss., Tech. 1986 (Pub. **1987**), (1N) 245–260 (German)

Gomes de Oliveira, A.
Bueno, J. H. F.
Dias de Andrade, S.
Saab, S. A.
Macroscopic and Microscopic Study of Oil/Water Emulsions, in Regard to HLB and Emulsifier Systems Used.
An. Farm. Quim., **1988**, *26–28*, 8–16 (Portuguese).

Graacia, A.
Lachaise, J.
Marion, G.
Schechter, R. S.
A Study of the Required Hydrophile–Lipophile Balance for Emulsification.
Langmuir, **1989**, *5(6)*, 1315–1318.

Grossiord, J. L. — See Depraetère, P.

Gu, X.
Shi, N.
Determination of the HLB Value of Fatty Acid Sucrose Esters.
Huaxue Gongye, **1988**, *(2)*, 81–82 (Chinese).

Guerrero, A. — See Gallegos, C.

Hahn, L.
Sucker, H.
HLB Determination by HPLC.
Tenside, Surfactants, Deterg., **1989**, *26(3)*, 192–194.

Hayashi, Y. — See Sagitani, H. (3)

Hebrant, M.
Tondre, C.
Partitioning of Homologous Complexing Agents with Varying HLB Between Water and Cationic or Nonionic Micelles.
J. Colloid Interface Sci., **1992**, *154*, 378–384.

Holtzscherer, C.
Candau, F.
(1) Application of the Cohesive Energy Ratio Concept (CER) to the Formation of Polymerizable Microemulsions.
Colloids Surfaces, **1988**, *29*, 411–423.

Holtzscherer, C.
Candau, F.
(2) Salt Effect on Solutions of Nonionic Surfactants and Its Influence on the Stability of Polymerized Microemulsions.
J. Colloid Interface Sci., **1988**, *125(1)*, 97–110.

Hreczuch, W.
Refractive Index–Hydrophilic Lipophilic Balance Relationship for Alcohol Ethoxylates.
Ind. Eng. Chem. Res., **1995**, *34*, 410–412.

Hyakutake, M. See Yamaguchi, S.

Itoh, K. See Sagitani, H. (1)

James, A. D.
Wates, J. M.
Wyn-Jones, E.
Determination of the Hydrophilicities of Nitrogen-Based Surfactants by Measurement of Partition Coefficients Between Heptane and Water.
J. Colloid Interface Sci., **1993**, *160*, 158–165.

Janas, J. See Voelkel, A. (1)

Jiang, C.
Wang, C.
Determination of Hydrophile–Lipophile Balance of the Nonionic Emulsifier by Reversed Phase Gas Chromatography.
Heching Xianjiao Gongye, **1991**, *14(6)*, 399–401 (Chinese).

Jonstroemer, M. See Lindman, B.

Jovanovic, S. See Rasulic, G.

Kalashinskaya, N. Yu. See Nikolaeva, N. M.

Kaneko, T. See Shinoda, K. (8)

Kern, F.	See Candau, F. (1)
Khan, A.	See Shinoda, K. (9)
Khutoryanskii, F. M.	See Nikolaeva, N. M.
Komatsu, H.	See Miyajima, K.
Kruglyakov, P. M.	(1) Hydrophilic–Lipophilic Balance: Empirical Systems and Hydrodynamic Approaches. *Kolloidn. Zh.*, **1990**, *52(5)*, 1027–1031 (Russian).
Kruglyakov, P. M. Sviridov, V. V.	(2) The Hydrophile–Lipophile Balance Value of a Solid. *Kolloidn. Zh.*, **1993**, *52(2)*, 181–187 (Russian).
Kunieda, H.	(1) Correlation Between the Formation of a Lamellar Liquid Crystalline Phase and the Hydrophile–Lipophile Balance (HLB) of Surfactants. *J. Colloid Interface Sci.*, **1986**, *114(2)*, 378–385.
Kunieda, H. Ohyama, K.	(2) Three-Phase Behavior and HLB Numbers of Bile Salts and Lecithin in a Water–Oil System. *J. Colloids Interface Sci.*, **1990**, *136(2)*, 431–439.
Kunieda, H. Evans, D. F. Solans, C. Yoshida, M.	(3) The Structure of Gel-Emulsions in a Water/Nonionic Surfactant/Oil System. *Colloids Surfaces*, **1990**, *47*, 35–43.
Kunieda, H. Nakamura, K. Davis, H. T. Evans, D. F.	(4) Formation of Vesicles and Microemulsions in a Water/Tetraethylene Glycol Dodecyl Ether/Dodecane System. *Langmuir*, **1991**, *7*, 1915–1919.
Kunieda, H. Sato, Y.	(5) Microemulsions of Bile Acid Salt and Hydrophile–Lipophile Balance Number. *Hyomen*, **1992**, *30*, 360–366 (Japanese).
Kunieda, H.	See Shinoda, K. (1)(7)(8)

Kusz, P. — See Szymanowski, J. (7)

Lachaise, J. — See Graacia, A.

Laurent, A. — See Treiner, C.

Levchenko, D. N. — See Nikolaeva, N. M.

Lin, I. — Critical Micelle Concentration. Hydrophile–Lipophile Balance, Effective Chain Length and Hydrophobicity Index of Ionic Surfactants Containing Two Long-Chain Alkyl Groups.
Tenside Deterg., **1980**, *17(3)*, 119–123.

Lindman, B.
Jonstroemer, M.
Shinoda, K. — Changes of Organized Solution (Microemulsion) Structure with Small Change in Amphiphile Composition.
Yukagaku, **1989**, *38(2)*, 157–160 (Japanese).

Lindman, B. — See Shinoda, K. (5)(9)

Ljubic, R. M.
Manojlovic, V. S. — (1) Measure of the Hydrophilicity of the Oil Phase and the Hydrophile–Lipophile Balance.
J. Serb. Chem. Soc., **1986**, *51(6)*, 317–325.

Ljubic, R. M.
Gasic, S. M. — (2) Notes on the Possibility of Mathematical Interpretation of Shinoda's PIT Curves and Hydrophilic–Lyophilic Balance of Nonionic Emulsifiers.
J. Serb. Chem. Soc., **1987**, *52(10)*, 615–616.

Ljubic, R. M.
Gasic, S. M. — (3) Notes on the Mathematical Relationship Between Hydrophilic–Lyophilic Balance According to Griffin and Expressions Obtained from the Phase Inversion Temperature.
J. Serb. Chem. Soc., **1989**, *54(2)*, 115–117.

Ljubic, R. M.
Gasic, S. M. — (4) Temperature Region of Water Solubilization in Some Chlorinated Hydrocarbons in the Presence of Nonyl Phenyl Polyglycol Ethers.
J. Serb. Chem. Soc., **1988**, *53(5)*, 265–270.

Machida, K. — See Miyajima, K.

Maekawa, M. — See Shinoda, K. (3)

Magdassi, S. — Formation of Oil-in-Polyethylene Glycol/Water Emulsions. *J. Dispersion Sci. Technol.*, **1988**, *9(4)*, 391–399.

Mohamad, A. I. — See Barakat, Y.

Manojlovic, V. S. — See Ljubic, R. M. (1)

Marion, G. — See Graacia, A.

Marszall, L. — (1) Effect of the Type of Nonionic Surfactants on the Emulsion Inversion Point of Aliphatic Hydrocarbons. *Kolor. Ert.*, **1985**, *27(5/6)*, 115–119.

Marszall, L. — (2) The Effect of Nonionic Surfactant Properties on the Phase Inversion Temperature. *Pollena: TSPK*, **1986**, *30(1/2)*, 1–9 (Polish).

Marszall, L. — (3) Evidence for the Enhanced Emulsion Stability in the Presence of Ionic Surfactant–Long Chain Alcohol Mixtures Through Emulsion Inversion Point. *J. Colloid Interface Sci.*, **1986**, *122*, 296–297.

Matveenko, V. N. — See Sineva, A. V.

McGowan, J. C. — A New Approach for the Calculation of Hydrophile–Lipophile Balance Values of Surfactants. *Tenside Sur. Det.*, **1990**, *27(4)*, 229–330.

McGowan, J. C. — See Sowada, R.

Meissner, E. — See Voelkel, A.

Merkwitz, H. — See Miesiac, I.

Merkwitz, H. — See Szymanowksi, J. (6)

Miesiac, I., Merkwitz, H., Szymanowski, J., Beger, J. — The Hydrophilicity of Polyoxyethylene Glycol Dialkyl Ethers and Some of Their Sulfur Analogs. *J. Colloid Interface Sci.*, **1986**, *114(2)*, 425–431.

Miyajima, K.
Machida, K.
Taga, T.
Komatsu, H.
Nakagaki, M.
Correlation Between the Hydrophobic Nature of Monosaccharides and Cholates, and Their Hydrophobic Indices.
J. Chem. Soc., Faraday Trans. I, **1988**, *84(8)*, 2537–2544.

Mohamad, A. I. See Gendy, T. S.

Munoz, J. See Gallegos, C.

Myszkowski, J. See Voelkel, A. (2)

Nakagaki, M. See Miyajima, K.

Nakamura, K. See Kunieda, H. (4)

Nikolaeva, N. M.
Levchenko, D. N.
Kalashinskaya, N. Yu.
Khutoryanskii, F. M.
Study of the Effect of Synthesis Conditions on the Polarity and Hydrophilic–Lipophilic Balance of Surfactants.
Sb. Nauchn. Tr.—Vses. Nauchno-Issled. Inst. Pererab. Nefti, **1980**, *35*, 42–46 (Russian).

Oberthur, R. See Ravey, J. C. (2)

Ochiai, M. See Sagitani, H. (1)(3)

Ohyama, K. See Kunieda, H. (2)

Orecchioni, A. M. See Treiner, C.

Parker, W. O., Jr.
Albonico, P.
Borgarello, E.
Blute, I.
Acidic Microemulsions of Dodecane/*p*-Dodecylbenzenesulfonic Acid/Water/HCl. Part 2. Structure by Proton NMR Self-Diffusion and Relaxation Studies.
Colloids Surfaces, **1990**, *48*, 21–38.

Parra, J. L. See Azemar, N.

Parra, J. L. See Solans, C. (1)(2)

Pilosof, A. M. R. See De Kanterewicz, R. J.

Prochaska, K. See Szymanowksi, J. (2)

Puisieux, F. See Rabaron, A.

Rabaron, A.
Cavé, G.
Puisieux, F.
Seiller, M.

Physical Methods for Measurement of the HLB of Ether and Ester Nonionic Surface-Active Agents—H-NMR and Dielectric Constant.
Int. J. Pharm., **1993**, *99(1)*, 29–36.

Raney, K. H.
Benson, H. L.

The Effect of Polar Soil Compounds on the Phase Inversion Temperature and Optimum Detergency Conditions.
JAOCS, **1990**, *67(11)*, 722–729.

Rashid, Z. A.

See Szymanowski, J. (2)(4)

Rasulic, G.
Jovanovic, S.
Dokic, P.

Study of the Effect of Emulsifier on Thermal Behavior of Concentrated Oil/Water Emulsions.
Thermochim. Acta, **1988**, *134*, 91–96.

Ravey, J. C.

(1) Nonionic Surfactant Systems Above the Phase Inversion Temperature. I. Phase Behavior.
Prog. Colloid Polym. Sci., **1987**, *73*, 107–112.

Ravey, J. C.
Buzier, M.
Oberthur, R.

(2) Nonionic Surfactant Systems Above the Phase Inversion Temperature. II. Molecular Aggregate Structure.
Prog. Colloid Polym. Sci., **1987**, *73*, 113–126.

Richmond, H. N.

See Brooks, B. W. (1)(2)(3)(4)

Richter, L.
Vollhardt, D.

Determination of Wetting Tensions as a Basis of HLB Values of Water-Soluble Emulsifiers.
Tenside Surf. Det., **1993**, *30(3)*, 179–185.

Robbins, M. L.
Bock, J.

(1) Model for Microemulsions. I. Effect of Sulfonate Surfactant Cation and Chain Size and Concentration on Phase Behavior.
J. Colloid Interface Sci., **1988**, *124(2)*, 462–485.

Robbins, M. L.
Bock, J.

(2) Model for Microemulsions. II. Hydrophile–Lipophile Balance-Salinity-Oil Molar Volume Phase Maps.
J. Colloid Interface Sci., **1988**, *124(6)*, 1318–1321.

Robu, C.
Effect of Surface Active Agents on the Properties of Emulsion Copolymers of Vinyl Acetate and 2-Ethylhexyl Acrylate.
Bul. Tehn.-Inf. Centrul Cerc. Pent. Mat. Prot., **1970**, *(2)*, 3–39 (Romanian).

Rolland, J. P.
See Depraetère, P.

Romero Sambrano, Y.
Garcia Fadrique, J.
HLBP: New Scale for the Hydrophilic–Lipophilic Balance.
Commun. Jorn. Com. Esp. Deterg., **1994**, *25*, 437–442 (Spanish).

Ruckenstein, E.
Interfacial Free Energy, Surface Excess, and Stability of Macroemulsions.
Langmuir, **1988**, *4(6)*, 1318–1321.

Saab, S. A.
See Gomes de Oliveira, A.

Sadaghiani, A.
See Shinoda, K. (9)

Sagitani, H.
Ochiai, M.
Itoh, H.
(1) Characteristics of Poly(oxytetramethylene) Poly(glyceryl) Alkyl Ether Nonionic Surfactants as Emulsifiers.
J. Soc. Cosmetic Chemists, **1987**, *38(6)*, 419–433.

Sagitani, H.
(2) Formation of O/W Emulsions by Surfactant Phase Emulsification and the Solution Behavior of Nonionic Surfactant System in the Emulsification Process.
J. Dispersion Sci. Technol., **1988**, *9(2)*, 115–129.

Sagitani, H.
Hayashi, Y.
Ochiai, M.
(3) Solution Properties of Homogeneous Polyglycerol Dodecyl Ether Nonionic Surfactants.
J. Am. Oil Chemists' Soc., **1989**, *66(1)*, 146–152.

Saijo, H.
See Shinoda, K. (1)

Saito, Y.
Sato, T.
Anazawa, I.
Effects of Molecular Weight Distribution of Nonionic Surfactants on Stability of O/W Emulsions.
J. Am. Oil Chemists' Soc., **1990**, *67(3)*, 145–148.

Sato, T.
See Saito, Y.

Sato, Y.
See Kunieda, H. (5)

Schambil, F.
See Förster, Th.

Schechter, R. S.
See Graacia, A.

Schott, H.
(1) Comments on Hydrophile–Lipophile Balance Systems.
J. Colloid Interface Sci., **1989**, *133(2)*, 527–529.

Schott, H.
(2) Comments on Hydrophile–Lipophile Balance Systems.
J. Pharm. Sci., **1990**, *79(1)*, 87–88.

Seiller, M.
See Depraetère, P.

Seiller, M.
See Rabaron, A.

Selle, M. H.
Sjöblom, J.
Skurtveit, R.
Emulsions Under Elevated Pressure and Temperature Conditions. II. The Model System Water(Electrolyte)-Octanoic Acid–Sodium Octanoate-*n*-Heptane at 20°C.
J. Colloid Interface Sci., **1991**, *144(1)*, 36–44.

Shchukin, E. D.
See Sineva, A. V.

Shi, M.
Wang, D.
Wang, H.
Sun, X.
Relation of the Antifoaming Effects of Polyoxyalkylene Ether Block Polymers and Their Stearates to the Content Percent of the Hydrophilic Group and Hydrophilic–Lipophilic Balance.
Gaodeng Xuexiao Huaxue Xuebo, **1988**, *9(6)*, 640–642 (Chinese).

Shi, N.
See Gu, X.

Shibata, Y.
See Shinoda, K. (3)(4)

Shinoda, K.
Kunieda, H.
Arai, T.
Saijo, H.
(1) Principles of Very Large Solubilization (Microemulsion): Inclusive Understanding of the Solubilization of Oil and Water in Aqueous and Hydrocarbon Media.
J. Phys. Chem., **1984**, *88*, 5126.

Shinoda, K.
(2) The Significance and Characteristics of Organized Solutions.
J. Phys. Chem., **1985**, *89*, 2429–2431.

Shinoda, K.
Maekawa, M.
Shibata, Y.
(3) Ionic Surfactants Soluble in Hard Water and in Hydrocarbons: Behavior of Organized Surfactant Solutions as a Function of the Hydrophilic–Lipophilic Balance.
J. Phys. Chem., **1986**, *90*, 1228–1230.

Shinoda, K.
Shibata, Y.
(4) Principles for the Attainment of Minimum Oil–Water Interfacial Tension by Surfactants: The Characteristics of Organized Surfactant Phase.
Colloids Surfaces, **1986**, 185–196.

Shinoda, K.
Lindman, B.
(5) Organized Surfactant Systems: Microemulsions.
Langmuir, **1987**, *3(2)*, 135–149.

Shinoda, K.
Kunieda, H.
(6) The Effect of Salt Concentration, Temperature, and Additives on the Solvent Property of Aerosol OT Solution.
J. Colloid Surface Sci., **1987**, *118(2)*, 586–589.

Shinoda, K.
Kunieda, H.
(7) Advance in Emulsification by Phase Inversion Temperature Method.
Fragrance J., **1989**, *17(4)*, 24–28 (Japanese).

Shinoda, K.
Kaneko, T.
(8) Characteristic Properties of Lecithin as a Surfactant.
J. Dispersion Sci. Technol., **1988/1989**, *9(5/6)*, 555–559.

Shinoda K.
Araki, M.
Sadaghiani, A.
Khan, A.
Lindman, B.
(9) Lecithin-Based Microemulsions: Phase Behavior and Microstructure.
J. Phys. Chem., **1991**, *95*, 989.

Shinoda, K.
Fukuda, M.
Carlsson, A.
(10) Characteristic Solution Properties of Mono-, Di-, and Triglyceryl Alkyl Ethers: Lipophobicity of Hydrophilic Groups.
Langmuir, **1990**, *6*, 334.

Shinoda, K.
See Lindman, B.

Shinoda, K.
See Yago, K.

Silha, J.
Bares, M.
Zeman, I.
Smidrkal, J.
The HLB Number Determination of Polyoxyethylene Surfactants.
Czech. Chem. Commun., **1989**, *52*, 945–952.

Sineva, A. V.
Matveenko, V. N.
Shchukin, E. D.
Interrelation Between the Hydrophile–Lipophile Balance, Viscosity, and Solubilizing Capacity of Microemulsions.
Isv. Vyssh. Uchebn. Zaved., Khim. Khim. Tekhnol., **1988**, *31(11)*, 62–65 (Russian).

Sjöblom, J.
See Selle, M. H.

Skrylev, L. D.
Sazanova, V. F.
Skryleva, T. L.
Evaluation of Hydrophilic and Lipophilic Properties of Surfactants by Their Critical Micelle Formation Concentration.
Izv. Vyssh. Unchebn. Zaved., Khim. Khim. Tekhnol., **1987**, *30(5)*, 72–76 (Russian).

Skryleva, T. L.
See Skrylev, L. D.

Skurtveit, R.
See Selle, M. H.

Smidrkal, J.
See Silha, J.

Solans C.
Azemar, N.
Parra, J. L.
Calbet, J.
Camp, S. A.
(1) Phase Behaviour and Detergency in Water/Nonionic Surfactant/Hydrocarbon Systems.
Proc. World Surfactants Congr., 2nd, **1988**, *I*, 421–429.

Solans, C.
Azemar, N.
Parra, J. L.
(2) High-Internal-Phase-Volume Emulsions in Water/Nonionic Surfactant/Hydrocarbon Systems.
Progr. Colloid Polym. Sci., **1988**, *76*, 224–227.

Solans, C.
See Azemar, N.

Solans, C.
See Kunieda, H. (3)

Somasundaran, P.
A New Method of Determining the Effective Hydrophobicity of Surfactants and Chelating Agents by HPLC.

J. Colloid Interface Sci., **1988**, *126(2)*, 634–637.

Sowada, R.
McGowan, J. C.
Calculation of Hydrophile–Lipophile Balance Group Numbers for Some Structural Units of Emulsifying Agents.
Tenside, Surfactants, Deterg., **1992**, *29(2)*, 109–113.

Stanislawska, A. — See Dabrowksi, J.

Stepniak-Biniakiewicz, D. — See Szymanowski, J. (2)

Sucker, H. — See Hahn, L.

Sun, X. — See Shi, M.

Sviridov, V. V. — See Kruglyakov, P. M. (2)

Szewczyk, H. — See Szymanowksi, J. (7)

Szymanowksi, J.
Beger, J.
(1) The Estimation of Hydrophile–Lipophile Balance from Solubility and Polarity Data.
J. Colloid Interface Sci., **1987**, *119(1)*, 286–288.

Szymanowski, J.
Stepniak-Biniakiewicz, D.
Prochaska, K.
Rashid, Z. A.
(2) The Surface Activity and the Hydrophile–Lipophile Balance of Hydroxyoximines and Mechanism of Copper Extraction.
Abh. Akad. Wiss. DDR, Abt Math., Naturwiss, Tech., **1987**(1N), 437–441. *Int. Congr. Surface Active Subst., 6th*, **1985**, 437–441.

Szymanowski, J.
(3) The Polarity of Nonionic Surfactants as Determined by Reversed-Phase Gas Chromatography.
Zesz. Nauk. Politechn. Pozn., **1986**, *(16)*, 267–287.

Szymanowski, J.
Voelkel, A.
Rashid, Z. A.
(4) Polarity and Hydrophile–Lipophile Balance of Model Hydroxyoximes and Their Intermediates as Determined by Reversed-Phase Gas Chromatography.
J. Chromatogr., **1987**, *402*, 55–64.

Szymanowski, J.
(5) The Polarity of Nonionics. The Polarity of Nonionic Surfactants as Determined by Gas-Chromatography.
Tenside, Surfactants, Deterg., **1989**, *26(3)*, 198–210.

Szymanowski, J.; Voelkel, A.; Beger, J.; Merkwitz, H.
(6) The Hydrophilic–Hydrophobic Balance of Some Nonionics.
Tenside, Surfactants, Deterg., **1989**, *26(4)*, 260–265.

Szymanowski, J.; Kusz, P.; Szewczyk, H.
(7) Increments of the Arithmetic Retention Index for Polyoxyethylene Glycol Monoalkyl Ethers and Their Degradation Products.
J. Chromatogr., **1989**, *477*, 407–412.

Szymanowski, J.
(8) Polarity of Nonionic Surfactants as Determined by Gas Chromatography.
Crit. Rev. Anal. Chem., **1990**, *21(6)*, 407–451.

Szymanowski, J. — See Miesiac, I.

Szymanowski, J. — See Wisniewski, M.

Szymanowski, J. — See Voelkel, A.

Taga, T. — See Miyajima, K.

Tesmann, H. — See Förster, Th.

Tondre, C. — See Hebrant, M.

Treiner, C.; Laurent, A.; Fessi, H.; Orecchioni, A. M.
Oil-in-Water Emulsion Investigation in the System Oil/Brij 35/Phenoxy,[sic]2-Ethanol/ Sea Water; Comparison of Partitioning and Droplet Size Data.
J. Colloid and Interface Sci., **1989**, *129(1)*, 201–207.

Turai, L. L.; Williams, L. D.
Effect of HLB Factor of Nonionic Surfactants on Deinking Efficiency.
TAPPI, **1977**, *60(11)*, 167–168.

Ugri, H. E.
Interfacial Properties of Surfactants.
Olaj, Szappan, Kozmet., **1991**, *40(3)*, 82–87 (Hungarian).

Vaution, C. — See Depraetère, P.

Voelkel, A.
Janas, J.
Garcia-Dominguez, J. A. — (1) Inverse Gas Chromatography in Characterization of Surfactants. Determination of Binary Parameter.
J. Chromatogr., **1993**, *654*, 135–141.

Voelkel, A.
Szymanowski, J.
Meissner, E.
Myszowski, J. — (2) Polarity of Fluorine-Containing Oxyethylates as Determined by Inverse Gas Chromatography.
J. Fluorine Chem., **1993**, *64*, 177–186.

Voelkel, A. — See Szymanowski, J. (4)(6)

Vollhardt, D. — See Richter, L.

Vucinic-Milankovic, N.
Milankovic, R.
Stupar, M. — Effect of the Phase Volume and Emulsifier Location on Phase Inversion.
Arh. Farm., **1990**, *40(3)*, 81–87 (Serbo-Croatian).

Wang, C. — See Jiang, C.

Wang, D. — See Shi, M.

Wang, H. — See Shi, M.

Wärnheim, T. — See Alander, J.

Wates, J. M. — See James, A. D.

Williams, J. M. — High Internal Phase Water-in-Oil Emulsions: Influence of Surfactants on Emulsion Stability and Foam Quality.
Langmuir, **1991**, *7(7)*, 1370–1377.

Wisniewski, M.
Szymanowski, J. — Polarity and Hydrophile–Lipophile Balance of Polydisperse Poly(oxyethylene) 4-Alkylphenylamines as Determined by Gas Chromatography.
Colloid Polymer Sci., 1989, *267(1)*, 59–64.

Wyn-Jones, E. — See James, A. D.

Xikui, J. — Hydrophobic–Lipophilic Interactions. Aggregation and Self-Coiling of Organic Molecules.
Acc. Chem. Res., **1988**, *21(10)*, 362–367.

Yago, K.
Shinoda, K.

Emulsification and Interfacial Tension of Lipid in Brine in the Presence of Cholates and Monoglyceride. Surface Science of Emulsification and Absorption of Lipids in Intestine.
Yukagaku, **1989**, *38(3)*, 231–235 (Japanese).

Yamaguchi, S.
Hyakutake, M.

Change of Emulsion Type with the Mixture of Hydrophilic Ionic and Lipophilic Nonionic Surfactants.
Yukagaku, **1990**, *39(3)*, 180–183 (Japanese).

Yoshida, M.

See Kunieda, H. (3)

Youssef, A. M.

See Gendy, T. S.

Zarev, E. E.

HLB Numbers and Geometric Parameters of Molecules of Noniongenic Surfactants.
Colloid J. Russ. Acad. Sci., **1993**, *55(1)*, 49–51.

Zeman, I.

See Silha, J.

Zgoda, M. M.

(1) Surfactants in the Group of Propylene and Ethylene Oxide Copolymers. II. Application of Infrared and Proton NMR Spectroscopy to the Determination of the Hydrophile–Lipophile Balance in Polyetherdiols.
Acta Pol. Pharm., **1988**, *45(1)*, 63–70 (Polish).

Zgoda, M. M.

(2) Tensides from the Group of Propylene Oxide–Ethylene Oxide Copolymers. VI. Use of the Critical Micelle Concentration (CMC) in the Estimation of the Hydrophile–Lipophile Balance (HLB), the HI Index and the Thermodynamic Potential of Micelle Formation (ΔG^0_m) of Polyetherdiols.
Acta Pol. Pharm., **1990**, *47(5–6)*, 55–60 (Polish).

Index